CARBON NANOTUBES - POLYMER NANOCOMPOSITES

CARBON NANOTUBES - POLYMER NANOCOMPOSITES

Editor

Felix Ryler

scitus academics

Carbon Nanotubes - Polymer Nanocomposites
Edited by **Felix Ryler**

ISBN: 978-1-68117-667-3
Library of Congress Control Number: 2016946533

SCITUS Academics LLC,
616, Corporate Way, Suite 2, 4766,
Valley Cottage, NY 10989

www.scitusacademics.com

Notice

Printed in United States of America on Acid Free Paper

PREFACE

Polymer nanocomposites are promising potential applications in a variety of industries from construction to aerospace. These are high end materials with greater concentrations of properties and unique design options and are considered to be materials of the 21st century. With a projected growth rate of approximately 25 percent per annum and a high demand for engineering polymers, their potential is so promising that applications range from packaging to bio-medical applications benefit. Reinforcement filler is the main difference between conventional composites and polymer nano-composites. The nanocomposite reinforcement is in the order of the nanometer, which leads to a very different macroscopic property. As a result of this special feature, polymer nanocomposites have only been studied in the last decade using different nanofillers such as minerals, fibers or sheets. A carbon nanotube (CNT) is a cylindrical miniature carbon structure fitted with hexagonal graphite molecules at the edges. Nanotubes look like a black soot or a powder, but are actually rolled up graphene sheets that form hollow strands with one atom thick walls. Nanotubes often referred to as buckytubes, have been created from Fullerene, a structure similar to geodesic dome. They take the form of cylindrical carbon molecules and have new properties that make them useful in a wide range of applications in nanotechnology, electronics, other fields and materials science optics. They exhibit extraordinary strength and unique electrical properties and are efficient heat conductors.

Carbon Nanotubes – Polymer Nanocomposites focus on the planning and property analysis of CNT (fibers) polymer nanocomposites as nano-fillers. Research in academia and industry has concentrated on the development of advanced polymer nanocomposites utilizing nano-scale fillers to accomplish desirable polymer properties like ionic conductivity, mechanical and thermal properties.

TABLE OF CONTENTS

1

POLYMER/CARBON NANOTUBE NANOCOMPOSITES

Veena Choudhary[1] and Anju Gupta

[1] Centre for Polymer Science and EngineeringIndian Institute of Technology Delhi, India

1. INTRODUCTION

Polymer is a versatile material having many unique properties like low density, reasonable strength, flexibility, easy processibilty, etc. However, the mechanical properties of these materials are inadequate for many engineering applications. Hence, there is a continuous search towards new polymeric materials with improved properties. Initially, blending of different class of polymer was used to fabricate new materials with unique properties. However, blending lead to only marginal improvement in physical properties which were still inadequate for engineering applications. So to improve the strength and stiffness of polymer materials different kinds of organic and inorganic fillers were used. It was observed that strength and stiffness of long fibers reinforced thermosetting polymer is comparable to metals at a fraction of their weight. As a result of which these material were used in aircraft and in sport equipment. However, processing of these materials is very difficult; therefore small fiber or particle reinforced composites were developed. The common particle fillers used were silica, carbon black, metal particles etc. But significantly high filler loading was required to achieve desired mechanical property, which thus increased cost and made processibility difficult. So to achieve high mechanical properties at lower filler loading, nanofillers were used. The nanofiller reinforced polymer matrix is known as polymer nanocomposite.

Polymer nanocomposites are a new class of composite materials, which is receiving significant attention both in academia and industry. As nano fillers are only a few nanometers (~10,000 times finer than a human hair) in dimension, it offers ultra-large interfacial area per volume between the nano-element and polymer matrix. As a result, the nanofiller reinforced composites exhibit enhanced toughness without sacrificing stiffness or optical clarity. It also possesses greater thermal and oxidative stability, better barrier, mechanical properties as well as unique properties like self-extinguishing behavior. Compared to different range of nanofillers, carbon nanotubes (CNTs) have emerged as the most promising nanofiller for polymer composites due to their remarkable mechanical and electrical properties (Ishikawa, 2001; Kracke & Damaschke, 2000). Currently, one of the most intriguing applications of CNTs is the CNT/polymer nanocomposites (Cai, 2000; Fiege, 1999; Gomes, 1999;Hersam, 1998; Ruiz, 1998). For the last two decades, a lot of research work has been done on evaluating the potential of CNTs as filler for polymer nanocomposites. In the present chapter, we will briefly discuss on CNTs and their properties, different fabrication methods of polymer nanocomposites and their mechanical, electrical and thermal properties.

2. CARBON NANOTUBES (CNTS)

CNTs are long cylinders of covalently bonded carbon atoms which possess extraordinary electronic and mechanical properties. There are two basic types of CNTs: single-wall carbon nanotubes (SWCNTs) which are the fundamental cylindrical structure and multi-wall carbon nanotubes (MWCNTs) which are made of coaxial cylinders (Fig. 1), having interlayer spacing close to that of the interlayer distance in graphite (0.34 nm). These cylindrical structures are only few nanometre in diameter, but the cylinder can be tens of microns long, with most end capped with half of a fullerene molecule. It was first discovered by M. Endo in 1978, as part of his Ph.D. work at the University of Orleans in France, but real interest in CNTs started when Iijima (1991) first reported it in 1991. The field thrives after that and the first polymer composites using CNT as filler was reported by Ajayan et al (1994).

2.1. Synthesis Of Carbon Nanotubes

CNTs can be prepared using three methods which includes arc discharge, laser ablation and chemical vapor deposition (CVD). Most of these processes take place in vacuum or with process gases. CVD growth of CNTs can occur in vacuum or at atmospheric pressure. High quality and large quantities of nanotubes can be synthesized by these methods.

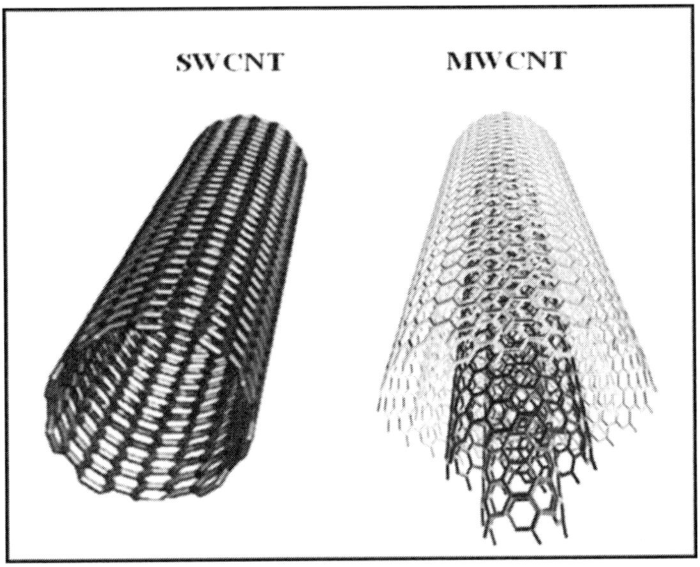

Figure 1. Schematic diagrams of single-wall carbon nanotube (SWCNT) and multi-wall carbon nanotube (MWCNT)

i. Arc discharge

The carbon arc discharge method, initially used for producing C_{60} fullerenes, is the most common and perhaps easiest way to produce CNTs. But this technique produces graphitic impurities such as carbon soot containing amorphous carbon, anions and fullerens. In this method an inert gas atmosphere is created in the reaction vessel by passing an inert gas at controlled pressure. Two graphitic rods constitute the electrodes, between which a potential difference is applied. As the rods are brought closer, a discharge occurs, resulting in formation of plasma (Fig. 2). The deposit, which contains CNTs, forms on the large negative electrode (cathode) while the smaller positive electrode (anode) is consumed. When a metal catalyst is used along with graphite, a hole is drilled in the carbon anode and it is filled with a mixture of metal and graphite powder. In this case, most nanotubes are found in soot deposited on the arc chamber wall.

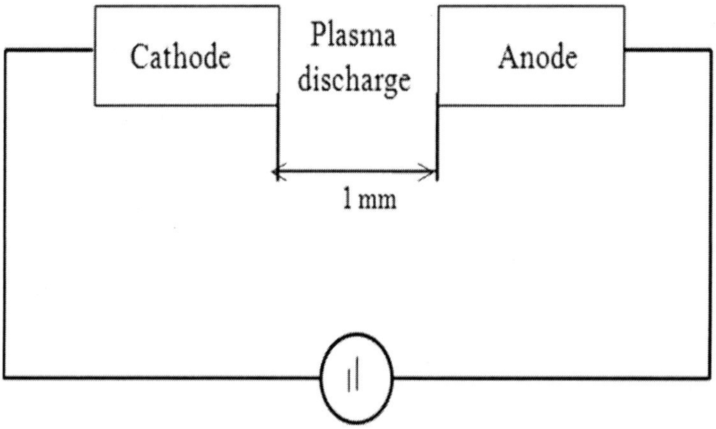

Figure 2. Plasma arc discharge setup

ii. Laser ablation

Laser ablation uses an intense laser pulse to vaporize a carbon target, which also contains small amount of metals such as nickel and cobalt and is placed in a tube furnace at 1200º C. As the target is ablated, inert gas is passed through the chamber carrying the grown nanotubes on a cold finger for collection (Fig. 3). This method mainly produces SWCNT in the form of ropes.

iii. Chemical vapor deposition

In this process a mixture of hydrocarbon, metal catalyst along with inert gas is introduced into the reaction chamber (Fig. 4a). During the reaction, nanotubes form on the substrate by the decomposition of hydrocarbon at temperatures 700–900º C and atmospheric pressure. The diameter of nanotubes that are to be grown are related to the size of the metal particles. This mechanism of CNT growth is still being studied. In Figure 4b two growth modes can be seen. The first 'tip growth mode' where, the catalyst particles can stay at the tips of the growing nanotube during the growth process and second 'base growth mode' where, catalyst particles remain at the nanotube base, depending on the adhesion between the catalyst particle and the substrate (Fig. 4b). This technique offers more control over the length and structure of the produced nanotubes compared to arc and laser methods. This process can also be scaled up to produce industrial quantities of CNTs. A number of reviews (Awasthi, 2005;Monthioux, 2002; Thostenson, 2001) are available which briefly discusses on these production techniques.

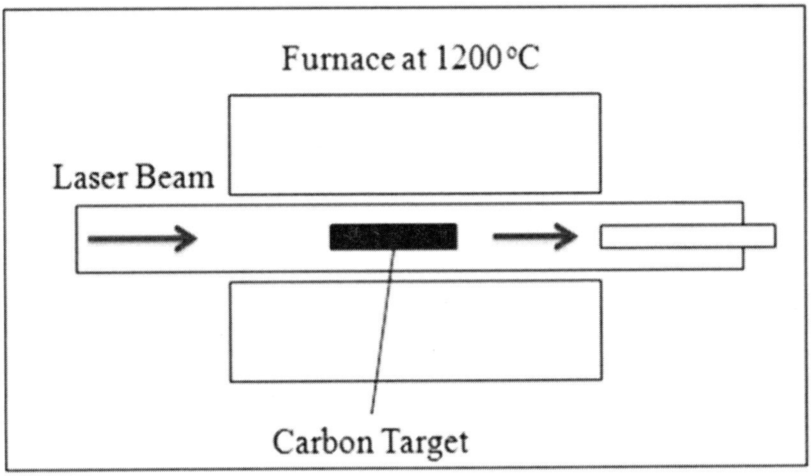

Figure 3. Laser ablation setup

2.2. Properties of carbon nanotubes

The determination of physical properties of CNTs is relatively more difficult compared to other fillers due to very small size of CNTs. However a number of experimental studies have been carried out on the direct determination of mechanical properties of individual CNTs (Yu et al., 2000a). The stiffness of CNTs was first determined by observing the amplitude of thermal vibrations in transmission electron microscopy (TEM) and the average stiffness values of 1.8 TPa and 1.25 TPa (Yu et al., 2000b) were reported for MWCNTs and SWCNTs, respectively. The in-situ tensile tests on individual MWCNTs and ropes of SWCNTs was performed by carrying out a stress-strain measurement using a "nanostressing stage" operating inside a scanning electron microscope (SEM). Experimental results showed that strength of outer shell of MWCNT ranged from 11 to 63 GPa at fracture strains of up to 12% and modulus values ranged from 270 to 950 GPa (Yu et al., 2000b). It was observed that strength of nanotubes depends on the number of defects, as well as interlayer interactions in MWCNTs and bundles of SWCNTs. Structural defects as well as bends or twists significantly affect mechanical strength of CNTs (Kane & Mele, 1997). Theoretical studies of the electronic properties of SWCNTs, suggest that nanotube shells can be either metallic or semiconducting, depend critically on helicity (Fig. 5) (Saito, 1992; Hamada, 1992; Mintmire, 1992).Tans et al. (1997) first, experimentally showed that there are indeed metallic and semiconducting SWCNTs, which verified the theoretical predictions. It was observed that due to poor control on synthesis, on average, approximately 1/3 of SWCNTs formed are metallic and 2/3 semiconductors (Odom et al., 1998). The room temperature conductivity of metallic SWCNT was found to be 10^5 to 10^6 S/m and for semiconducting tubes about 10 S/m. The conductivity of SWCNT is close to the in-plane conductivity of graphite [10^6 S/m (Charlier & Issi 1996)]. Conductivities of individual MWCNTs have been reported in the range of 10^7 to 10^8 S/m (Ebbesen et al., 1996), depending on the helicities of the outermost shells or the presence of defects (Dai et al., 1996). The axial thermal conductivity of individual, perfect CNTs is reported to be as high as 3300 W/m/K (Kim et al., 2001).

3. FUNCTIONALIZATION OF CARBON NANOTUBES

For a nanocomposite, a good dispersion of the filler within the host matrix is very important. At the same time it is also important to stabilize the dispersion to prevent re-aggregation of the filler. These tasks are particularly very challenging in case of nanofillers since the extremely large surface area lead to strong tendency to form agglomerates. CNTs are very well known to form aggregates during compounding and hence various techniques have been used to overcome this problem like use of sonication or mechanical mixing during the fabrication of nanocomposite which generally help in dispersing CNTs. But the most effective way to resolve this problem is surface functionalization of CNTs. Surface functionalization helps in stabilizing the dispersion, since it can prevent re-aggregation of nanotubes and also leads to coupling of CNT with polymeric matrix. Coupling between CNT and polymer matrix is also very important for efficient transfer of external stress to nanotube. In recent years, various methods have been developed for surface functionalization of CNT which includes functionalization of defect groups, covalent functionalization of sidewalls, non-covalent functionalization, e.g., formation of supramolecular adducts with surfactants or polymers (Fig. 6). Although surface functionalization leads to significant improvement in CNT dispersion and stress transfer but this method also causes deterioration of intrinsic properties of CNTs. Covalent functionalization often lead to tube fragmentation and the non-covalent functionalization results in poor exfoliation. Alteration of CNT properties lead to poor reinforcement and conductivity. Hence, it becomes obvious that dispersion and stabilization are not simple issues and compromises have to be made depending on the applications (Hirsch, 2002).

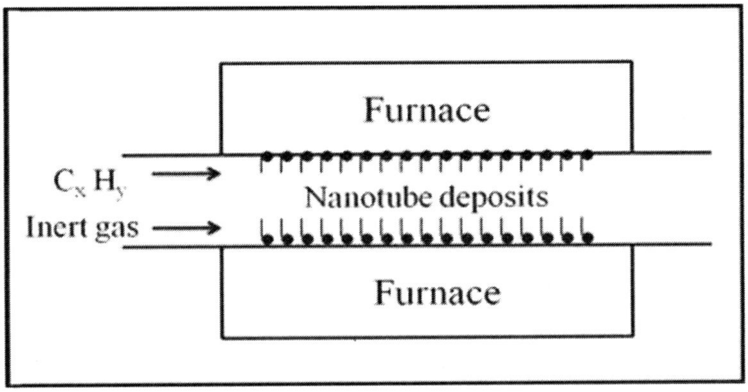

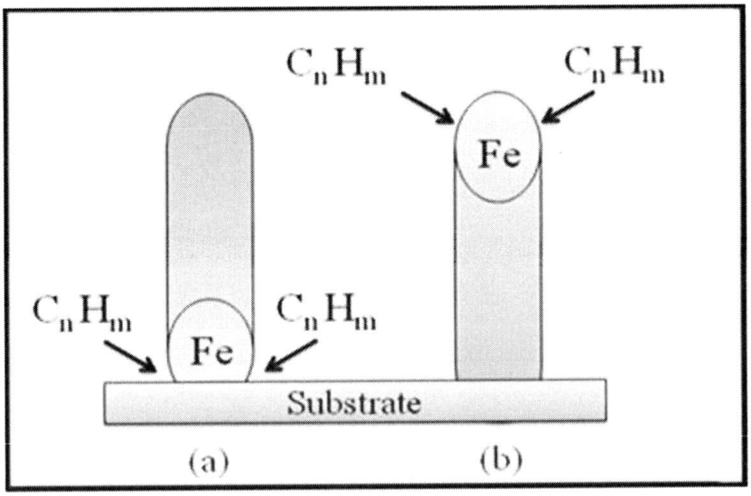

Figure 4. a. Chemical vapor deposition setup; b. Growth modes for CVD (a) base growth mode and (b) tip growth mode

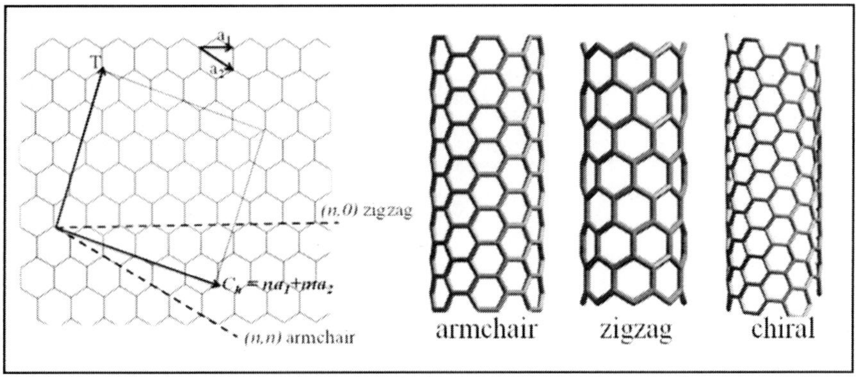

Figure 5. A sheet of graphene rolled to show formation of different types of single walled carbon nanotube

3.1. Covalent Functionalization Of Sidewall

Local strain in CNTs, which arises from either pyramidalization or misalignment of π-orbitals of the sp^2-hybridized carbon atoms, makes nanotubes more reactive than a flat graphene sheet, thereby paving the path to covalently attach chemical species to nanotubes (Banerjee et al., 2005). Covalent functionalization of CNTs can be achieved by introducing some functional groups on defect sites of CNTs (Fig. 7) by using oxidizing agents such as strong acids, which results in the formation of carboxylic or hydroxyl groups (-COOH, -OH) on the surface of nanotubes (Coleman, 2000, 2006, Singh, 2009). This type of functionalization is known as defect group functionalization. Such functionalization improves nanotube dispersion in solvents and polymers and imparts high stability in polar solvents. For example, Feng et al. (2008) reported that oxidation of MWCNTs with HNO_3/H_2O_2 and HNO_3/H_2SO_4 introduces some carboxylic groups on CNTs, which enhanced their stability in water at room temperature for more than 100 days. As a result, the water-stable nanotubes can easily be embedded in water soluble polymers such as poly(vinyl alcohol) (PVA), giving CNT/polymer nanocomposites with homogeneous dispersion of CNTs (Zhao et al., 2008). Oxidized nanotubes also show excellent stability in other solvents such as caprolactam, which is used in the production of polyamide (PA6) (Gao et al., 2005). Studies on acid functionalization of CNTs have shown significant improvement in interfacial bonding between CNTs and polymer matrices, which have been shown to give a stronger nanotube–polymer interaction, leading to improved Young's modulus and mechanical strength (Gao, 2006; Sui, 2008; Yuen, 2008a, 2008b, 2008c; Luo, 2008;Rasheed, 2006a, 2006b; Sahoo, 2006; Wong, 2007).

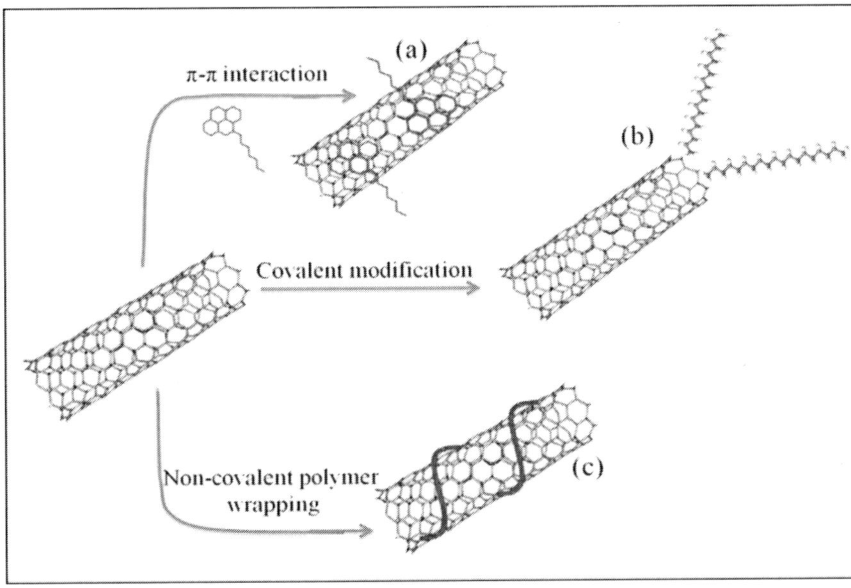

Figure 6. Possibilities for the functionalization of SWCNTs a) л-л interaction; b) defect group functionalization ; c) non-covalent functionalization with polymers

Polymer molecules can also be grafted on the surface of CNTs in presence of these active functional groups (–COOH, -NH_2, -OH). Grafting of polymer chain on CNTs can be carried out either by 'grafting from' or 'grafting to' technique (Coleman, 2000, 2006; Liu, 2005). The grafting-from approach is based on the initial immobilization of initiators onto the nanotube surface, followed by in-situ polymerization with the formation of polymer molecules attached to CNTs. The advantage of this approach is that polymer-functionalized nanotubes with high grafting density can be prepared. However, this process needs a strict control of the amounts of initiator and substrate. The grafting from technique is widely used for the preparation of poly(methyl methacrylate) (PMMA) and related polymer grafted nanotubes. For example, Qin et al. (2004) reported the preparation of poly(n-butyl methacrylate) grafted SWCNTs by attaching n-butyl methacrylate (nBMA) to the ends and sidewalls of SWCNT via atom transfer radical polymerization (ATRP) using methyl 2-bromopropionate as the free radical initiator. A similar approach was reported by Hwang et al. (2004) for the synthesis of PMMA grafted MWCNTs by potassium persulfate initiated emulsion polymerization reactions and use of the grafted nanotubes as a reinforcement for commercial PMMA by solution casting.

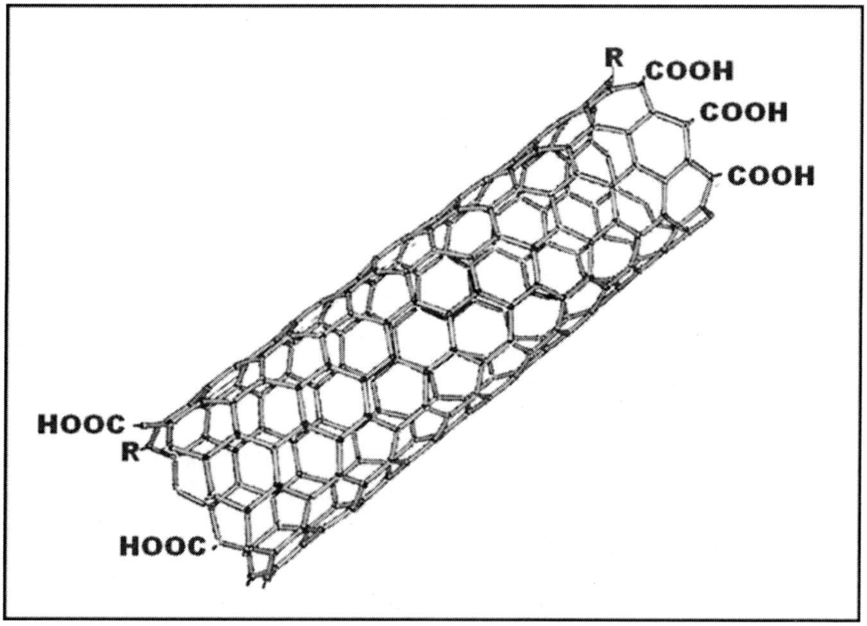

Figure 7. Covalent functionalization of carbon nanotubes on defects sites

On the other hand, in case of grafting-to approach, attachment of already functionalized polymer molecules to functionalized nanotube surface takes place via appropriate chemical reactions. One of the first examples of 'grafting to' approach was published by Fu et al (2001). In this work carboxylic acid groups on the nanotube surface were converted to acyl chlorides by refluxing the samples with thionyl chloride. Then the acyl chloride functionalized CNTs were reacted with hydroxyl groups of dendritic poly(ethylene glycol) (PEG) polymers via the esterification reactions. The grafting-to method was applied for the preparation of epoxy-polyamidoamine–SWCNT composites (Sun et al., 2008). An advantage of this method is that, commercially available polymers containing reactive functional groups can be utilized.

3.2. Non-Covalent Functionalization With Surfactant Or Polymer

The non-covalent functionalization has unique ability of preserving the intrinsic properties of nanotube which is very important for its electrical and thermal conductivity. Various studies have shown that surfactant or wrapping with polymer can lead to individualization of nanotube in aqueous or organic solvent. (Moore, 2003; Matarredona, 2003; Vigolo, 2000; Regev, 2004; Grossiord, 2005; Curran, 1998; Coleman, 2000; O'Connell, 2001). Moore *et al.* studied various anionic, cationic and non-ionic surfactants and polymers to determine their ability to disperse nanotube in aqueous media. They reported that size of the hydrophilic group of surfactant or polymer play a key role in nanotube dispersion. It was also observed that surfactant alone is not capable of suspending nanotubes effectively and vigorous sonication is required (Matarredona et al., 2003). Polymers such as poly(*m*-phenylene-*co*- 2,5-dioctoxy-*p*-phenylenevinylene) (PmPV) can be used to wrap around nanotube in organic solvents such as $CHCl_3$ (Coleman et al., 2000). Polar side chain containing polymer, such as poly(vinyl pyrrolidone) [PVP] or poly(styrene sulfonate) [PSS] gave stable solutions of SWCNT/polymer complexes in water (O'Connell et al., 2001). The thermodynamic driving force for wrapping of polar polymer on nanotube is the need to avoid unfavorable interactions between the non-polar tube walls and the polar solvent (water). To disperse the nanotube in non-polar polymer matrices, such as polyolefins, polymerization-filling technique (PFT) was used. In this process nanotube is first dispersed with catalyst and co-catalyst and followed by polymerization (Dubois & Alexandre, 2006). This technique leads to individualization of nanotube and allows the homogeneous dispersion of nanotubes upon melt blending.

4. CNT/POLYMER NANOCOMPOSITES

In order to utilize CNTs and their extra ordinary properties in real-world applications, CNT/polymer nanocomposites were developed. Currently, polymer composite is the biggest application area for CNTs. These nanocomposites are being utilized in different fields including transportation, automotive, aerospace, defense, sporting goods, energy and infrastructure sectors. Such wide range applications of such materials are due to their high durability, high strength, light weight, design and process flexibility. CNT/polymer nanocomposites are also used as electrostatic discharge (ESD) and electromagnetic interference (EMI) shielding material because of high electrical conductivity of this material. However, the effective utilization of CNTs for fabricating nanocomposites strongly depends on the homogeneous dispersion of CNTs throughout the matrix without destroying their integrity. Furthermore, good interfacial bonding is also required to achieve significant load transfer across the CNT–matrix interface, a necessary condition for improving the mechanical properties of composites. So it is very important to achieve high degree of CNT dispersion during processing without affecting its property. In the following section we will discuss about the different processing techniques and properties of CNT/polymer nanocomposite.

4.1. Processing Of Cnt/Polymer Nanocomposites

From the above discussion, it is very clear that CNTs have strong tendency to form aggregates due to their large surface area. These aggregates persist unless high shear forces are applied e.g., vigorous mixing of the polymer. But such mixing often damages nanotube structures, compromising their properties. Therefore, the biggest challenge is to fully disperse individual nanotubes in the matrices to realize full potential of CNTs. Surface modification of CNTs have somewhat helped in dispersing CNT but long term stability is still a real challenge. Nevertheless, several approaches have been successfully adopted to obtain intimate mixing of nanotubes with polymer matrices, including dry powder mixing, solution blending, melt mixing, in-situ polymerization and surfactant-assisted mixing.

i. Solution processing

The most common method for preparing CNT/polymer nanocomposites involves mixing of CNT and polymer in a suitable solvent. The benefit of solution blending is rigorous mixing of CNTs with polymer in a solvent which facilitates nanotube de-aggregation and dispersion. This method consists of three steps: dispersion of nanotubes in a suitable solvent, mixing with the polymer (at room temperature or elevated temperature) and recovery of the nanocomposite by precipitating or casting a film. Both organic and aqueous medium have been used to produce CNT/polymer nanocomposites [Bandyopadhyaya, 2002; Pei, 2008; Wu, 2007; Cheng, 2008]. In this method dispersion of nanotube can be achieved by magnetic stirring, shear mixing, reflux or most commonly, ultrasonication. Sonication can be provided in two forms, mild sonication in a bath or high-power sonication. The use of high-power ultrasonication for a long period of time can shorten the nanotube length, i.e. reduces the aspect ratio, which is detrimental to the composite properties (Badaire et al., 2004).

To minimize this problem, surfactants have been used to disperse higher loadings of nanotubes (Islam, 2003; Barrau, 2003; Bryning, 2005a). Islam et al. (2003) reported that SWCNT (20 mg/mL) can be dispersed in water by using 1% sodium dodecylbenzene sulfonate as surfactant and low power, high-frequency (12 W, 55 kHz) sonication for 16-24 h. However, it has a major drawback of retaining surfactant in the nanocomposites which deteriorate the transport properties of nanocomposites. Bryning et al. (2005a) prepared SWCNT/epoxy nanocomposites and showed that the thermal conductivity of composite is much lower if surfactant is used for SWCNT dispersion.

In solvent blending, slow evaporation step often lead to CNT aggregation. To overcome this problem, CNT/polymer suspension can be kept on a rotating substrate [spin-casting (Lamy de la Chapelle et al., 1999)] or can be dropped on a hot substrate [drop-casting (Benoit, et al., 2001)] to expedite the evaporation step. Coagulation [developed by Du et al. (Du et al., 2003)] is another method, which involves pouring of a CNT/polymer suspension into an excess of non-solvent. This lead to entrapment of SWCNT by precipitating polymer chains which inturn prevents the SWCNT from bundling. The method is very successful in case of PMMA and polyethylene [PE] nanocomposites (Haggenmueller et al., 2006).

ii. Melt blending

While solution processing is a valuable technique for both nanotube dispersion and nanocomposite formation, it is less suitable for industrial scale processes. For industrial applications, melt processing is a preferred choice because of its low cost and simplicity to facilitate large scale production for commercial applications. In general, melt processing involves the melting of polymer pellets to form a viscous liquid and application of high shear force to disperse the nanofillers such as CNTs. Successful examples of melt blending include MWCNT/polycarbonate (Poetschke et al., 2003) MWCNT/ nylon-6, (Liu, 2004; Zhang, 2004) SWCNT/polypropylene, (Bhattacharya et al., 2003) and SWCNT/polyimide (Siochi et al., 2004) nanocomposites. Although melt blending is very simple, but high shear force and high temperature can deteriorate nanocomposite property, as high shear force which is required to achieve CNT dispersion can also lead to CNT fragmentation. So an optimum shear stress is required to achieve desired dispersion at lowest possible damage of CNTs.

The use of high temperature is also very critical, as high temperature enhances CNT dispersion by lowering the viscosity but too high temperature can degrade the polymer intrinsic properties. So optimization of temperature is also very important. To overcome these challenges many modification in melt compounding has been made like : Haggenmueller et al. (2000) combined the solution and melt blending by subjecting a solvent cast SWCNT/polymer film to several cycles of melt pressing. An approach developed by Jin et al (2002) introduces polymer-coated MWCNT (rather than pristine MWCNT) into the polymer melt to promote compatibilization. However optimization of processing conditions is an important issue, not just for different nanotube types, but for the whole range of polymer–nanotube combinations (Dubois & Alexandre, 2006).

iii. In-situ polymerization

In recent years, in-situ polymerization has been extensively explored for the preparation of polymer grafted nanotubes and processing of corresponding polymer composite materials. The main advantage of this method is that it enables grafting of polymer macromolecules onto the walls of CNTs. In addition, it is a very convenient processing technique, which allows the preparation of nanocomposites with high nanotube loading and very good miscibility with almost each polymer matrix. This technique is particularly important for the preparation of insoluble and thermally unstable polymers, which cannot be processed by solution or melt processing. Depending on required molecular weight and molecular weight distribution of polymers, chain transfer, radical, anionic, and ring-opening metathesis polymerizations can be used for in-situ polymerization processing. Initially, in-situ radical polymerization was applied for the synthesis of PMMA/MWCNT nanocomposites (Jia, 1999; Velasco-Santos, 2003; Putz, 2004). More recently (Wu, 2009) studied the mechanical and thermal properties of hydroxyl functionalized MWCNTs/acrylic acid grafted PTT nanocomposites and showed a significant enhancement in thermal and mechanical properties of PTT matrix due to the formation of ester bonds between –COOH groups of acrylic acid grafted PTT and –OH groups of MWCNTs.

In-situ polymerization was also very useful for the preparation of polyamide/CNT polymer nanocomposites. Park et al. (2002) also reported the synthesis of SWCNT reinforced polyimide nanocomposites by in-situ polymerization of diamine and dianhydride under sonication. Epoxy nanocomposites comprise the majority of reports using in-situ polymerization methods, (Schadler, 1998; Zhu, 2003, 2004; Gong, 2000; Ajayan, 2000; Moniruzzaman, 2006a) where the nanotubes are first dispersed in the resin followed by curing the resin with the hardener. Zhu et al. (2003) prepared epoxy nanocomposites by this technique using carboxylated end-cap SWCNT and an esterification reaction to produce a composite with improved tensile modulus. It is important to note that as polymerization progresses and the viscosity of the reaction medium increases, the extent of in-situ polymerization reactions might be limited.

In general, in -situ polymerization can be applied for the preparation of almost any polymer nanocomposites containing CNTs which can be non-covalently or covalently bound to polymer matrix. Non-covalent binding between polymer and nanotube involves physical adsorption and wrapping of polymer molecules through van der Waals and л–л interactions. The role of covalently functionalized and polymer grafted nanotubes will be considered in more detail below.

5. ALIGNMENT OF CARBON NANOTUBES IN NANOCOMPOSITES

The superior properties of CNTs offer exciting opportunities for new nanocomposites, but the important limitation for some potential applications of CNTs come from the fact that randomly oriented nanotubes embedded in polymer matrices have exhibited substantially lower electrical and thermal conductivities than expected (Fischer, 1997; Hone, 1999). Nanotube alignment can be obtained prior to composite fabrication or during composite fabrication or after composite fabrication by in-situ polymerization (Raravikar, 2005; Feng,

2003), mechanical stretching (Jin et al., 1998), melt fiber spinning (Haggenmueller, 2000, 2003), electrospinning (Gao, 2004; Hou, 2005; Ko, 2003) and application of magnetic or electric field (Ma, 2008; Camponeschi, 2007). Haggenmueller et al. (2000) have tried a combination of solvent casting and melt mixing methods to disperse single-walled CNTs in PMMA films and subsequently melt spun into fibers. However, only the melt mixing method was found to be successful in forming continuous fibers. Ma et al. (2008) studied alignment and dispersion of functionalized nanotube composites of PMMA induced by electric field and obtained significant enhancement in dispersion quality and alignment stability for oxidized MWCNTs as compared to pristine MWCNTs. Camponeschi et al. (2007) found that orientation and alignment of CNTs embedded in the epoxy under a magnetic field increased and showed improvement in mechanical properties of the resulting nanocomposites. Gao et al. (2004) prepared SWCNT/poly(vinyl pyrrolidone) fibres by electrospinning using electrostatic forces and found SWCNT exhibit good alignment and dispersion.Xie et al. (2005) showed that enhanced dispersion and alignment of CNTs in polymer matrices greatly improve mechanical, electrical, thermal, electrochemical, optical and super-hydrophobic properties of CNT/polymer nanocomposites. Safadi et al. (2002) prepared PS/MWCNT nanocomposite films by spin casting at high speed (2200 rpm) and found that MWCNTs were aligned in specific angles relative to the radial direction: 45º and 135º on average. The presence of ~2.5 vol.% MWCNTs doubles the tensile modulus and transforms the film from insulating to conducting. Thostenson & Chou (2002) found that tensile strength and modulus of melt drawn PS/MWCNT composite films increased by 137 and 49% respectively, compared to the undrawn PS film.

6. PROPERTIES OF CNT/POLYMER NANOCOMPOSITES

Incorporation of CNT in polymer matrix resulted in a significant change in mechanical, electrical and thermal properties of polymer matrices. Various factors that influence property modification are processing techniques, type of CNT, aspect ratio and CNT content. It is generally observed that a particular processing method which is good for one property may not be good for another. One such example is surface modification of CNT which generally enhances the mechanical properties but deteriorates the electronic properties. So it is very important to optimize the various conditions to obtain the nanocomposite with desired properties. A number of studies have been aimed at evaluating the mechanical, electrical and thermal properties of CNT/polymer nanocomposite under different conditions and filler loading.

6.1. Mechanical Properties Of Cnt/Polymer Nanocomposites

The excellent mechanical properties of CNTs, as discussed above, suggest that incorporation of very small amount of CNTs into a polymer matrix can lead to structural materials with significantly high modulus and strength. Significant advancement has been made in improving the mechanical properties of polymer matrix by mixing small fraction of CNTs. Qian et al. (2000) reported that adding 1 wt.% MWCNTs in the PS by solution-evaporation method, results in 36–42 and ~25% improvements in tensile modulus and tensile strength, respectively. Biercuk et al. (2002) have also reported increase of indentation resistance (Vickers hardness) by 3.5 times on adding 2 wt. % SWCNTs in epoxy resin.Cadek et al. (2002) also found significant improvement in the modulus and hardness (1.8 times and 1.6 times) on addition of 1 wt% MWCNTs in PVA matrix. Homogeneous dispersion and alignment of CNTs in polymer matrix had a significant effect on the properties of resulting composites. Velasco-Santos et al. (2003) reported that by enhancing the dispersion of CNT by using an in-situ polymerization, the storage modulus of PMMA/MWCNT nanocomposites at 1 wt.% of MWCNTs at 90º C increased by 1135%.

Although, addition of CNTs lead to enhancement of mechanical properties of the polymer matrix but the improvement is still well below the expected value. At current stage, the extraordinary properties of CNTs are still not fully utilized in polymer composites. Many research works have indicated that poor adhesion between the matrix and nanotube is the limiting factor in imparting the excellent mechanical properties of nanotubes in composites. As load transfer from matrix to CNTs play a key role in mechanical properties of the nanocomposite, good interfacial bonding is very important. Load transfer between the matrix and filler depends on the interfacial shear stress between the two (Schadler et al., 1998). A high interfacial shear stress will transfer the applied load to the filler over a short distance and a low interfacial shear stress will require a long distance. There are three main load transfer mechanisms operating between a matrix and filler.

a. Micro-mechanical interlocking

This is the amount of load transfer due to mechanical interlocking which is very poor in nanotube composites because of the atomically smooth surface of nanotubes. As CNTs has some surface defects like varying diameter and bends/twist due to non-hexagonal defects, along a CNT, mechanical interlocking do play a role in CNT–polymer adhesion.

b. Chemical bonding between filler and matrix

A chemical bond either ionic or covalent significantly improves the interfacial interaction between matrix and filler that enables a stress transfer.

c. Weak van der Waals bonding between filler and matrix

The van der Waals interaction arises from the molecular proximity and is the only mode of interaction between CNTs and the matrix in absence of chemical bonding.

Hence, formation of chemical bonding between CNT and polymer can significantly improve the mechanical properties of nanocomposites. Recently, Blake et al. (2004) developed butyl-lithium-functionalized MWCNTs which can be covalently bonded to chlorinated polypropylene (CPP). The CPP/MWCNT was then compounded with CPP/tetrahydrofuran (THF) solution to obtain CPP/MWCNT nanocomposites. They showed that on addition of 0.6 vol% MWCNT, the modulus increased by three times and both tensile strength and toughness (measured by the area under the stress–strain curve) increased by 3.8 times (from 13 to 49 MPa) and 4 times (from 27 to 108 J/g), respectively. Bal & Samal (2007) showed that the amine functionalized CNTs get completely dispersed in polymer matrix in comparison to unmodified CNTs. Telescopic pull-out was also observed in case of functionalized MWCNTs (Gojny et al., 2005). It was observed that although CNTs get pulled out from the matrix the outer wall still remained in the matrix. This is possible because only weak van der Waals forces are present between the various concentric tubes of the MWCNT where as the outer tube is covalently bonded to the matrix. Such a pull-out process suggests that efficient load transfer occurs from matrix to the outer tube, due to strong covalent bonding between epoxy matrix and CNT.

These observations suggest that the efficiency of property improvement depends on the type of CNT, processing techniques and the compatibility between CNT and host matrix. Although chemical functionalization of CNT can improve the compatibility between CNT and polymer which inturn improves the mechanical properties but it has a deteriorating effect on the other properties of nanocomposites such as electrical and thermal conductivity. However, the rapid growth of this field suggests the solution to these problems are not very far and in coming few years' desire of obtaining super strong polymer material will be realized.

6.2. Electrical properties of CNT/polymer nanocomposites

With exceptional mechanical properties, CNTs also possess very high intrinsic electrical conductivity. The electrical conductivity of individual CNTs ranged between 10^7 to 10^8 S/m that is comparable to metals (Ebbesen et al., 1996). Very high electrical conductivity of CNTs have helped to impart conductivity in highly insulating material like polymer by fabricating polymer nanocomposites. The enhancement in electrical conductivity of insulating polymer by several orders of magnitude has been achieved with a very small loading (0.021 wt %) of nanotubes in the polymer matrices, which helped in preserving other performance aspects of the polymers such as optical clarity, mechanical properties, low melt flow viscosities, etc. As a result, these conducting materials are in growing demand in different application area such as transparent conductive coatings, electrostatic dissipation, electrostatic painting and electromagnetic interference shielding applications.

The electrical conductivity of CNT reinforced polymer nanocomposites depends on many factors including type of CNTs, aspect ratio, surface functionalization and CNT content. The electrical conductivity of nanocomposite increases with increasing CNT loading till a critical filler concentration where a dramatic increase in conductivity is observed. This critical filler concentration is called electrical percolation threshold concentration. At percolation threshold concentration, filler forms a three-dimensional conductive network within the matrix, hence electron can tunnel from one filler to another, and in doing so it overcomes the high resistance offered by insulating polymer matrix. The percolation threshold is typically determined by plotting the electrical conductivity as a function of the reduced mass fraction of filler and fitting with a power law function (Fig. 8). As formation of percolating networks depends on both intrinsic conductivity and aspect ratio of the filler, the nanotube/polymer nanocomposites exhibit very low percolation threshold because of high conductivity and high aspect ratio of CNTs. Bryning et al. (2005b) studied the effect of aspect ratio on percolation threshold concentration by preparing SWCNT/epoxy nanocomposites with nanotubes from two different sources, HiPco and laser oven, having aspect ratios of ~150 and ~380 respectively. They reported a smaller percolation threshold with the higher aspect ratio nanotubes. Similar observation was also made by Bai & Allaoui (2003); they found more than 8-fold decrease in threshold concentration in MWCNT/epoxy composites when the MWCNT length was increased from 1 to 50 μm. In general, the minimum percolation threshold concentration for SWCNT/polymer nanocomposite is 0.0021wt % in epoxy matrix (Martin et al., 2004). For different polymer systems it ranged from 0.0021 to 15 wt% CNT loading (Bauhofer & Kovacs, 2009). These studies show that the percolation threshold concentration and nanocomposite conductivity also depends on polymer type, synthesis method, aspect ratio of CNTs, disentanglement of CNT agglomerates, uniform spatial distribution of individual CNTs and degree of alignment [Bryning, 2005b; Moniruzzaman, 2006b; Du, 2005].

Another factor which significantly affects the electrical conductivity of nanocomposite is the chemical functionalization of CNTs because it disrupts the extended π-conjugation of nanotubes and thereby reduces the electrical conductivity of isolated nanotubes. Sulong et al. (2009) showed that incorporation of acid and octadecylated functionalized MWCNT in epoxy matrix decreased the electrical conductivity of nanocomposites. Similar results have also been reported by Park et al (2009). Hence, it is important to optimize the modification condition or reagent to achieve minimum deterioration of electronic properties of CNTs.

Nevertheless, significant improvement in electrical conductivity of polymer on CNT addition lead to the development of CNT/polymer conductive nanocomposites for electronics, automotive and aerospace applications with uses such as electrostatic dissipation, electromagnetic interference (EMI) shielding, multilayer printed circuits, and conductive coatings (Baughman, 2002; Dresselhaus, 2004; Holt, 2006; Martel, 1998; Minoux, 2005;Zhang, 2005). The rapid development of electrical industry, demands fabrication of light weight and effective EMI shielding material for the protection of workspace and environment from radiation coming from computers and telecommunication equipment as well as for protection of sensitive circuits. Thus electrically conducting polymer nanocomposites have received much attention recently compared to conventional metal-based EMI shielding materials (Chung, 2001; Joo & Epstein, 1994;Kim, 2003; Yang, 2005a), because of their light weight, resistance to corrosion, flexibility and processing advantages. The use of CNT have significantly reduced the filler loading required to achieve desired EMI SE, thus reduced the cost and weight of the material (Chung, 2001; Micheli, 2009; Sundararaj & Al-Saleh, 2009; Yang, 2005b).

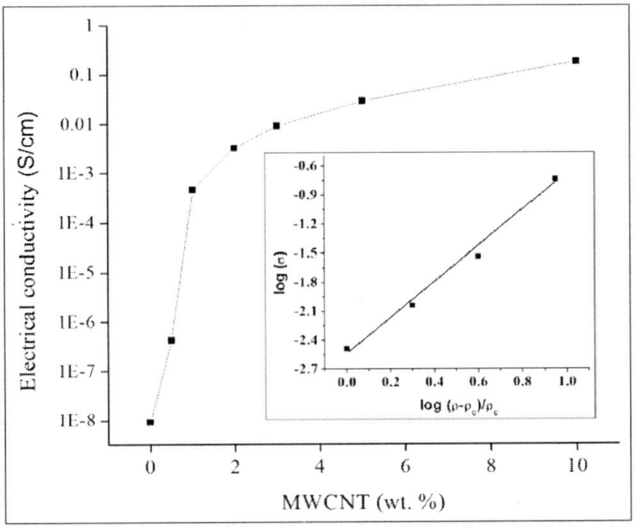

Figure 8. Plot of dc electrical conductivity (σ) vs weight fraction of MWCNT in PTT/MWCNT composites [inset shows the log-log plot of σ vs [(ρ- ρ_c)/ρ_c]

The effect of aspect ratio on EMI SE was shown by Sundararaj & Al-Saleh. They reported that SE of 1 mm thick shielding plate made of 7.5 vol% MWCNT/polypropylene (PP) nanocomposite was much higher (35 dB) than 7.5 vol% (HS-CB)/PP composite (18 dB) in the X-band frequency range. Yang et al. (2005b) studied the EMI shielding behavior of MWCNT/PS nanocomposites and achieved SE of ~20 dB at 7 wt % MWCNT loading. Although, lower value of SE of CNT composites have also been reported, e.g., Liu et al. (2007) achieved only ~17 dB at 20 wt% MWCNTs loading in polyurethane (PU) whereas Kim et al. (2004) reported ~27 dB SE at 40 wt% CNT loading for MWCNT/PMMA films. These observations suggest that SE of CNT filled polymer nanocomposite depends on many factors including fabrication techniques, purity of CNT, dispersion etc. Recently significant efforts have been made in understanding the EMI shielding mechanisms of polymer nanocomposite as it is very important for the best utilization of shielding capabilities of material and for designing nanocomposite at lowest possible filler loading and cost. Three types of EMI shielding mechanisms have been proposed, namely: reflection, absorption and multiple reflections [Sundararaj & Al-Saleh, 2009; Chung, 2001; Liu, 2007]. Some previous studies have shown that in MWCNT/polymer nanocomposites, SE is mainly absorption dominated where as SWCNT/polymer nanocomposites are mainly reflection dominated material. The application area of EMI shielding materials depends on the dominant shielding mechanism like absorption based EMI shielding materials are used in radar, microwave communication technology, stealth (self concealing) technology, microwave darkroom and anti-EMI coating application.

6.3. Thermal properties

Thermal properties of a composite are equally important as its mechanical and electrical properties because it gives more freedom in selecting processing conditions and also application area. It is observed that incorporation of CNTs in polymer matrices results in increase of glass transition, melting and thermal decomposition temperatures due to hindered chain and segmental mobility of the polymers. To improve the thermal endurance of CNT/polymer nanocomposites, surfactant as the wetting agent were also incorporated. Gong (2000) and Velasco-Santos et al. (2003) reported that addition of 1 wt.% CNTs with a surfactant in epoxy and PMMA matrix increased the glass transition temperature by ~25 and ~40º C respectively. Incorporation of CNTs in polymer matrices also enhances the rate of crystallization by acting as nucleating sites [Deshpande & Jape, 2009; Zhang, 2008] which in-turn reduces the processing time and enhances mechanical strength. There are reports according to which addition of CNTs in polymer matrix can hinder the crystallization of polymer matrices (Jin, 2007; Diez-Pascual, 2009). Our recent study on MWCNT/PTT nanocomposite suggested that presence of MWCNTs in PTT matrix delays the crystallization and lead to formation of bigger crystallites (Gupta & Choudhary, 2010).

Thermal stability and melting temperature of polymer matrices were improved in presence of CNTs.Kashiwagi et al. (2000) found that the thermal decomposition temperature of polypropylene in nitrogen increases by ~12º C on 2 vol. % MWCNTs loading. Because of the excellent thermal conductivity of CNTs, incorporation of CNTs significantly improves the thermal transport properties of polymer nanocomposites which lead to its usage as printed circuit boards, connectors, thermal interface materials, heat sinks, and other high-performance thermal management systems. Choi et al. (2003) reported 300% increase in thermal conductivity of epoxy matrix at room temperature on 3 wt % SWCNT loading and an additional increase (10%) when aligned magnetically. Biercuk et al. (2002) prepared an epoxy nanocomposite with 1 wt % raw (not purified) laser-oven SWCNT and showed a 125% increase in thermal conductivity at room temperature. Alignment of CNTs also plays an important role in improving the transport properties of a material. Choi et al. (2003) found 10% increase in thermal conductivity of epoxy composite with aligned MWCNTs in comparison to non-aligned MWCNTs.

7. APPLICATION OF CNT/POLYMER NANOCOMPOSITES

With their excellent range of properties, CNTs have opened up a new age of advanced multifunctional materials. Incorporation of CNTs in polymer matrices provides materials that could be used for many high performance engineering applications. Currently, the most widespread use of CNT nanocomposites is in electronics. These nanocomposites could be used to shield electromagnetic interference and as electrostatic-discharge components. The microwave-absorbing capability of nanotubes could be exploited to heat temporary housing structures and may have applications in space exploration. Thin layers of nanotubes on plastics might also be used in transparent conducting composites. High mechanical strength of these nanocomposites could be utilized to make some high-end sporting goods such as tennis rackets, baseball bat etc, and thus delivering superior performance. In short, the biggest market for CNT nanocomposites will undoubtedly be for high-value applications that can absorb the added costs, which includes commercial sectors such as electronics especially aerospace (which requires lightweight, high-strength, high-temperature-resistant composites) and energy (for example, in nanotube-reinforced rubber seals for large oil recovery platforms). Once the cost of nanotubes becomes comparable to that of carbon fibre (or even to that of the much cheaper reinforcing agent, carbon black), commodities such as nanotube-filled rubber tyres could become a reality.

8. CONCLUSION AND FUTURE SCOPE

Studies on CNT/polymer nanocomposite suggests that CNT has great potential in altering the properties of polymer matrices. The quality of CNT/polymer nanocomposites depends on many factors including type of CNTs, chirality, purity, defect density, aspect ratio, % loading, dispersion, alignment and interfacial adhesion between the nanotube and polymer matrix. A lot of research work has been carried out to improve the quality of CNTs and processing techniques. The biggest challenge in realizing the full potential of CNTs is to achieve homogeneous dispersion of CNTs so that maximum filler surface area will be available for load-transfer between filler and matrix. The functionalization of nanotubes provides a convenient route to improve dispersion and stress transfer between CNT and polymer matrix, but more improvement in this field is required to preserve the intrinsic properties of CNTs. It is important to focus on different methods of noncovalent functionalization of nanotubes and discover a route which can improve the dispersion and compatibility without negatively affecting the composite properties. The actual task of dispersing the CNTs in polymer matrix is performed during its manufacturing. The three major processing techniques namely, solution, melt and in-situ polymerization have their unique advantages in fabricating CNT/polymer nanocomposite. Although solution blending produces high quality composite but melt compounding is much simpler and provide option for large scale production. Recently, in-situ polymerization is also showing great potential in fabricating CNT/polymer nanocomposite. The success of processing technique is directly related to performance of composite. The maximum improvement in mechanical properties of polymer matrix is observed in case of in-situ polymerization which produces covalent bond between CNT and polymer matrix. However it negatively influences the electronic properties of the composite. The increase in electrical conductivity of polymer material on CNT addition is the biggest advantage of fabricating CNT/polymer nanocomposite. As significant improvement in electrical conductivity is observed at very low CNT loading, CNT/polymer nanocomposite is finding application as light weight, low cost and highly effective ESD and EMI shielding material.

The thermal properties of the polymer matrix also modified by CNT addition like CNT increases the glass transition, melting and thermal decomposition temperatures. CNT also influences the crystallization rate and percentage crystallinity by acting as nucleating agent. Improvement in both crystallization rate and percentage crystallinity enhances its mechanical and processing properties. So finally we can conclude that CNT is ideal filler for fabricating polymer composite but some serious challenges need to be addressed before fully realizing the extraordinary properties of CNT in polymer nanocomposite.

REFERENCES

1. P. M. Ajayan, O. Stephan, C. Colliex, D. Trauth, 1994 Aligned carbon nanotube arrays formed by cutting a polymer resin-nanotube composite. Science, 265 12121214 , 0036-8075

2. P. M. Ajayan, L. S. Schadler, C. Giannaris, A. Rubio, 2000 Single-walled carbon nanotube-polymer composites: strength and weakness. Adv. Mater., 12 750753 , 0935-9648

3. K. Awasthi, A. Srivastava, O. N. Srivastava, 2005 Synthesis of carbon nanotubes. J Nanosci Nanotechnol, 5 16161636 , 1550-7033

4. S. Badaire, P. Poulin, M. Maugey, C. Zakri, 2004 In Situ Measurements of Nanotube Dimensions in Suspensions by Depolarized Dynamic Light Scattering. Langmuir, 20 1036710370 , 0743-7463

5. J. B. Bai, A. Allaoui, 2003 Effect of the length and the aggregate size of MWNTs on the improvement efficiency of the mechanical and electrical properties of composites- experimental investigation. Composites, Part A, 34A 689694 , 0135-9835X

6. S. Bal, S. S. Samal, 2007 Carbon nanotube reinforced polymer composites- a state of the art. Bull. Mater. Sci., 30 379386 , 0250-4707

7. R. Bandyopadhyaya, E. Nativ-Roth, O. Regev, R. Yerushalmi-Rozen, 2002 Stabilization of individual carbon nanotubes in aqueous solutions. Nano Lett., 2 2528 , 1530-6984

8. S. Banerjee, T. Hemraj-Benny, S. S. Wong, 2005 Covalent surface chemistry of single-walled carbon nanotubes. Adv. Mater., 17 1729 , 0935-9648

9. S. Barrau, P. Demont, E. Perez, A. Peigney, C. Laurent, C. Lacabanne, 2003 Effect of palmitic acid on the electrical conductivity of carbon nanotubes-epoxy resin composites. Macromolecules, 36 96789680 , 0024-9297

10. R. H. Baughman, A. A. Zakhidov, W. A. de Heer, 2002 Carbon nanotubes-the route toward applications. Science, 297 787792 , 0036-8075

11. W. Bauhofer, J. Z. Kovacs, 2009 A review and analysis of electrical percolation in carbon nanotube polymer composites. Compos. Sci. Technol., 69 14861498 , 0000-0266- 3538

12. J. M. Benoit, B. Corraze, S. Lefrant, W. J. Blau, P. Bernier, O. Chauvet, 2001 Transport properties of PMMA-Carbon Nanotubes composites. Synth. Met., 121 12151216 , 0379-6779

13. A. R. Bhattacharyya, T. V. Sreekumar, T. Liu, S. Kumar, L. M. Ericson, R. H. Hauge, 2003 Crystallization and orientation studies in polypropylene/single wall carbon nanotube composite. Polymer,44 23732377 , 0032-3861

14. M. J. Biercuk, M. C. Llaguno, M. Radosavljevic, J. K. Hyun, A. T. Johnson, J. E. Fischer, 2002 Carbon nanotube composites for thermal management. Appl. Phys. Lett., 80 27672769 , 0003-6951

15. R. Blake, Y. K. Gun'ko, J. Coleman, M. Cadek, A. Fonseca, J. B. Nagy, 2004 A Generic Organometallic Approach toward Ultra-Strong Carbon Nanotube Polymer Composites. J. Am. Chem. Soc., 126 1022610227 , 0002-7863

16. Bryning. M. B. (a, D. E. Milkie, M. F. Islam, J. M. Kikkawa, A. G. Yodh, 2005 Thermal conductivity and interfacial resistance in single-wall carbon nanotube epoxy composites. Appl. Phys. Lett., 87 161909 161901-161909/161903, 0003-6951

17. Bryning. M. B. (b, M. F. Islam, J. M. Kikkawa, A. G. Yodh, 2005 Very low conductivity threshold in bulk isotropic single-walled carbon nanotube-epoxy composites. Adv. Mater., 17 11861191 , 0935-9648

18. M. Cadek, J. N. Coleman, V. Barron, K. Hedicke, W. J. Blau, 2002 Morphological and mechanical properties of carbon-nanotube-reinforced semicrystalline and amorphous polymer composites. Appl. Phys. Lett., 81 51235125 , 0003-6951

19. L. Cai, H. Tabata, T. Kawai, 2000 Self-assembled DNA networks and their electrical conductivity. Appl. Phys. Lett., 77 31053106 , 0003-6951

20. J. C. Charlier, J. P. Issi, 1996 Electrical conductivity of novel forms of carbon. J. Phys. Chem. Solids, 57 957965 , 0022-3697

21. F. Cheng, P. Imin, C. Maunders, G. Botton, A. Adronov, 2008 Soluble, discrete supramolecular complexes of single-walled carbon nanotubes with fluorene-based conjugated polymers. Macromolecules, 41 23042308 , 0024-9297

22. E. S. Choi, J. S. Brooks, D. L. Eaton, M. S. Al-Haik, M. Y. Hussaini, H. Garmestani, 2003 Enhancement of thermal and electrical properties of carbon nanotube polymer composites by magnetic field processing. J. Appl. Phys., 94 60346039 , 0021-8979

23. D. D. L. Chung, 2001 Electromagnetic interference shielding effectiveness of carbon materials. Carbon, 39 279285 , 0008-6223

24. J. N. Coleman, A. B. Dalton, S. Curran, A. Rubio, A. P. Davey, A. Drury, 2000 Phase separation of carbon nanotubes and turbostratic graphite using a functional organic polymer. Adv. Mater., 12 213216 , 0935-9648

25. J. N. Coleman, U. Khan, W. J. Blau, Y. K. Gunko, 2006 Small but strong: A review of the mechanical properties of carbon nanotube-polymer composites. Carbon, 44 16241652 , 0008-6223

26. E. Camponeschi, R. Vance, M. Al-Haik, H. Garmestani, R. Tannenbaum, 2007 Properties of carbon nanotube-polymer composites aligned in a magnetic field. Carbon, 45 20372046 , 0008-6223

27. S. A. Curran, P. M. Ajayan, W. J. Blau, D. L. Carroll, J. N. Coleman, A. B. Dalton, 1998 A composite from poly(m-phenylenevinylene-co-2,5-dioctoxy-p-phenylenevinylene) and carbon nanotubes. A novel material for molecular optoelectronics. Adv. Mater., 10 10911093 , 0935-9648

28. H. Dai, E. W. Wong, C. M. Lieber, 1996 Probing electrical transport in nanomaterials: conductivity of individual carbon nanotubes. Science, 272 523526 , 0036-8075

29. V. D. Deshpande, S. Jape, 2009 Nonisothermal crystallization kinetics of nucleated poly(ethylene terephthalate). J. Appl. Polym. Sci., 111 13181327 , 0021-8995

30. A. M. Dıez-Pascual, M. Naffakh, M. A. Gomez, C. Marco, G. Ellis, B. Simard, 2009 2009). Development and characterization of PEEK/ carbon nanotube composites.

31. Vol. Carbon, 47 7, 30793090 , 0008-6223

32. M. S. Dresselhaus, 2004 Applied physics: Nanotube antennas. Nature, 432 959960 , 0028-0836

33. F. Du, J. E. Fischer, K. I. Winey, 2005 Effect of nanotube alignment on percolation conductivity in carbon nanotube/polymer composites. Phys. Rev. B: Condens. Matter Mater. Phys., 72 121404 121401-121404/121404, 1098-0121

34. F. Du, J. E. Fischer, K. I. Winey, 2003 Coagulation method for preparing single-walled carbon nanotube/poly(methyl methacrylate) composites and their modulus, electrical conductivity, and thermal stability. J. Polym. Sci., Part B: Polym. Phys., 41 33333338 , 0887-6266

35. P. Dubois, M. Alexandre, 2006 Performant clay/carbon nanotube polymer composites. Adv. Eng. Mater., 8 147154 , 1438-1656

36. T. W. Ebbesen, H. J. Lezec, H. Hiura, J. W. Bennett, H. F. Ghaemi, T. Thio, 1996 Electrical conductivity of individual carbon nanotubes. Nature, 382 54-56, 0028-0836

37. J. Feng, J. Sui, W. Cai, Z. Gao, 2008 Microstructure and mechanical properties of carboxylated carbon nanotubes/poly(L-lactic acid) composite. J. Compos. Mater., 42 15871595 , 0021-9983

38. W. Feng, X. D. Bai, Y. Q. Lian, J. Liang, X. G. Wang, K. Yoshino, 2003 Well-aligned polyaniline/carbon-nanotube composite films grown by in-situ aniline polymerization. Carbon, 41 15511557 , 0008-6223

39. G. B. M. Fiege, A. Altes, R. Heiderhoff, L. J. Balk, 1999 Quantitative thermal conductivity measurements with nanometer resolution. J. Phys. D: Appl. Phys., 32 L13L17 , 0022-3727

40. J. E. Fischer, H. Dai, A. Thess, R. Lee, N. M. Hanjani, D. L. Dehaas, 1997 Metallic resistivity in crystalline ropes of single-wall carbon nanotubes. Phys. Rev. B: Condens. Matter, 55 R4921R4924 , 1098-0121

41. K. Fu, W. Huang, Y. Lin, L. A. Riddle, D. L. Carroll, Y. Sun, P. , 2001 Defunctionalization of Functionalized Carbon Nanotubes. Nano Lett., 1 439441 , 1530-6984

42. J. Gao, B. Zhao, M. E. Itkis, E. Bekyarova, H. Hu, V. Kranak, 2006 Chemical engineering of the single-walled carbon nanotube-nylon 6 interface. J. Am. Chem. Soc., 128 74927496 , 0002-7863

43. J. Gao, M. E. Itkis, A. Yu, E. Bekyarova, B. Zhao, R. C. Haddon, 2005 Continuous Spinning of a Single-Walled Carbon Nanotube-Nylon Composite Fiber. J. Am. Chem. Soc., 127 38473854 , 0002-7863

44. J. Gao, A. Yu, M. E. Itkis, E. Bekyarova, B. Zhao, S. Niyogi, 2004 Large-scale fabrication of aligned single-walled carbon nanotube array and hierarchical single-walled carbon nanotube assembly. J. Am. Chem. Soc., 126 1669816699 , 0002-7863

45. F. H. Gojny, M. H. G. Wichmann, B. Fiedler, W. Bauhofer, K. Schulte, 2005 Influence of nano-modification on the mechanical and electrical properties of conventional fibre-reinforced composites. Composites, Part A, 36A 15251535 , 0135-9835X

46. S. Gomes, N. Trannoy, P. Grossel, 1999 DC thermal microscopy: study of the thermal exchange between a probe and a sample. Meas. Sci. Technol., 10 805811 , 0957-0233

47. X. Gong, J. Liu, S. Baskaran, R. D. Voise, J. S. Young, 2000 Surfactant-assisted processing of carbon nanotube/polymer composites. Chem. Mater., 12 10491052 , 0897-4756

48. N. Grossiord, J. Loos, C. E. Koning, 2005 Strategies for dispersing carbon nanotubes in highly viscous polymers. J. Mater. Chem., 15 23492352 , 0959-9428

49. A. Gupta, V. Choudhary, 2010 Isothermal and non-Isothermal crystallization kinetics and morphology of poly(trimethylene terephthalate)/multiwalled carbon nanotube composites. Macromole. Symp., 290 5669 , 1022-1360

50. R. Haggenmueller, J. E. Fischer, K. I. Winey, 2006 Single wall carbon nanotube/polyethylene composites: nucleating and templating polyethylene crystallites. Macromolecules, 39 29642971 , 0024-9297

51. R. Haggenmueller, H. H. Gommans, A. G. Rinzler, J. E. Fischer, K. I. Winey, 2000 Aligned single-wall carbon nanotubes in composites by melt processing methods. Chem. Phys. Lett., 330 219225 , 0009-2614

52. R. Haggenmueller, W. Zhou, J. E. Fischer, K. I. Winey, 2003 Production and characterization of highly aligned single-walled carbon nanotubes in polymer composites. J. Nanosci.Nanotechnol., 3 105110 , 1550-7033

53. N. Hamada, S. Sawada, A. Oshiyama, 1992 New one-dimensional conductors: graphitic microtubules. Phys. Rev. Lett., 68 15791581 , 0031-9007

54. M. C. Hersam, A. C. F. Hoole, S. J. O'Shea, M. E. Welland, 1998 Potentiometry and repair of electrically stressed nanowires using atomic force microscopy. Appl. Phys. Lett., 72 915917 , 0003-6951

55. A. Hirsch, 2002 Functionalization of single-walled carbon nanotubes. Angewandte Chemie International Edition, 41 18531859 , 0570-0833

56. J. K. Holt, H. G. Park, Y. Wang, M. Stadermann, A. B. Artyukhin, C. P. Grigoropoulos, 2006 Fast mass transport through sub-2-nanometer carbon nanotubes. Science, 312 10341037 , 0036-8075

57. J. Hone, M. Whitney, A. Zettl, 1999 Thermal conductivity of single-walled carbon nanotubes. Synth. Met., 103 24982499 , 0379-6779

58. H. Hou, J. J. Ge, J. Zeng, Q. Li, D. H. Reneker, A. Greiner, 2005 Electrospun polyacrylonitrile nanofibers containing a high concentration of well-aligned multiwall carbon nanotubes. Chem. Mater., 17 967973 , 0897-4756

59. G. L. Hwang, Y. Shieh, T. , K. C. Hwang, 2004 Efficient load transfer to polymer-grafted multiwalled carbon nanotubes in polymer composites. Adv. Funct. Mater., 14 487491 , 0161-6301X

60. M. F. Islam, E. Rojas, D. M. Bergey, A. T. Johnson, A. G. Yodh, 2003 High weight fraction surfactant solubilization of single-wall carbon nanotubes in water. Nano Lett., 3 269273 , 1530-6984

61. S. Iijima, 1991 Helical microtubules of graphitic carbon. Nature, 354 5658 , 0028-0836

62. H. Ishikawa, S. Fudetani, M. Hirohashi, 2001 Mechanical properties of thin films measured by nanoindenters. Appl. Surf. Sci., 178 5662 , 0169-4332

63. Z. Jia, Z. Wang, C. Xu, J. Liang, B. Wei, D. Wu, 1999 Study on poly(methyl methacrylate)/carbon nanotube composites. Mater. Sci. Eng., A, A271 395400 , 0921-5093

64. J. Jin, M. Song, F. Pan, 2007 A DSC study of effect of carbon nanotubes on crystallization behaviour of poly(ethylene oxide). Thermo. Acta, 45 2531 , 0040-6031

65. Z. Jin, K. P. Pramoda, S. H. Goh, G. Xu, 2002 Poly(vinylidene fluoride)-assisted melt-blending of multi-walled carbon nanotube/poly(methyl methacrylate) composites. Mater. Res. Bull., 37 271278 , 0025-5408

66. L. Jin, C. Bower, O. Zhoua, 1998 Alignment of carbon nanotubes in a polymer matrix by mechanical stretching. Appl. Phys. Lett., 73 11971202 , 0003-6951

67. J. Joo, A. J. Epstein, 1994 Electromagnetic radiation shielding by intrinsically conducting polymers. Appl. Phys. Lett., 65 22782280 , 0003-6951

68. T. Kashiwagi, J. W. Gilman, K. Butler, R. H. Harris, J. R. Shields, A. Asano, 2000 Flame retardant mechanism of silica gel/silica. Fire Mater., 24 277289 , 0308-0501

69. C. L. Kane, E. J. Mele, 1997 Size, shape, and low energy electronic structure of carbon nanotubes. Phys. Rev. Lett., 78 19321935 , 0031-9007

70. P. Kim, L. Shi, A. Majumdar, P. L. Mc Euen, 2001 Thermal transport measurements of individual multiwalled nanotubes. Phys. Rev. Lett., 87 215502 215501-215502/215504, 0031-9007

71. H. M. Kim, K. Kim, C. Y. Lee, J. Joo, S. J. Cho, H. S. Yoon, 2004 Electrical conductivity and electromagnetic interference shielding of multiwalled carbon nanotube composites containing Fe catalyst. Appl. Phys. Lett., 84 589591 , 0003-6951

72. S. H. Kim, S. H. Jang, S. W. Byun, J. Y. Lee, J. S. Joo, S. H. Jeong, 2003 Electrical properties and EMI shielding characteristics of polypyrrole-nylon 6 composite fabrics. J. Appl. Polym. Sci., 87 19691974 , 0021-8995

73. F. Ko, Y. Gogotsi, A. Ali, N. Naguib, H. Ye, G. Yang, 2003 Electrospinning of continuous carbon nanotube-filled nanofiber yarns. Adv. Mater., 15 11611165 , 0935-9648

74. B. Kracke, B. Damaschke, 2000 Measurement of nanohardness and nanoelasticity of thin gold films with scanning force microscope. Appl. Phys. Lett., 77 361363 , 0003-6951

75. Chapelle. M. Lamy la, C. Stephan, T. P. Nguyen, S. Lefrant, C. Journet, P. Bernier, 1999 Raman characterization of singlewalled carbon nanotubes-PMMA composites. Synth. Met., 103 25102512 , 0379-6779

76. J. Luo, T. , H. Wen, C. , W. Wu, F. , C. Chou, P. , 2008 Mechanical research of carbon nanotubes/PMMA composite films. Polym. Compos., 29 12851290 , 0272-8397

77. P. Liu, 2005 Modifications of carbon nanotubes with polymers. Eur. Polym. J., 41 26932703 , 0014-3057

78. Z. Liu, G. Bai, Y. Huang, Y. , F. Du, F. Li, 2007 Reflection and absorption contributions to the electromagnetic interference shielding of single-walled carbon nanotube/polyurethane composites. Carbon, 45 821827 , 0008-6223

79. T. Liu, I. Y. Phang, L. Shen, S. Y. Chow, W. Zhang-D, 2004 Morphology and Mechanical Properties of Multiwalled Carbon Nanotubes Reinforced Nylon-6 Composites. Macromolecules, 37 72147222 , 0024-9297

80. C. , W. Zhang, Y. Zhu, L. Ji, R. Zhang, N.. Koratkar, J. Liang, 2006 Alignment and dispersion of functionalized carbon nanotubes in polymer composites induced by an electric field. Carbon, 46 706720 , 0008-6223

81. R. Martel, T. Schmidt, H. R. Shea, T. Hertel, P. Avouris, 1998 Single- and multi-wall carbon nanotube field-effect transistors. Appl. Phys. Lett., 73 24472449 , 0003-6951

82. C. A. Martin, J. K. W. Sandler, M. S. P. Shaffer, M. K. Schwarz, W. Bauhofer, K. Schulte, 2004 Formation of percolating networks in multi-wall carbon-nanotube-epoxy composites. Compos. Sci. Technol., 64 23092316 , 0000-0266- 3538

83. O. Matarredona, H. Rhoads, Z. Li, J. H. Harwell, L. Balzano, D. E. Resasco, 2003 Dispersion of single-walled carbon nanotubes in aqueous solutions of the anionic surfactant NaDDBS. J. Phys. Chem. B, 107 1335713367 , 1089-5647

84. D. Micheli, R. Pastore, C. Apollo, M. Marchetti, G. Gradoni, F. Moglie, 2009 Carbon based nanomaterial composites in RAM and microwave shielding applications, 9th IEEE Conf. Nanotechnol., 226235 , 978-1-42444-832-6 Genoa, 26-30 July, 2009

85. E. Minoux, O. Groening, K. B. K. Teo, S. H. Dalal, L. Gangloff, J. Schnell, P. , 2005 Achieving high-current carbon nanotube emitters. Nano Lett., 5 21352138 , 1530-6984

86. J. W. Mintmire, B. I. Dunlap, C. T. White, 1992 Are fullerene tubules metallic? Phys. Rev. Lett., 68 631634 , 0031-9007

87. Moniruzzaman. M. (a, F. Du, N. Romero, K. I. Winey, 2006 Increased flexural modulus and strength in SWNT/epoxy composites by a new fabrication method. Polymer, 47 293298 , 0032-3861

88. Moniruzzaman. M. (b, K. I. Winey, 2006 Polymer composites containing carbon nanotubes. Macromolecules, 39 51945205 , 0024-9297

89. M. Monthioux, 2002 Filling single-wall carbon nanotubes. Carbon, 40 18091823 , 0008-6223

90. V. C. Moore, M. S. Strano, E. H. Haroz, R. H. Hauge, R. E. Smalley, J. Schmidt, 2003 Individually suspended single-walled carbon

nanotubes in various surfactants. Nano Lett., 3 13791382 , 1530-6984

91. M. J. O'Connell, P. Boul, L. M. Ericson, C. Huffman, Y. Wang, E. Haroz, 2001 Reversible water-solubilization of single-walled carbon nanotubes by polymer wrapping. Chem. Phys. Lett., 342 265271 , 0009-2614

92. T. W. Odom, J. Huang, P. Kim, C. M. Lieber, 1998 Atomic structure and electronic properties of single-walled carbon nanotubes. Nature, 391 6264 , 0028-0836

93. C. Park, Z. Ounaies, K. A. Watson, R. E. Crooks, J. Smith, S. E. Lowther, 2002 Dispersion of single wall carbon nanotubes by in-situ polymerization under sonication. Chem. Phys. Lett., 364 303308 , 0009-2614

94. O. Park-K, T. Jeevananda, N. H. Kim, S.-i. Kim, J. H. Lee, 2009 Effects of surface modification on the dispersion and electrical conductivity of carbon nanotube/polyaniline composites. Scr. Mater., 60 551554 , 1359-6462

95. X. Pei, L. Hu, W. Liu, J. Hao, 2008 Synthesis of water-soluble carbon nanotubes via surface initiated redox polymerization and their tribological properties as water-based lubricant additive. Eur. Polym. J., 44 24582464 , 0014-3057

96. P. Poetschke, A. R. Bhattacharyya, A. Janke, H. Goering, 2003 Melt mixing of polycarbonate/multi-wall carbon nanotube composites. Compos. Interfaces, 389404 , 1568-5543

97. K. W. Putz, C. A. Mitchell, R. Krishnamoorti, P. F. Green, 2004 Elastic modulus of single-walled carbon nanotube/poly(methyl methacrylate) composites. J. Polym. Sci., Part B: Polym. Phys., 42 22862293 , 0887-6266

98. D. Qian, E. C. Dickey, R. Andrews, T. Rantell, 2000 Load transfer and deformation mechanisms in carbon nanotube-polystyrene composites. Appl. Phys. Lett., 76 28682870 , 0003-6951

99. S. Qin, D. Qin, W. T. Ford, D. E. Resasco, J. E. Herrera, 2004 Functionalization of single-walled carbon nanotubes with polystyrene via grafting to and grafting from methods. Macromolecules, 37 752757 , 0024-9297

100. N. R. Raravikar, L. S. Schadler, A. Vijayaraghavan, Y. Zhao, B. Wei, P. M. Ajayan, 2005 Synthesis and characterization of thickness-aligned carbon nanotube-polymer composite films. Chem. Mater., 17 974983 , 0897-4756

101. Rasheed. A. (a, M. D. Dadmun, P. F. Britt, 2006 Polymer-nanofiber composites: enhancing composite properties by nanofiber oxidation. J. Polym. Sci., Part B: Polym. Phys., 44 30533061 , 0887-6266

102. Rasheed. A. (b, H. G. Chae, S. Kumar, M. D. Dadmun, 2006 Polymer nanotube composites: Correlating intermolecular interaction to ultimate properties. Polymer, 47 47344741 , 0032-3861

103. O. Regev, P. N. B. El Kati, J. Loos, C. E. Koning, 2004 Preparation of conductive nanotube-polymer composites using latex technology. Adv. Mater., 16 248251 , 0935-9648

104. F. Ruiz, W. D. Sun, F. H. Pollak, C. Venkatraman, 1998 Determination of the thermal conductivity of diamond-like composite films using a scanning thermal microscope. Appl. Phys. Lett., 73 18021804 , 0003-6951

105. B. Safadi, R. Andrews, E. A. Grulke, 2002 Multiwalled carbon nanotube polymer composites: synthesis and characterization of thin films. J. Appl. Polym. Sci., 84 26602669 , 0021-8995

106. N. G. Sahoo, Y. C. Jung, H. J. Yoo, J. W. Cho, 2006 Effect of functionalized carbon nanotubes on molecular interaction and properties of polyurethane composites. Macromol. Chem. Phys., 207 17731780 , 1521-3935

107. R. Saito, M. Fujita, G. Dresselhaus, M. S. Dresselhaus, 1992 Electronic structure of chiral graphene tubules. Appl. Phys. Lett., 60 22042206 , 0003-6951

108. L. S. Schadler, S. C. Giannaris, P. M. Ajayan, 1998 Load transfer in carbon nanotube epoxy composites. Appl. Phys. Lett., 73 38423844 , 0003-6951

109. P. Singh, S. Campidelli, S. Giordani, D. Bonifazi, A. Bianco, M. Prato, 2009 Organic functionalisation and characterisation of single-walled carbon nanotubes. Chem Soc Rev, 38 22142230 , 0306-0012

110. E. J. Siochi, D. C. Working, C. Park, P. T. Lillehei, J. H. Rouse, C. C. Topping, 2004 Melt processing of SWCNT-polyimide composite fibers. Composites, Part B, 35B 439446 , 0000-1359- 8368

111. G. Sui, W. H. Zhong, X. P. Yang, Y. H. Yu, S. H. Zhao, 2008 Preparation and properties of natural rubber composites reinforced with pretreated carbon nanotubes. Polym. Adv. Technol., 19 15431549 , 0000-1042- 7147

112. A. B. Sulong, N. Muhamad, J. Sahari, R. Ramli, B. Deros, J. Park, 2009 Electrical conductivity behavior of chemical functionalized MWCNTs epoxy composites. European Journal of Scientific Research. 29 1321 , 0145-0216X

113. L. Sun, G. L. Warren, J. Y. O'Reilly, W. N. Everett, S. M. Lee, D. Davis, 2008 Mechanical properties of surface-functionalized SWCNT/epoxy composites. Carbon, 46 320328 , 0008-6223

114. U. Sundararaj, M. H. Al-Saleh, 2009 Electromagnetic interference shielding mechanisms of CNT/polymer composites. Carbon, 47 17381746 , 0008-6223

115. S. J. Tans, M. H. Devoret, H. Dal, A. Thess, R. E. Smalley, L. J. Geerligs, 1997 Individual single-wall carbon nanotubes as quantum wires. Nature, 386 474477 , 0028-0836

116. E. T. Thostenson, Z. Ren, T. W. Chou, 2001 Advances in the science and technology of carbon nanotubes and their composites: a review. Compos. Sci. Technol., 61 18991912 , 0000-0266- 3538

117. E. T. Thostenson, T. Chou, W. , 2002 Aligned multi-walled carbon nanotube-reinforced composites: processing and mechanical characterization. J. Phys. D: Appl. Phys., 35 L77L80 , 0022-3727

118. C. Velasco-Santos, A. L. Martinez-Hernandez, F. T. Fisher, R. Ruoff, V. M. Castano, 2003 Improvement of thermal and mechanical properties of carbon nanotube composites through chemical functionalization. Chem. Mater., 15 44704475 , 0897-4756

119. B. Vigolo, A. Penicaud, C. Coulon, C. Sauder, R. Pailler, C. Journet, 2000 Macroscopic fibers and ribbons of oriented carbon nanotubes. Science, 290 13311334 , 0036-8075

120. K. Wong, K. , S. Shi, Q. , K. Lau, T. , 2007 Mechanical and thermal behavior of a polymer composite reinforced with functionalized carbon nanotubes. Key Eng. Mater., 334-335 , 705708 , 1013-9826

121. C. Wu, S. , 2009 Synthesis and characterization of poly(trimethylene terephthalate) composites incorporating multi-walled carbon nanotubes. J. Appl. Polym. Sci., 114 16331642 , 0021-8995

122. H. Wu, X. , X. Qiu, Q. , W. Cao, M. , Y. Lin, H. , R. Cai, F. , S. Qian, X. , 2007 Polymer-wrapped multiwalled carbon nanotubes synthesized via microwave-assisted in situ emulsion polymerization and their optical limiting properties. Carbon, 45 28662872 , 0008-6223

123. X. Xie, L. , Y. Mai, W. , X. Zhou, P. , 2005 Dispersion and alignment of carbon nanotubes in polymer matrix: A review. Mater. Sci. Eng., R, R49 89112 , 0092-7796X

124. Yang. Y. (a, M. C. Gupta, K. L. Dudley, R. W. Lawrence, 2005 Conductive carbon nanofiber-polymer foam structures. Adv. Mater., 17 19992003 , 0935-9648

125. Yang. Y. (b, M. C. Gupta, K. L. Dudley, R. W. Lawrence, 2005 Novel carbon nanotube-polystyrene foam composites for electromagnetic interference shielding. Nano Lett., 5 21312134 , 1530-6984

126. Yu. M. (a, F. , B. S. Files, S. Arepalli, R. S. Ruoff, 2000 Tensile loading of ropes of single wall carbon nanotubes and their mechanical properties. Phys. Rev. Lett., 84 55525555 , 0031-9007

127. Yu. M. (b, F. , O. Lourie, M. J. Dyer, K. Moloni, T. F. Kelly, R. S. Ruoff, 2000 Strength and breaking mechanism of multiwalled carbon nanotubes under tensile load. Science, 287 637640 , 0036-8075

128. Yuen. S. (a, M. , C. , C. M. , 2008 Morphological, electrical, and mechanical properties of multiwall carbon nanotube/polysilsesquioxane composite. J. Appl. Polym. Sci., 109 20002007 , 0021-8995

129. Yuen. S. (b, M. , C. , C. M. , C. Chuang, Y. , K. Yu, C. , S. Wu, Y. , C. Yang, C. , 2008 Effect of processing method on the shielding

effectiveness of electromagnetic interference of MWCNT/PMMA composites. Compos. Sci. Technol., 68 963968 , 0000-0266- 3538

130. Yuen. S. (c, M. , C. , C. M. , C. Teng, C. , H. Wu, H. , H. Kuan, C. , C. Chiang, L. , 2008 Molecular motion, morphology, and thermal properties of multiwall carbon nanotube/polysilsesquioxane composite. J. Polym. Sci., Part B: Polym. Phys., 46 472482 , 0887-6266

131. W. D. Zhang, L. Shen, I. Y. Phang, T. Liu, 2004 Carbon nanotubes reinforced nylon-6 composite prepared by simple melt-compounding. Macromolecules, 37 256259 , 0024-9297

132. M. Zhang, S. Fang, A. A. Zakhidov, S. B. Lee, A. E. Aliev, C. D. Williams, 2005 Strong, transparent, multifunctional, carbon nanotube sheets. Science, 309 12151219 , 0036-8075

133. S. Zhang, M. L. Minus, L. Zhu, C. Wong, P. , S. Kumar, 2008 Polymer transcrystallinity induced by carbon nanotubes. Polymer, 49 13561364 , 0032-3861

134. B. Zhao, J. Wang, Z. Li, P. Liu, D. Chen, Y. Zhang, 2008 Mechanical strength improvement of polypropylene threads modified by PVA/CNT composite coatings. Mater. Lett., 62 43804382 , 0016-7577X

135. J. Zhu, J. Kim, H. Peng, J. L. Margrave, V. N. Khabashesku, E. V. Barrera, 2003 Improving the dispersion and integration of single-walled carbon nanotubes in epoxy composites through functionalization. Nano Lett., 3 11071113 , 1530-6984

136. J. Zhu, H. Peng, F. Rodriguez-Macias, J. L. Margrave, V. N. Khabashesku, A. M. Imam, 2004 Reinforcing epoxy polymer composites through covalent integration of functionalized nanotubes. Adv. Funct. Mater., 14 643648 , 0161-6301X

2

CARBON NANOTUBE-REINFORCED THERMOTROPIC LIQUID CRYSTAL POLYMER NANOCOMPOSITES

Jun Young Kim [1,2]

[1]Material Laboratory, Corporate R&D Center, Samsung SDI Co. Ltd., 575 Shin-dong, Yeongtong-gu, Suwon-si, Gyeonggi-do, 443-731, Korea

[2]Department of Materials Science and Engineering, Massachusetts Institute of Technology, 77 Massachusetts Avenue, Cambridge, MA 02139, USA

ABSTRACT

This paper focuses on the fabrication via simple melt blending of thermotropic liquid crystal polyester (TLCP) nanocomposites reinforced with a very small quantity of modified carbon nanotube (CNT) and the unique effects of the modified CNT on the physical properties of the nanocomposites. The thermal, mechanical, and rheological properties of modified CNT-reinforced TLCP nanocomposites are highly dependent on the uniform dispersion of CNT and the interactions between the CNT and TLCP, which can be enhanced by chemical modification of the CNT, providing a design guide of CNT-reinforced TLCP nanocomposites with great potential for industrial uses.

KEYWORDS

carbon nanotube; nanocomposites; thermotropic liquid crystal polymer

1. INTRODUCTION

Liquid-crystalline polymers (LCPs) have a phase of matter intermediate between the isotropic liquid and crystalline solid states termed as a mesophase or mesomorphic phase, and this LC state offers broad perspectives for various application materials that exhibit anisotropic but liquid properties [1]. In general, LCPs can be distinguished as thermotropic LCPs or lyotropic LCPs, depending on whether they are in a LC state in the melt or in solution. Lyotropic LCPs exhibit liquid crystallinity in the solution, being controlled by the characteristics and temperature of solvent and the concentration of polymer, and they cannot exhibit liquid crystallinity in the melt because they degrade before melting and show a phase transition through the addition or removal of solvents [2,3]. For thermotropic LCPs, the phase transition to LC state is induced by a thermal process, being controlled by the melt temperature and thermal history, and they form thermally activated mesogenic phases that extend from the crystal melting temperature up to the isotropic temperature [2,3].

Thermotropic LCPs, that are a class of the materials used to produce anisotropic melts of relatively low viscosity, have been extensively investigated because of their excellent mechanical properties, superior chemical resistance, low gas and liquid permeability, low controllable coefficient of thermal expansion, excellent dimensional stability under temperature or in severe environment, and easy processability with high precision by extrusion and injection molding [2,3,4,5]. Thermotropic LCPs with high strength and stiffness due to their rigid-rod- like molecules can be preferentially oriented to form fibrils under elongational or shear flow during melt processing, and oriented fibrous structures are developed in the extruded thermotropic LCPs, resulting in self-reinforcing characteristics [6,7,8]. For this reason, thermotropic LCPs have received considerable attention both in the neat state and as reinforcing fillers for thermoplastic polymers, and much research has been performed to date both to displace conventional thermoplastic polymers and to develop commercial applications of the thermotropic LCPs and their composites such a high performance engineering plastics or fibers [9,10,11,12].

Carbon nanotubes (CNTs) have attracted a great deal of scientific interest as advanced materials since they were first reported by Iijima [13]. In particular, the excellent electronic properties, thermal conductivity and mechanical strength of CNTs have created a high level of activity in materials research for practical applications in a broad range of industries such as hydrogen storage, lithium batteries, fuel cells, field-emitting-diodes, gas sensors, and polymer nanocomposites [14,15,16,17,18,19,20,21,22]. Furthermore, a number of efforts have been made to develop high performance polymeric materials based on CNTs, with the benefit of nanotechnology. Fundamental research progress on the potential applications of CNTs suggests that CNTs can be regarded as promising reinforcements in polymer nanocomposites due to the combination of their unique excellent properties with high aspect ratio and small size [23,24,25,26,27,28]. However, due to their high cost and limited availability, only a few practical applications have been realized to date in the various industrial fields.

Polymer nanocomposites that constitute a new class of the materials based on the reinforcement of polymers using nanofillers have attracted a great deal of interest in fields ranging from the purely scientific to industrial applications due to the remarkable improvements in the overall properties achieved at low filler loadings. For the fabrication of polymer nanocomposites containing CNTs, some major goals to realize the potential application of CNTs as nanoreinforcing fillers are: (a) homogeneous dispersion of the nanoutbes in the polymer matrix and (b) good interfacial adhesion between the nanotubes and the polymer matrix. Typically, CNTs tend to bundle together and to form agglomerations due to intrinsic van der Waals attractions between the individual tubes [24,25]. Weak interfacial bonding between CNTs and polymer matrix has limited efficient load transfer to polymers, playing a limited reinforcement role in the polymer nanocomposites [29,30,31]. The functionalization of CNTs, which can be considered as an effective method to achieve the uniform dispersion of CNTs and their compatibility with the polymers, can lead to the enhancement of the interfacial adhesion between CNTs and polymer matrix, thus improving physical properties of polymer nanocomposites containing CNTs [32,33,34]. In addition, one of major challenges for the fabrication of CNT/polymer nanocomposites is to optimize the processing operations for cheap manufacturing costs. Currently, four processing techniques are in common use to fabricate CNT/polymer nanocomposites [35,36,37,38,39,40]: direct mixing, solution method, *in situ* polymerization, and melt blending. Among these processing techniques, melt blending process has been accepted as the simplest and the most effective method from an industrial perspective, because this process makes it possible to fabricate high performance polymer nanocomposites at low process costs, and facilitates commercial scale-up [40,41,42,43,44,45,46,47,48,49]. Furthermore, the combination of a small quantity of CNTs with thermotropic LCPs provides attractive possibilities for improving the physical properties of polymer nanocomposites using a cost-effective method.

This paper focused on the fabrication and characterization of thermotropic liquid crystal copolyester (TLCP) nanocomposites reinforced with a very small quantity of CNTs using simple melt blending in a twin-screw extruder to create high performance composite materials with low processing cost. The effects of the unique characteristics of CNTs on the physical properties of TLCP/CNT nanocomposites are discussed. The attempt to disperse CNTs in TLCP matrix and to fabricate TLCP/CNT nanocomposites by simple melt compounding has been rarely investigated to date, and the effects of CNTs on the mechanical, rheological, and thermal properties of TLCP/CNT nanocomposites have not yet been reported in the literature. Thus, it is expected that this paper will be helpful in providing the preliminary understanding on the physical properties of TLCP nanocomposites reinforced with a very small quantity of CNTs. In addition, this paper suggests a simple and cost-effective method that will facilitate the industrial realization of TLCP nanocomposite with enhanced physical properties, providing a design guide of TLCP/CNT nanocomposites with a great industrial use potential. The exhaustive description of all LCP composites containing CNTs is beyond of the scope of this paper, and for this readers are referred to recent, excellent review papers [50,51,52].

2. RESULTS AND DISCUSSION

2.1. Acid oxidation of carbon nanotubes

For polymer nanocomposites containing CNTs, homogeneous dispersion of CNTs in the polymer matrix and good interfacial adhesion between CNTs and the polymer matrix is required in order to overcome the limitations of efficient load transfer to polymers, playing a limited reinforcement role in polymer nanocomposites. In general, acid oxidation of CNTs is considered as an effective method to achieve the uniform dispersion of CNTs and their compatibility with the polymers, leading to the enhancement of the interfacial adhesion between CNTs and polymer matrix [32]. The carboxylic acid groups effectively introduced on the surface of CNTs increase the chemical affinity of the thus modified CNT with the polymer matrix. As shown in Figure 1(a), the characteristic peak at ~1,570 cm^{-1} was attributed to the IR-phonon mode of multi-walled CNTs [53]. For the modified CNTs, characteristic peaks were observed at 1,090, 1,189, 1,399, and 1,723 cm^{-1}, corresponding to the stretching vibrations of the carboxylic acid groups

[54]. Similar patterns observed in the Raman spectra of pristine and modified CNTs indicated that this chemical modification did not affect the graphite structure of CNTs (see Figure 1c). Two characteristic peaks observed at around 1,350 and 1,570 cm^{-1} can be termed as *D*-band and *G*-band, respectively [55,56].

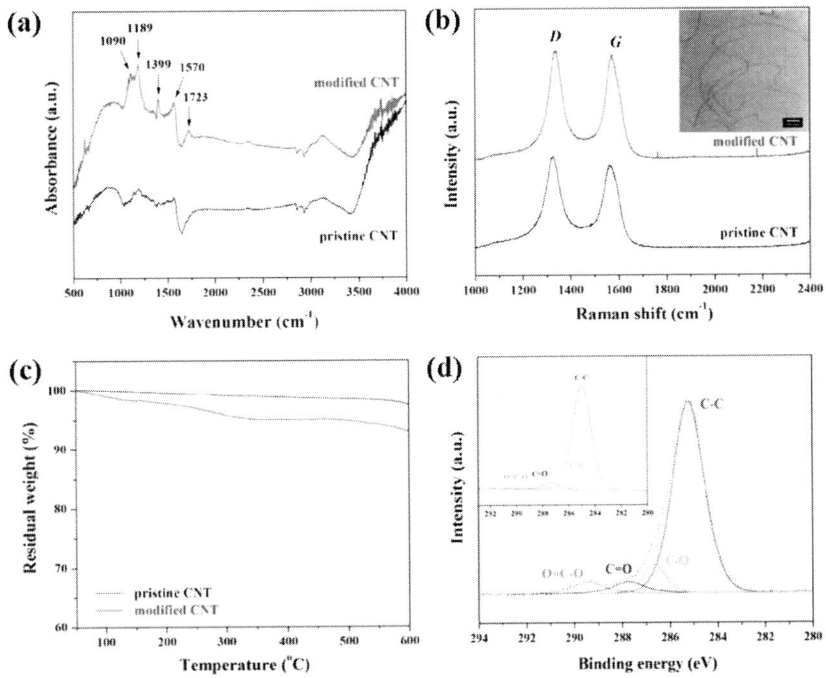

Figure 1. Characterization of modified CNT and TLCP nanocomposites. (a) FT-IR spectra of pristine and modified CNT. (b) Raman spectra of pristine and modified CNT. The decrease in the relative intensity ratio of *G*-band to *D*-band (I_G/I_D) of modified CNT indicates the increase in the degree of disorder and the presence of the defects on the surface of modified CNT by chemical modification. The inset shows TEM image of modified CNT. (c) TGA thermograms of pristine and modified CNT under N_2. (d) High resolution C_{1s} XPS spectra of TLCP nanocomposite containing 1.5 wt% of modified CNT. The inset shows high resolution C_{1s} XPS spectra of neat TLCP. The shifting of characteristic peaks indicates the interactions between modified CNT and TLCP macromolecular chains in TLCP nanocomposites. Reproduced with permission from Ref. [43]. © 2008 Elsevier Ltd.

The *G*-band, *i.e.*, the Raman-allowed phonon high-frequency mode, was related to the structural intensity of the sp^2-hybridized carbon atoms of CNTs, and the *D*-band reflected the disorder-induced carbon atoms, resulting from the defects in CNTs and their ends, and its intensity decreases with the degree of the graphitization of CNTs. The intensity of *D*-band was slightly increased in the modified CNT and the relative *G*-band to *D*-band ratio (I_G/I_D) of the modified CNT was slightly increased as compared to pristine CNT, which was attributed to the increase in the degree of disorder and the presence of the defects on the surface of modified CNT as a result of the chemical modification. As shown in Figure 1(c), for the modified CNT, the first weight loss observed below 100 °C in the TGA thermograms was attributed to the loss of water molecules in the nanocomposite samples [57]. In addition, the further gradual decrease in the residual weight was attributed to organic decomposition of thermally unstable functional groups formed on the surface of modified CNT [58].

This result demonstrates that carboxylic acid groups were effectively introduced on the surface of modified CNT via chemical modification. In general, CNTs often tend to bundle together and form agglomerations due to the intrinsic van der Waal attractions between the individual tubes in combination with their high aspect ratio and large surface area, thus leading to the obstruction of efficient load transfer to the polymer matrix [24,25]. The inset of Figure 1(b) shows that the modified CNT exhibits less entangled organization after chemical modification, instead of showing the agglomerates or bundles observed in pristine CNT. This feature makes it possible to enhance the dispersion of modified CNT relative to pristine CNT. Therefore, it is expected that the functional groups induced on the surface of CNT will be helpful for enhancing the dispersion state of CNT and the interfacial interaction in TLCP nanocomposites through hydrogen bonding formation, as illustrated in Figure 2. However, it was reported that there was a possibility to cause slight damage and a decreased length in the CNT structure induced by strong acid treatment, perhaps having a negative effect on the physical properties of the resulting nanocomposites [59]. In the XPS survey spectra, two characteristic peaks corresponding to $C_{1s}$and O_{1s} at around 285 and 523 eV were observed, which were assigned to the sp^2 carbon and the ester-type groups with oxygen bonded to carbon [60]. High resolution C_{1s} XPS spectra of TLCP and TLCP nanocomposite containing 1.5 wt% of modified CNT are shown in Figure 1(d). The C_{1s} spectra of neat TLCP exhibited four characteristic peaks at 285.0, 286.5, 287.5, and 289.0 eV, corresponding to the C–C, C–O, C=O, and O=C–O groups, respectively. However, the C_{1s} spectra of TLCP nanocomposites shifted to higher binding energy, and also revealed the presence of four peaks, corresponding to the C–C (285.3 eV), C–O (286.8 eV), C=O (287.8 eV), and O=C–O (289.6 eV) groups, respectively. This slight shifting of the characteristic peaks to higher bonding energy for TLCP nanocomposites was attributed to some interactions of the functional groups induced on the surface of modified CNT via chemical modification with TLCP macromolecular chains in TLCP nanocomposites.

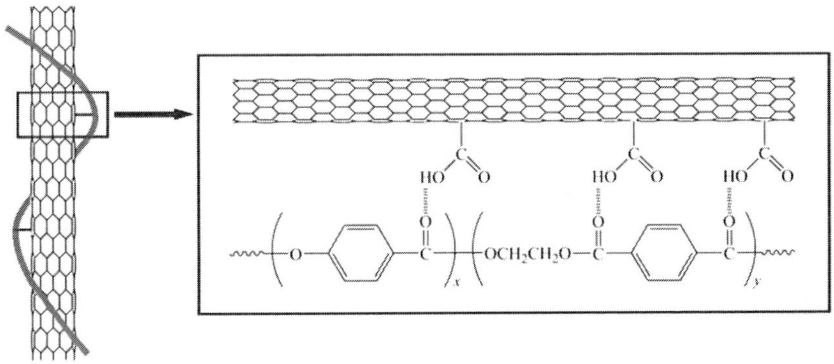

Figure 2. Schematic of possible interactions between modified CNT and TLCP though hydrogen bond formation.

2.2. Thermal stability

The thermal stability of polymer nanocomposites plays an important role in determining their working temperature limits and the environmental conditions for their use, which are related to their thermal decomposition temperature and decomposition rate. The patterns of TGA curves of TLCP nanocomposites are similar to those of pure TLCP, indicating that the thermal decomposition of TLCP nanocomposites must mostly stem from the TLCP. As shown in Figure 3, the incorporation of CNTs into TLCP matrix increased the thermal decomposition temperatures and the residual yields of TLCP nanocomposites. The presence of CNT could lead to the stabilization of TLCP matrix, resulting in the enhancement of the thermal stability of TLCP nanocomposites. The introduced CNT could induce protective barriers against thermal decomposition and retard the thermal decomposition of TLCP nanocomposites, resulting from the effective function of CNTs acting as physical barriers to hinder the transport of volatile decomposed products out of TLCP nanocomposites during thermal decomposition [61]. In addition, the incorporation of pristine and modified CNT into TLCP matrix increased the thermal stability of TLCP nanocomposites, and this enhancing effect was more significant in TLCP nanocomposites containing modified CNTs. For TLCP nanocomposites containing modified CNTs, good interfacial adhesion between modified CNT and TLCP matrix restricted the thermal motion of TLCP macromolecules [62], resulting in further enhancement of the thermal stability. TGA kinetic analysis was also conducted on TLCP nanocomposites to clarify the effect of CNT on thermal stability of TLCP nanocomposites. The decomposition temperatures and decomposition kinetic parameters such as the initial decomposition temperatures, the integral procedure decomposition temperature, the temperature at maximum rate of the weight loss, and the activation energy for thermal decomposition are commonly used to determine the thermal stability of polymers or composites. The activation energy for thermal decomposition (E_a) of TLCP nanocomposites can be calculated from TGA thermograms by using the Horowits-Metzger integral kinetic method [63] as follows:

$$\ln\left[\ln(1-\alpha)^{-1}\right] = \frac{E_a \theta}{RT_{dm}^2} \quad (1)$$

where α is the weigh loss; E_a is the activation energy for thermal decomposition; T_{dm} is the temperature at the maximum rate of weight loss; θ is the variable auxiliary temperature defined as $\theta = T - T_{dm}$, and R is the universal gas constant. The $E_a$value can be estimated from the slope of the plot of $\ln[\ln(1 - \alpha)^{-1}]$ versus θ as shown in Figure 3(b). The E_a values of pure TLCP and TLCP/CNT nanocomposites were estimated to be 290.8, 293.7, 303.8, and 322.6 kJ/mol, respectively. As compared to pure TLCP, higher E_a value of TLCP nanocomposites suggested that they were more thermally stable than pure TLCP. In addition, TLCP nanocomposites containing modified CNT showed higher E_a values as compared to TLCP nanocomposites containing pristine CNT, suggesting that TLCP nanocomposites with more uniform dispersion of modified CNT were more thermally stable than those containing pristine CNT. This feature was also attributed to the interactions between modified CNT and TLCP matrix, which increase the activation of the thermal decomposition of TLCP matrix relative to pristine CNT. Similar observation has been reported that the enhanced thermal stability of CNT-reinforced poly(ethylene 2,6-naphthalate) (PEN) nanocomposites was associated with the increased interfacial interactions between modified CNT and PEN [41]. Because of excellent thermal conductivity of CNTs [64], the enhanced interfacial interactions between modified CNT and TLCP matrix resulted in the increased thermal conductivity of

TLCP nanocomposites, leading to the improvement in the thermal stability.

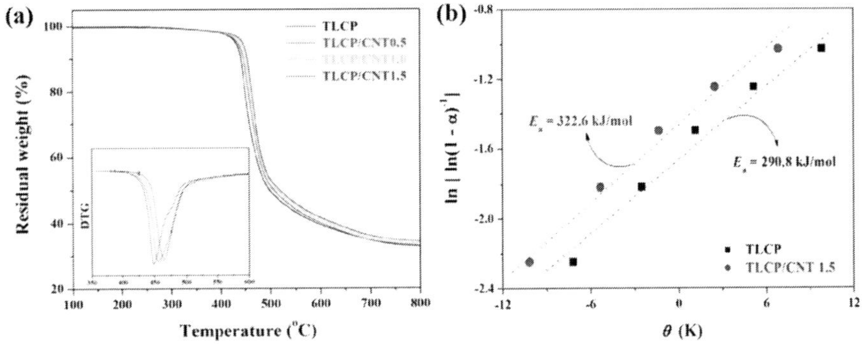

Figure 3. Thermal stability of TLCP nanocomposites. (a) TGA curves and (b) the plots of $\ln[\ln(1\text{-}\alpha)^{-1}]$ versus θ for TLCP nanocomposites. By Eq. (1), the slope provides an estimate of activation energy (E_a) for thermal decomposition of TLCP nanocomposites. Reproduced with permission from Ref. [43]. © 2008 Elsevier Ltd.

2.3. Rheological behavior

The rheological properties of TLCP nanocomposites are shown in Figure 4. The complex viscosities ($|\eta^*|$) of pure TLCP and TLCP nanocomposites decreased with increasing frequency, indicating a non-Newtonian behavior over the whole frequency range studied. The shear thinning behavior of TLCP was attributed to the random orientation of rigid molecular chains during shearing force. The TLCP nanocomposites exhibited higher $|\eta^*|$ than that of pure TLCP at low frequency, indicating that the interconnected or network-like structures of CNTs were formed in TLCP nanocomposites because of the nanotube-nanotube or nanotube-polymer interactions. In addition, TLCP nanocomposites exhibited strong shear thinning behavior, resulting from the break-down of these structures with increasing frequency. As shown in Figure 4(a), the enhancing effect of $|\eta^*|$ of TLCP nanocomposites with increasing CNT content was more pronounced at low frequency than that at high frequency. The irregular decrease in the $|\eta^*|$ with increasing frequency indicated the pseudo-plastic characteristics of TLCP nanocomposites due to random orientation and entangled molecules in

TLCP nanocomposites. As compared to TLCP nanocomposites containing pristine CNT, the complex viscosity of TLCP nanocomposites containing modified CNT was further increased due to the increased interactions between modified CNT and TLCP matrix. Furthermore, TLCP nanocomposites containing modified CNT exhibited higher $|\eta^*|$ and more distinct shear thinning behavior over the whole frequency range studied relative to pure TLCP and pristine CNT, suggesting either better dispersion of modified CNT or stronger nanotube-polymer interactions. Similar observation has been reported that higher viscosity and more distinct shear thinning behavior were indicative of stronger interfacial interactions in the CNT/epoxy nanocomposites with more uniform dispersion of CNTs [65]. The increase in the $|\eta^*|$ of TLCP nanocomposites was closely related to the large increase in the storage modulus. The shear thinning exponent (n) of TLCP nanocomposites can be estimated from the relationship of $|\eta^*| \approx \omega^{n}$ [66]. There is a significant dependence of shear thinning behavior of TLCP nanocomposites on the presence of CNT. The shear thinning exponents of TLCP nanocomposites slightly decreased by incorporating pristine and modified CNTs, and this effect was more pronounced in the case of modified CNT. As shown inFigure 4(d), the n values of TLCP nanocomposites significantly decreased with the introduction of modified CNT, which may be related to the enhancement of the interfacial interactions between modified CNT and TLCP matrix as well as the uniform dispersion of modified CNT. In addition, TLCP composites containing pristine and modified CNTs exhibited slightly different values of shear thinning exponent even at the same CNT content, indicating the influence of the modification on the extent of CNT dispersion. In this study, TLCP nanocomposites exhibited the relationship between shear thinning behavior and mechanical properties: the more the shear thinning behavior, the better reinforcing effect on the mechanical properties of TLCP nanocomposites. Wagner and Reisinger [67] reported that the tensile modulus of poly(butylene terephthalate) nanocomposites containing modified clay prepared via melt compounding was significantly enhanced with lower values of the shear thinning exponent. This result will be described in the following discussion of the mechanical properties of TLCP nanocomposites.

The storage modulus (G') and loss modulus (G'') of TLCP nanocomposites as a function of frequency are shown in Figures 4(b) and (c). The values of G' and G'' of TLCP nanocomposites significantly increased with increasing frequency and CNT content, and this enhancing effect was more pronounced at low frequency region. This feature is similar to the relaxation behavior of typical filled-polymer composite system [68,69]: if polymer chains are fully relaxed and exhibit a characteristic homopolymer-like terminal behavior, the flow curves of polymers can be expressed by a power law of $G' \propto \omega^2$ and $G'' \propto \omega$. Krisnamoorti *et al.* [70,71] reported that the slopes of $G'(\omega)$ and $G''(\omega)$ for layered silicate and polymer nanocomposite system were much smaller than 2 and 1, respectively, and they suggested that large deviations in the presence of a small quantity of layered silicate resulted from the formation of a network structure in polymer molten state. As shown in Figure 4(d), the variations of terminal zone slopes of TLCP nanocomposites indicated the non-terminal behavior with the power law dependence for $G'(\omega)$ and $G''(\omega)$ of TLCP nanocomposites. Similar non-terminal low-frequency behaviors have been also observed in the ordered block copolymers and the smectic liquid-crystalline small molecules [72,73]. The decrease in the slope of for TLCP nanocomposites with the introduction of CNT can be explained by the fact that the nanotube-nanotube or nanotube-polymer interactions can induce the formation of the interconnected or network-like structures, leading to the pseudo solid-like behavior more elasticity of TLCP nanocomposites relative to TLCP matrix. As the applied frequency increased, the interconnected or network-like structures were collapsed by high levels of shear force and TLCP nanocomposites exhibited almost similar or slightly higher G' and G'' values than that of pure TLCP at high frequency region. As compared to pure TLCP and TLCP nanocomposites containing pristine CNT, higher values of G' and G'' of TLCP nanocomposites containing modified CNT also suggested the increased interactions between modified CNT and TLCP matrix.

2.4. Morphology

The morphologies of TLCP/CNT nanocomposites are shown in Figure 5. In general, the drawbacks related to the uniform dispersion of CNT in the polymer matrix were caused by intrinsic van der Walls attractions between the individual nanotubes in combination with high aspect ratio and large surface area. This feature could make it difficult for CNT to disperse well in the polymer matrix and lead to more aggregated bundles of CNT, causing severe stress concentration phenomena and preventing efficient load transfer to the polymer matrix, thus resulting in some deterioration of the overall performance of the CNT/polymer nanocomposites. In this study, a chemical modification was performed in order to achieve both uniform dispersion of CNT in TLCP matrix and a stronger interfacial adhesion between CNT and TLCP. After chemical modification, CNT exhibited less entangled organization due to the functional groups formed on the nanotube surfaces instead of showing more agglomerated structures or bundles observed in the pristine CNT.

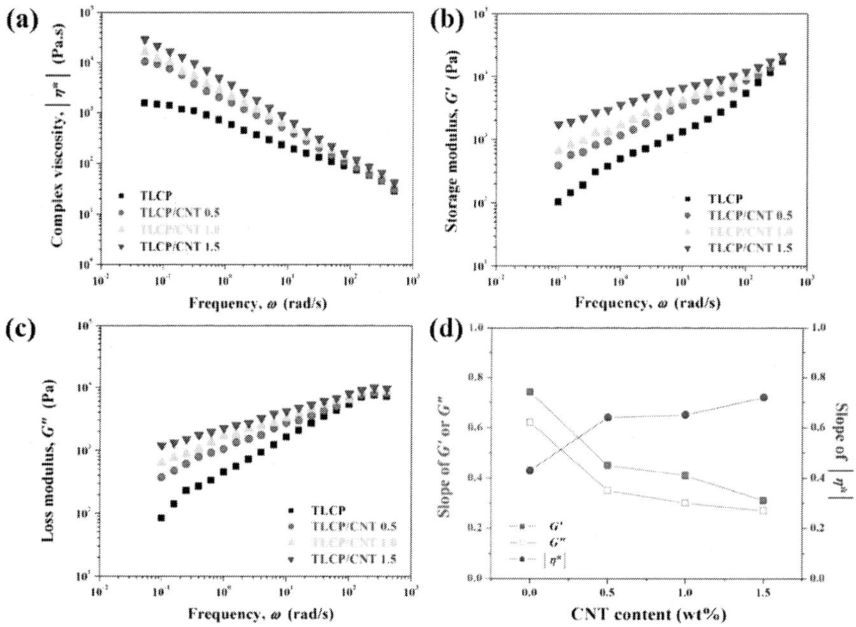

Figure 4. Rheological properties of TLCP nanocomposites. (a) Complex viscosity ($|\eta^*|$), (b) storage modulus (G'), and (c) loss modulus (G'') of TLCP nanocomposites as a function of applied frequency. Reproduced with permission from Ref. [43]. © 2008 Elsevier Ltd. (d) Variations of slopes of G', G'', and $|\eta^*|$ versus ω for TLCP nanocomposites with CNT content. Data from Ref. [43].

The modified CNT was, on a large scale, dispersed well in TLCP/CNT nanocomposites. Morphological observation was explained by the fact that CNT stabilized its dispersion by good interactions with TLCP, resulting from the interfacial interactions of –COOH groups of CNT with C=O groups of TLCP matrix. This interfacial adhesion between CNT and TLCP plays an important role in improving the overall mechanical properties of TLCP/CNT nanocomposites. It can be seen that two ends of CNT were covered by TLCP matrix despite a few CNT being pulled out from TLCP matrix. Interestingly, some of CNTs were broken with their two ends still strongly embedded in TLCP matrix, and bridging the local microcracks, which may delay the failure of polymer nanocomposites [74]. This result indicates that CNT has a good wetting with TLCP matrix, suggesting the existence of strong interactions between them. Similar observation has been reported that the presence of the fractured nanotubes, along with the polymer matrix still adhered to the fractured tubes matrix in terms of a crack interacting with the nanotube reinforcement significantly increased the mechanical properties of CNT/polystyrene nanocomposites [75]. For TLCP/CNT nanocomposites, the presence of functional groups on the surfaces of CNT resulted in the interfacial interaction between CNT and TLCP: the hydrogen atoms at –COOH groups of CNT may form hydrogen boding with C=O groups of TLCP molecular chains (Figure 2). Similar effect has been reported in CNT/polyamide (PA) nanocomposite systems that chemically modified CNT was effectively dispersed in PA matrix, because the functional groups of CNT probably have reacted with PA molecular chains during melt-mixing process, resulting in better dispersion state and stronger interfacial adhesion [76]. This result can be considered as the evidence for efficient load transfer from TLCP matrix to CNT during tensile testing, leading to the reinforcing effect on the overall mechanical properties of TLCP/CNT nanocomposites.

2.5. Mechanical behavior

For TLCP nanocomposites, there is a significant dependence of the mechanical properties on CNT content. The incorporation of CNT can substantially improve the mechanical properties of TLCP nanocomposites due to the nanoreinforcing effect of CNT with high aspect ratio and relatively uniform dispersion in TLCP matrix, and this enhancing effect was more pronounced at lower CNT content as compared to that at higher CNT content. At lower CNT contents, the incorporated CNTs were, on a large scale, uniformly dispersed in TLCP matrix, despite some entangled CNT or bundles. However, as compared to those at low CNT content, the mechanical properties of TLCP nanocomposites was not as significantly increased at higher CNT content as expected. Less uniformly dispersed and more entangled bundles of pristine CNT were formed in TLCP matrix at higher CNT content, which may be the cause of the stress concentration. This feature can be explained by the characteristics of CNTs that tend to bundle together due to the intrinsic van der Waals attractions in combination with high aspect ratio and large surface area [24,25] and can lead to some agglomeration, causing stress concentration phenomenon and preventing efficient load transfer from the reinforcing phase to the polymer matrix phase [26,27,28]. For PEN nanocomposites containing pristine CNT, the dependence of their mechanical propertieson CNT content has been also reported by Kim *et al.* [46,47,48,49]. Thus, for achieving further enhanced mechanical properties of TLCP nanocomposites, the enhancement of both uniform dispersion of nanotubes and the interfacial adhesion between nanotubes and polymer matrix should be required via functionalization of CNT. The mechanical properties of TLCP nanocomposites containing modified CNT are shown in Figure 6. The tensile strength and modulus of TLCP nanocomposites were substantially improved with the introduction of modified CNT by efficiently transferring load from the polymers to the nanotubes, and this enhancing effect was more pronounced with lower CNT content. Higher mechanical properties of TLCP/CNT nanocomposites was attributed to good interfacial adhesion between CNT and TLCP, suggesting that functional groups on modified CNT surface are helpful for improving the interfacial interaction with TLCP, thus being favorable to more effective load transfer from TLCP matrix to CNT. Thus, the improvement in homogeneous dispersion and interfacial adhesion in TLCP/CNT nanocomposites, resulting from strong interactions between modified CNT and TLCP, can lead to the substantial enhancement of mechanical properties of TLCP nanocomposites.

For characterizing the effect of modified CNT on the mechanical properties of TLCP nanocomposites, it is also very instructive to compare the reinforcing efficiency of pristine and modified CNT for a given content in TLCP nanocomposites. In this paper, the reinforcing efficiency of CNT can be defined as the normalized mechanical properties of TLCP nanocomposites with respect to those of pure TLCP. The enhancing effect of the mechanical properties by incorporating CNT was more pronounced in TLCP nanocomposites containing modified CNT than in the case of pristine CNT, indicating that the incorporation of modified CNT into TLCP matrix was more effective in improving the mechanical properties of TLCP nanocomposites. The incorporation of modified CNT containing functional groups effectively induced via chemical modification into TLCP matrix resulted in the increased interfacial adhesion between CNT and TLCP matrix, thus being favorable to more efficient load transfer form the polymers to the nanotubes. Therefore, the enhanced interfacial adhesion between modified CNT and TLCP matrix as well as uniform dispersion of modified CNT in TLCP matrix results in significant improvement in the overall mechanical properties of TLCP nanocomposites.

Figure 5. SEM micrograph of the fracture surface for TLCP nanocomposites containing 1.5 wt% of modified CNT (Scale bar: 500 nm). The inset show uniform dispersion of CNT in TLCP matrix. The arrow indicates that CNT were to be broken with their ends still embedded in TLCP matrix or they were bridging the local microcracks in TLCP nanocomposites, suggesting good wetting with TLCP matrix and enhanced adhesion between them, thus being favorable to efficient load transfer from TLCP to CNT. Reproduced with permission from Ref. [43]. © 2008 Elsevier Ltd.

A comparative study of the experimental results with the values predicted from the theoretical models is also very instructive to characterize the effect of CNT on the mechanical properties of TLCP/CNT nanocomposites. It is known that the modified Halpin-Tsai equation has been used to predict the modulus of CNT-filled polymer nanocomposites [77,78]. By assuming the random-oriented discontinuous distribution of modified CNT in TLCP matrix, the modified Halpin-Tsai equation can be expressed as follows [77,78, 79,80]:

$$E_C = \left[\left(\frac{3}{8}\right)\left(\frac{1+2(l_{CNT}/d_{CNT})\eta_L V_{CNT}}{1-\eta_L V_{CNT}}\right)+\left(\frac{5}{8}\right)\left(\frac{1+2\eta_T V_{CNT}}{1-\eta_T V_{CNT}}\right)\right]E_{TLCP}$$
$$\eta_L = \frac{(E_{CNT}/E_{TLCP})-1}{(E_{CNT}/E_{TLCP})+2(l_{CNT}/d_{CNT})} \quad and \quad \eta_T = \frac{(E_{CNT}/E_{TLCP})-1}{(E_{CNT}/E_{TLCP})+2} \tag{2}$$

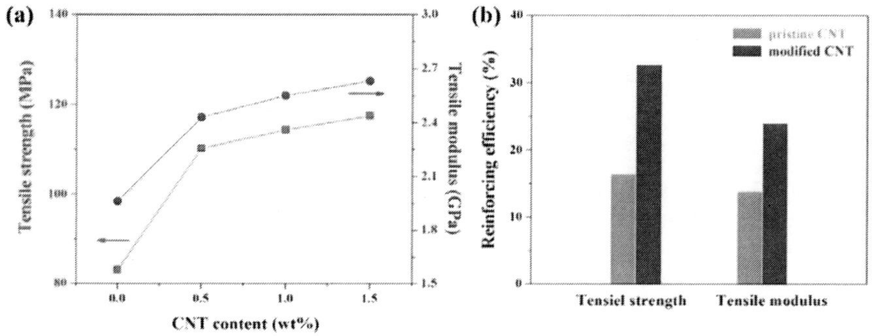

Figure 6. (a) Mechanical properties of TLCP nanocomposites with the CNT content. (b)Reinforcing efficiency of pristine and modified CNTs on mechanical properties of TLCP nanocomposites containing 0.5 wt% of CNT content. Reinforcing efficiency (%) = $[(M_c - M_m)/ M_m] \times 100$, where M_c and M_m represent the mechanical properties, such as tensile strength and tensile modulus, of TLCP nanocomposites and TLCP, respectively. Data from Ref. [43].

where E_C, E_{TLCP}, and E_{CNT} are the tensile modulus of TLCP nanocomposites, TLCP, and CNT, respectively; l_{CNT}/d_{CNT} is the ratio of length to diameter for CNT, and V_{CNT} is the volume fraction of CNT in the nanocomposites. To fit the Equation (2) to experimental results for TLCP nanocomposites, the weight fraction was transformed to the volume fraction, taking the densities of TLCP (1.41 g/cm^3) [81] and perfectly graphitized CNT (2.16 g/cm^3) [77,78]. The theoretically predicted values of TLCP nanocomposite modulus can be estimated assuming the aspect ratio of ~1,000 and E_{CNT} of ~450 GPa. The E_{CNT} values used in this paper represents a mid-range value in the modulus ranges of CNT previously measured [82]. The tensile strength of TLCP nanocomposites can be determined from the following equation [83]:

$$\sigma_C = \sigma_{CNT} V_{CNT} + \sigma_{TLCP} V_{TLCP} \tag{3}$$

where σ_C, σ_{CNT}, and σ_{TLCP} are the tensile strength of the nanocomposites, CNT, and TLCP, respectively. The theoretically predicted values of TLCP nanocomposite strength can be estimated based on the σ_{CNT} values of ~11 GPa based on previous literature [84]. The theoretically predicted values and experimental data for TLCP nanocomposites modulus and strength are compared in Figure 7. At lower CNT content, the experimental results for mechanical properties of TLCP nanocomposites were fitted well with the theoretical equations, while the large deviations were observed at higher CNT content as compared to the theoretically predicted values. The nanoreinforcing effect of CNT on the mechanical properties of TLCP nanocomposites appeared to be more pronounced at lower CNT content than at higher CNT content. This result suggested that the interfacial interaction was more effective in the enhancement of mechanical properties of TLCP nanocomposites at lower CNT content than at higher CNT content. Kim *et al.* [41] reported that for modified CNT/PEN nanocomposite system, the interfacial interactions were more effective in improving the mechanical properties of modified CNT/PEN nanocomposites at lower CNT concentration than at higher CNT concentration. Thus, the synergistic effect of the increased interfacial interactions between modified CNT and TLCP matrix in combination with uniform dispersion of modified CNT was more effective at lower CNT content for enhancing the overall mechanical properties of TLCP nanocomposites.

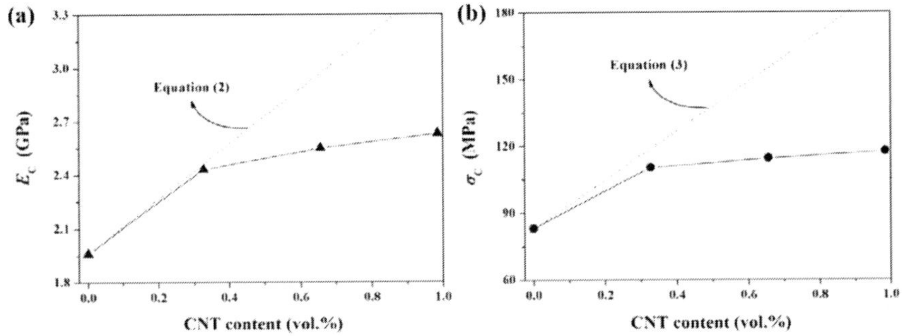

Figure 7. Comparative study for the mechanical properties of TLCP nanocomposites. Theoretically predicted values and the experimental results for (a) the nanocomposite modulus (E_C) and (b) the nanocomposite strength (σ_C) of TLCP nanocomposites. Dotted lines represent the theoretically predicted values based on the Equations (2) and (3). Reproduced with permission from Ref. [43]. © 2008 Elsevier Ltd.

However, the experimental results for mechanical properties of TLCP nanocomposites were lower than the theoretically predicted values, as shown in Figure 7. This result may be explained by the curvature of CNTs in TLCP nanocomposites: some CNTs embedded in TLCP matrix often exhibit curved morphology, and not a straight one (see Figure 5) and this curvature feature of CNT in TLCP matrix has a possibility to decrease the nanoreinforcing effect of CNT in TLCP nanocomposites, in comparison with the theoretical reinforcement provided by straight inclusions. Fisher *et al.* [85,86] have developed a model combining finite element results and micromechanical methods to determine the effective reinforcing modulus of wavy embedded CNT. They found that even slight curvature of CNT significantly reduced the effective reinforcement compared with straight CNT, and suggested that this feature may be an additional mechanism to limit the reinforcing effect of CNT on the mechanical properties of CNT/polymer nanocomposites, leading to the nanocomposite modulus and strength that were less than predicted by traditional theoretical models [85,86]. In addition, the orientation of CNT was an important factor in determining the mechanical properties of polymer nanocomposites, and the curvature morphology of CNTs was also related to the orientation of CNT in polymer nanocomposites. Gorga and Cohen [87] reported that aligned non-entangled CNTs should disperse and orient more readily in the polymer nanocomposites, thus improving their mechanical properties more significantly with optimal processing conditions. The nanoreinforcing effect of CNT will be more effective in improving the mechanical properties of TLCP nanocomposites when the introduced CNT exhibit straight morphology within TLCP matrix and they are preferentially aligned along their axial direction. Thus, the overall mechanical properties of TLCP nanocomposites is expect to be further improved by optimizing the unique geometric feature and alignment of CNT in TLCP matrix as well as the combination of the enhanced interfacial adhesion between CNT and TLCP matrix with the good dispersion of CNT in TLCP matrix during the melt processing.

3. EXPERIMENTAL SECTION

3.1. Materials

TLCP used was a flexible semi-aromatic copolyester synthesized from poly(*p*-hydroxybenzoate) (PHB) and poly(ethylene terephthalate) (PET) with a molar ratio of 80:20, purchased from Unitika Co. Ltd, Japan. The nanotubes used were multi-walled CNT (degree of purity >95%) synthesized by a thermal chemical vapor deposition process, purchased from Iljin Nanotech Co., Korea. The diameter and length of CNT were in the range of 10~30 nm and 10~50 μm, respectively. Concentrated nitric acid (HNO_3, 68%) and sulfuric acid (H_2SO_4, 98%)

were purchased from Sigma Aldrich Co., and they used as received without further purification.

3.2. Fabrication of TLCP/CNT nanocomposites

Pristine CNT was modified by the following steps: CNT was added to the mixture of concentrated HNO_3 and H_2SO_4 with a volumetric ratio of 1:3 and this mixture was sonicated at 80 °C for 4 h to create the carboxylic acid groups on the nanotube surface. This mixture was diluted with excess deionized water and then vacuum-filtered through 0.22 μm milipore PTFE membranes until the pH of the filtrate reached approximately 7. The filtrate solid was dried *in vacuo* at 100 °C for 24 h, yielding the modified CNT. The carboxylic acid groups on the surface of CNT were effectively induced via this chemical modification [32] to increase the chemical affinity of CNT with TLCP, yet to decrease the π–π stacking effect among the aromatic rings of the nanotubes, which often leads to the formation of their agglomeration [88]. All materials were dried at 120 °C *in vacuo* for at 24 h before use, to minimize the effects of moisture. The TLCP nanocomposites were prepared by a melt blending in a Haake rheometer (Haake Technik GmbH, Germany) equipped with a twin-screw (intermeshing co-rotating type). The temperature of the heating zone, from the hopper to the die, was set to 290, 300, 305, and 295 °C, and the screw speed was fixed at 40 rpm. Prior to melt blending, TLCP and CNT were physically premixed before being fed into hopper of the extruder to achieve better dispersion of CNT with TLCP matrix. For the fabrication of TLC nanocomposites, TLCP was melt blended with the addition of CNT content, specified as 0.5, 1.0, and 1.5 wt% in TLCP matrix, respectively. Upon completion of melt blending, the extruded strands were allowed to cool in water-bath, and then cut into pellets with constant diameter and length using a rate-controlled PP1 pelletizer (Haake Technik GmbH).

3.3. Characterization

The chemical structure of CNT was characterized by means of room temperature FT-IR measurements in the range of 400~4,000 cm^{-1} recorded using a Magna-IR 550 spectrometer (Nicolet Co.). The change in the structure of CNT by chemical modification was characterized by means of a JASCO NRS-3100 Raman spectrometer (JASCO Inc.) equipped with a 532 nm diode laser, and the laser beam with a nominal power of 30 mW was focused to a spot size of 5 μm diameter. Morphologies of CNT and TLCP nanocomposites were observed using a JEOL 2000 FX TEM and a JEOL JSM-6340F SEM. Rheological properties of TLCP nanocomposites were performed on an ARES (Advanced Rheometric Expansion System) rheometer (Rheometric Scientific, Inc.) in oscillation mode with the parallel-plate geometry using the plate diameter of 25 mm and the plate gap setting of ~1 mm at 305 °C, by applying a time-dependent strain, $\gamma(t) = \gamma_0 \sin(\omega t)$ and measuring resultant shear stress, $\gamma(t) = \gamma_0\ [G'\sin(\omega t) + G''\cos(\omega t)]$, where G' and G'' are storage and loss moduli, respectively. The frequency ranges were varied between 0.05 and 450 rad/s, and the strain amplitude was applied to be within the linear viscoelastic ranges. The mechanical properties of TLCP nanocomposites were measured using an Instron 4465 testing machine, according to the procedures in the ASTM D 638 standard. Thermogravimertic analysis of TLCP nanocomposite were performed with a TA instrument SDF 2960 TGA over the temperature rage of 30~800 °C at a heating rate of 10 °C/min under N_2.

4. CONCLUSIONS

This paper focused on the fabrication and characterization of thermotropic liquid crystal polymer (TLCP) nanocomposites reinforced with a very small quantity of modified carbon nanotube (CNT). The TLCP nanocomposites exhibited higher complex viscosity and more distinct shear thinning behavior than pure TLCP, resulting from strong interactions between CNT and TLCP matrix, which significantly influenced the relaxation behavior of polymer chains in the nanocomposites. The extent of the increase in the complex viscosity, storage modulus, and loss modulus of TLCP nanocomposites by the presence of modified CNT was more pronounced at low frequency as compared to those at high frequency. The decrease in the slopes of storage and loss moduli for TLCP/CNT nanocomposites with the introduction of modified CNT was attributed to the formation of the interconnected or network-like structures via the nanotube-nanotube and polymer-nanotube interactions, resulting in the pseudo solid-like behavior of TLCP nanocomposites. The incorporated CNT has a significant effect on the enhancement of thermal stability by acting as effective physical barriers against the thermal decomposition in TLCP nanocomposites. The improvement in the mechanical properties of TLCP nanocomposites resulted from the enhanced interfacial adhesion between modified CNT and TLCP as well as uniform dispersion of modified CNT in TLCP matrix, and their synergistic effect combining good interfacial interaction and uniform dispersion was more effective at lower CNT content relative to higher CNT content. This study demonstrates that the optimization of straight morphology and preferential alignment of CNT during melt processing as well as the enhanced interfacial adhesion between CNT and TLCP and uniform dispersion of CNT in TLCP matrix plays a key role in further improving the overall mechanical properties of TLCP nanocomposites.

REFERENCES AND NOTES

1. De Jeu, W.H. Physical Properties of Liquid Crystalline Materials; Gordon & Breach: London, UK, 1980.
2. Gray, G.W.; Winsor, P.A. Liquid Crystals and Plastic Crystals; John Wiley & Sons: New York, NY, USA, 1974.
3. Gray, G.W. Thermotropic Liquid Crystals; John Wiley & Sons: New York, NY, USA, 1987.
4. Chung, T.S. Thermotropic Liquid Crystal Polymers: solid thin Film Polymerization, Characteriza-tion, Blends, and Applications; CRC Press: London, UK, 2001.
5. Demus, D.; Gray, G.W.; Speiss, H.W.; Goodby, J.W.; Vill, V. Handbook of Liquid Crystals; Wiley-VCH: Weinheim, Germany, 1998.
6. Kiss, G. In situ composites: Blends of isotropic polymers and thermotropic liquid crystalline polymers. Polym. Eng. Sci.1987, 27, 410–423.
7. Dutta, D.; Fruitwala, H.; Kohli, A.; Weiss, R.A. Polymer blends containing liquid crystals: A review. Polym. Eng. Sci.1990, 30, 1005–1018.
8. He, J.; Bu, W. Microstructure formation in polyblends containing liquid crystalline poly-mers. Polymer 1994, 35, 5061–5066.

9. Kim, J.Y.; Kim, S.H. Influence of viscosity ratio on processing and morphology of thermo-tropic liquid crystal polymer-reinforced poly(ethylene 2,6-naphthalate) blends. Polym. Int. 2006, 55, 449–455.

10. Kim, J.Y.; Kim, S.H. In situ fibril formation of thermotropic liquid crystal polymer in polyes-ters blends. J. Polym. Sci. Part B: Polym. Phys. 2005, 43, 3600–3610.

11. Kim, J.Y.; Kim, S.H.; Kikutani, T. Fiber property and structure development of polyester blend fibers reinforced with a thermotropic liquid crystal polymer. J. Polym. Sci. Part B: Polym. Phys. 2004, 42, 395–403.

12. Kim, J.Y.; Kim, S.H. Structure and property relationship of thermotropic liquid crystal polymer and polyester composite fibers. J. Appl. Polym. Sci. 2006, 99, 2211–2219.

13. Iijima, S. Helical microtubules of graphitic carbon. Nature 1991, 354, 56–58.

14. Wong, E.W.; Sheehan, P.E.; Lieber, C.M. Nanobeam mechanics: elasticity, strength, and toughness of nanorods and nanotubes. Science 1997, 277, 1971–1975.

15. Liu, C.; Fan, Y.Y.; Liu, M.; Kong, H.T.; Cheng, H.M.; Dresselhaus, M.S. Hydrogen storage in single-walled carbon nanotubes at room temperature. Science 1999, 286, 1127–1129.

16. Ishihara, T.; Kawahara, A.; Nishiguchi, H.; Yoshio, M.; Takita, Y. Effects of synthesis condi-tion of graphitic nanocabon tube on anodic property of Li-ion rechargeable battery. J. Power Sources 2001, 97-98, 129–132.

17. Wu, M.; Shaw, L. A novel concept of carbon-filled polymer blends for applications in PEM fuel cell bipolar plates. Int. J. Hydrogen Energy 2005, 30, 373–380.

18. De Heer, W.A.; Chatelain, A.; Ugarte, D. A carbon nanotube field-emission electron source. Science 1995, 270, 1179–1180.

19. Fan, S.; Chapline, M.G.; Franklin, N.R.; Tombler, T.W.; Casell, A.M.; Dai, H. Self-oriented regular arrays of carbon nanotubes and their field emission properties. Science 1999, 283, 512–514.

20. Kim, P.; Lieber, C.M. Nanotube nanotweezers. Science 1999, 286, 2148–2150. [Google Schol-ar] [CrossRef]

21. Kong, J.; Franklin, N.R.; Zhou, C.; Peng, S.; Cho, J.J.; Dai, H. Nanotube molecular wires as chemical sensor. Science2000, 287, 622–625.

22. Alan, B.; Dalton, A.B.; Collins, S.; Munoz, E.; Razal, J.M.; Ebron, V.H.; Ferraris, J.P.; Cole-man, J.N.; Kim, B.G.; Baughman, R.H. Super-tough carbon nanotube fibres. Nature 2003, 423, 703–703.

23. Schadler, L.S.; Giannaris, S.C.; Ajayan, P.M. Load transfer in carbon nanotube epoxy composites. Appl. Phys. Lett. 1998,73, 3842–3844.

24. Ebbesen, T. Carbon Nanotubes: Preparation and Properties; CRC Press: New York, NY, USA, 1997.

25. Dresselhaus, M.S.; Dresselhaus, G.; Avouris, P.H. Carbon Nanotubes: Synthesis, Structure, Properties, and Applications; Springer: Berlin, Germany, 2001.

26. Bokobza, L. Multiwall carbon nanotube elastomeric composites. Polymer 2007, 48, 4907–4920.

27. Paul, D.R.; Robeson, L.M. Polymer nanotechnology: Nanocomposites. Polymer 2008, 49, 3187–3204.

28. Ajayan, P.M. Nanotubes from carbon. Chem. Rev. 1999, 99, 1787–1800.

29. Lourie, O.; Cox, D.M.; Wagner, H.D. Buckling and collapse of embedded carbon nano-tubes. Phys. Rev. Lett. 1998, 81, 1638–1641.

30. Shaffer, M.S.P.; Windle, A.H. Fabrication and characterization of carbon nano-tube/poly(vinyl alcohol) composites.Adv. Mater. 1999, 11, 937–941.

31. Gong, X.; Liu, J.; Baskara, S.; Voise, R.D.; Young, J.S. Surfactant-assisted processing of carbon nanotube/polymer composites. Chem. Mater. 2000, 12, 1049–1052.

32. Liu, J.; Rinzler, A.G.; Dai, H.; Hafner, J.H.; Bradley, R.K.; Boul, P.J.; Liu, A.; Iverson, T.; Shelimov, K.; Huffman, C.B.; Rodriguez-Macias, F.; Shon, Y.S.; Lee, T.R.; Colbert, D.T.; Smalley, R.E. Fullerene pipes. Science 1998, 280, 1253–1256.

33. Sun, Y.P.; Fu, K.; Lin, Y.; Huang, W. Functionalized carbon nanotubes: Properties and applications. Acc. Chem. Res.2002, 35, 1096–1104.

34. Bellayer, S.; Gilman, J.W.; Eidelman, N.; Bourbigot, S.; Flambard, X.; Fox, D.M.; De Long, H.C.; Trulove, P.C. Preparation of homogeneously dispersed multiwalled carbon nano-tube/poly-styrene nanocomposites via melt extrusion using trialkyl imidazolium compatibil-izer. Adv. Funct. Mater. 2005, 15, 910–916.

35. Pötschke, P.; Fornes, T.D.; Paul, D.R. Rheological behavior of multiwalled carbon nano-tube/poly-carbonate composites. Polymer 2002, 43, 3247–3255.

36. Pegel, S.; P.; Petzold, G.; Alig, I.; Dudkin, S.M.; Lellinger, D. Dispersion, agglomeration, and network formation of multiwalled carbon nanotubes in polycarbonate melts. Polymer 2008, 49, 974–984.

37. Mu, M.; Walker, A.M.; Torkelson, J.M.; Winey, K.I. Cellular structures of carbon nanotubes in a polymer matrix improve properties relative to composites with dispersed nano-tubes. Polymer 2008, 49, 1332–1337.

38. Jung, R.; Park, W.I.; Kwon, S.M.; Kim, H.S.; Jin, H.J. Location-selective incorporation of multiwalled carbon nanotubes in polycarbonate microspheres. Polymer 2008, 49, 2071–2076.

39. Moniruzzaman, M.; Winey, K.I. Polymer nanocomposites containing carbon nano-tubes. Macromolecules 2006, 39, 5194–5205.

40. Kim, J.Y.; Kim, S.H. Multiwall Carbon Nanotube Reinforced Polyester Nanocomposites. In Polyesters: Properties, Preparation and

Application; Yamashita, H., Nakano, Y., Eds.; Nova Science Publishers: New York, NY, USA, 2008; Chapter 2.

41. Kim, J.Y.; Han, S.I.; Hong, S. Effect of modified carbon nanotube on the properties of aro-matic polyester nanocomposites. Polymer 2008, 49, 3335–3345.

42. Kim, J.Y.; Han, S.I.; Kim, D.K.; Kim, S.H. Mechanical reinforcement and crystallization behavior of poly(ethylene 2,6-naphthalte) nanocomposites induced by modified carbon nano-tube. Composites: Part A 2009, 40, 45–53.

43. Kim, J.Y.; Kim, D.K.; Kim, S.H. Effect of modified carbon nanotube on physical properties of thermotropic liquid crystal polyester composites. Eur. Polym. J. 2009, 45, 316–324.

44. Kim, J.Y.; Park, H.S.; Kim, S.H. Thermal decomposition behavior of carbon nanotube-reinforced poly(ethylene 2,6-naphthalate) nanocomposites. J. Appl. Polym. Sci. 2009, 113, 2008–2017.

45. Kim, J.Y.; Park, H.S.; Kim, S. H. Unique nucleation of multiwalled carbon nanotube and poly(ethylene 2,6-naphthalate) nanocomposites during non-isothermal crystalliza-tion. Polymer 2006, 47, 1379–1389.

46. Kim, J.Y. The effect of carbon nanotube on the physical properties of poly(butylene tereph-thalate) nanocomposites by simple melt blending. J. Appl. Polym. Sci. 2009, 112, 2589–2600.

47. Kim, J.Y.; Kim, S.H. Influence of multiwall carbon nanotube on physical properties of poly(ethylene 2,6-naphthalate) nanocomposites. J. Polym. Sci. Part B: Polym. Phys. 2006, 44, 1062–1071.

48. Kim, J.Y.; Park, H.S.; Kim, S.H. Multiwall carbon nanotube-reinforced poly(ethylene tereph-thalate) nanocomposites by melt compounding. J. Appl. Polym. Sci. 2007, 103, 1450–1457.

49. Kim, J.Y.; Han, S.I.; Kim, S.H. Crystallization behavior and mechanical properties of poly(ethylene 2,6-naphthalate)/multiwall carbon nanotube nanocomposites. Polym. Eng. Sci. 2007, 47, 1715–1723.

50. Lagerwall, J.P.F.; Scalia, G. Carbon nanotubes in liquid crystals. J. Mater. Chem. 2008, 18, 2890–2898.

51. Zhang, S.; Kumar, S. Carbon nanotubes as liquid crystals. Small 2008, 4, 1270–1283.

52. Rahman, M.; Lee, W. Scientific duo of carbon nanotubes and nematic liquid crystals. J. Phys. D: Appl. Phys. 2009, 42, 063001:1–063001:12.

53. Liu, L.; Qin, Y.; Guo, Z.X.; Zhu, D. Reduction of solubilized multi-walled carbon nano-tubes. Carbon 2003, 41, 331–335.

54. Wu, C.S.; Liao, H.T. Study on the preparation and characterization of biodegradable polylac-tide/multi-walled carbon nanotubes nanocomposites. Polymer 2007, 48, 4449–4458.

55. Yuang, Y.; Young, R.J. Microstructure and mechanical properties of pitch-based carbon fibers. J. Mater. Sci. 1994, 29, 4027–4036.

56. Ruan, S.L.; Gao, P.; Yang, X.G.; Yu, T.X. Toughening high performance ultrahigh molecular weight polyethylene using multiwalled carbon nanotubes. Polymer 2003, 44, 5643–5654.

57. Gupta, M.C.; Umare, S.S. Studies on poly(o-methoxyaniline). Macromolecules 1992, 25, 138–142.

58. Watts, P.C.P.; Mureau, N.; Tang, Z.; Miyajima, Y.; Carey, J.D.; Silva, S.R.P. The importance of oxygen-containing defects on carbon nanotubes for the detection of polar and non-polar vapours through hydrogen bond formation.Nanotechnology 2007, 18, 175701:1–175701:6.

59. Grossiord, N.; Loss, J.; Regev, O.; Koning, C.E. Toolbox for dispersing carbon nanotubes into polymers to get conductive nanocomposites. Chem. Mater. 2006, 18, 1089–1099.

60. Zhao, L.P.; Gao, L. Novel in situ synthesis of MWNTs-hydroxyapatite compo-sites. Carbon 2004, 42, 423–426.

61. Kashiwagi, T.; Du, F.; Douglas, J.F.; Winey, K.I.; Harris, R.H.; Shields, J.R. Nanoparticle networks reduce the flammability of polymer nanocomposites. Nature Mater. 2005, 4, 928–933.

62. Tseng, C.H.; Wang, C.C.; Chen, C.Y. Functionalizing carbon nanotubes by plasma modifica-tion for the preparation of covalent-integrated epoxy composites. Chem. Mater. 2007, 19, 308–315.

63. Horowitz, H.H.; Metzger, G. A new analysis of thermogravimetric traces. Anal. Chem. 1963, 35, 1464–1468.

64. Berber, S.; Kwon, Y.K.; Tomanek, D. Unusually high thermal conductivity of carbon nano-tubes. Phys. Rev. Lett. 2000,84, 4613–4616.

65. Abdalla, M.; Derrick, D.; Adibempe, D.; Nyario, E.; Robinson, P.; Thompson, G. The effect of interfacial chemistry on molecular mobility and morphology of multiwalled carbon nanotubes epoxy nanocomposite. Polymer 2007, 48, 5662–5670.

66. Abdel-Goad, M.; Pötschke, P. Rheological characterization of melt processed polycarbonate-multiwalled carbon nanotube composites. J. Non-Newtonian Fluid Mech. 2005, 128, 2–6.

67. Wagener, R.; Reisinger, T.J.G. A rheological method to compare the degree of exfoliation of nanocomposites. Polymer2003, 44, 7513–7518.

68. Shenoy, A.V. Rheology of Filled Polymer Systems; Kluwer Academic Publisher: Dordrecht, The Netherlands, 1999.

69. Ferry, J. Viscoelastic Properties of Polymers; Wiley: New York, NY, USA, 1980.

70. Krishnamoorti, R.; Vaia, R.A.; Giannelis, E.P. Structure and dynamics of polymer-layered silicate nanocomposites.Chem. Mater. 1996, 8, 1728–1734.

71. Krisnamoorti, R.; Giannelis, E.P. Rheology of end-tethered polymer layered silicate nanocom-posites. Macromolecules1997, 30, 4097–4102.

72. Rosedalev, J.H.; Bates, F.S. Rheology of ordered and disordered symmetric poly(ethylene propylene)-poly(ethyl ethylene) diblock copolymers. Macromolecules 1990, 23, 2329–2338.

73. Larson, R.G.; Winey, K.I.; Patel, S.S.; Watanabe, H.; Bruinsma, R. The rheology of layered liquids: Lamellar block copolymers and smectic liquid crystals. Rheol. Acta 1993, 32, 245–253.

74. Cho, J.; Daniel, I.M. Reinforcement of carbon/epoxy composites with multiwall carbon nanotubes and dispersion enhancing block copolymers. Scripta Mater. 2008, 58, 533–536.

75. Thostenson, E.T.; Chou, T.W. Aligned multi-walled carbon nanotube-reinforced composites: Processing and mechanical characterization. J. Phys. D: Appl. Phys. 2002, 35, L77–L80.

76. Meng, H.; Sui, G.X.; Pang, P.F.; Yang, R. Effects of acid- and diamine-modified MWNTs on the mechanical properties and crystallization behavior of polyamide 6. Polymer 2008, 49, 610–620.

77. Qian, D.; Dickey, E.C.; Andrew, R.; Rantall, T. Load transfer and deformation mechanism in CNT-PS composites.Appl. Phys. Lett. 2000, 76, 2868–2870.

78. Cadek, M.; Coleman, J.N.; Barron, V.; Hedicke, K.; Blau, W.J. Morphological and mechanical properties of carbon-nanotube-reinforced semicrystalline and amorphous polymer compo-sites. Appl. Phys. Lett. 2002, 81, 5123–5125.

79. Mallick, P.K. Fiber-Reinforced Composites; Marcel Dekker: New York, NY, USA, 1993.

80. Zhang, X.; Liu, T.; Sreekumar, T.V.; Kumar, S.; Moore, V.C.; Hauge, R.H.; Smalley, R.E. Poly(vinyl alcohol)/SWNT composite film. Nano Lett. 2003, 3, 1285–1288.

81. Saengsuwan, S.; Bualek-Limcharoen, S.; Mitchell, G.R.; Olley, R.H. Thermotropic liquid crystalline polymer/polypropylene in situ composite films: rheology, morphology, molecular orientation and tensile properties.Polymer 2003, 44, 3407–3415.

82. Pan, Z.W.; Xie, S.S.; Lu, L.; Chang, B.H.; Sun, L.F.; Zhou, W.Y.; Wang, G.; Zhang, D.L. Tensile tests of ropes of very long aligned multiwall carbon nanotubes. Appl. Phys. Lett. 1999, 74, 3152–3154.

83. Agarwal, B.D.; Broutman, L.G. Analysis and Performance of Fiber Composites; Wiley: New York, NY, USA, 1980.

84. Yu, M.F.; Lourie, O.; Dyer, M.J.; Moloni, K.; Kelly, T.F.; Ruoff, R.S. Strength and breaking mechanism of multiwalled carbon nanotubes under tensile load. Science 2000, 287, 637–640.

85. Fisher, F.T.; Bradshow, R.D.; Brinson, L.C. Effects of nanotube waviness on the modulus of nanotube-reinforced polymers. Appl. Phys. Lett. 2002, 80, 4647–4649.

86. Fisher, F.T.; Bradshow, R.D.; Brinson, L.C. Fiber waviness in nanotube-reinforced polymer composites—I: Modulus predictions using effective nanotube properties. Compos. Sci. Tech-nol. 2003, 63, 1689–1703.

87. Gorga, R.E.; Cohen, R.E. Toughness enhancements in poly(methyl methacrylate) by addition of oriented multiwall carbon nanotubes. J. Polym. Sci. Part B: Polym. Phys. 2004, 42, 2690–2702.

88. Tzavalas, S.; Drakonakis, V.; Mouzakis, D.E.; Fisher, D.; Gregoriou, G. Effect of carboxy-functionalized multiwall carbon nanotube (MWNT-COOH) on crystallization and chain con-formations of poly(ethylene terephthalate) (PET) in PET-MWNT nanocompo-sites. Macromolecules 2006, 39, 9150–9156.

3

POLYMER NANOCOMPOSITES—A COMPARISON BETWEEN CARBON NANOTUBES, GRAPHENE, AND CLAY AS NANOFILLERS

Mrinal Bhattacharya

Department of Bioproducts and Biosystems Engineering, University of Minnesota, St. Paul, MN 55108, USA;

ABSTRACT

Nanofilled polymeric matrices have demonstrated remarkable mechanical, electrical, and thermal properties. In this article we review the processing of carbon nanotube, graphene, and clay montmorillonite platelet as potential nanofillers to form nanocomposites. The various functionalization techniques of modifying the nanofillers to enable interaction with polymers are summarized. The importance of filler dispersion in the polymeric matrix is highlighted. Finally, the challenges and future outlook for nanofilled polymeric composites are presented.

KEYWORDS

polymer nanocomposites; carbon nanotubes; graphene, clay; properties

1. INTRODUCTION

The use of fillers for the enhancement of polymer properties has been well documented. Initially, fillers were used to reduce the cost of the polymeric products. However, with time, fillers became an integral part in many applications, particularly for reinforcing the mechanical properties of the polymer. "Reinforced" polymers consist of a polymeric matrix and a relatively stiff inorganic filler that undergoes dramatic change in modulus or stress at given strain over the pure polymer. Traditional fillers include talc, glass fibers, carbon black, and calcium carbonate particles in the micrometer range. However, most micron sized traditional fillers require high loading for modest property enhancement, causing problems in melt flow and processing due to the high viscosity of the filled materials. Furthermore, the high density of traditional fillers also leads to heavier composites. Finally, the lack of interfacial interaction between the filler and the polymeric matrix leads to weak interfacial adhesion and results in failure.

A broad diversity of filler sizes has been used in reinforcing polymeric matrix. Edwards [1] in his review for filler reinforcement observed that "there is, nevertheless, good evidence that small particle size is a necessary requirement, and very likely the predominant requirement, for the reinforcement effect in rubber". High degree of reinforcement is observed in the particle size range of 100 nm and below [1]. It has been reported that nylon 6 required three times more mass of glass fibers than clay montmorillonite (MMT) platelets to cause a doubling in the modulus [2]. Nanofillers in the range of 3%–5% by weight achieve the same reinforcement as 20%–30% of microsized fillers. Thus, nanocomposites have a weight advantage over conventional composites and, nanoscale materials have emerged as an attractive candidate as fillers as their increased specific interfacial area enables potentially higher interfacial interactions and hence, higher modulus.

Nanofillers can be categorized on the basis of their dimensions—one dimensional which include nanotubes and nanowires [3,4], two dimensional such as nanoclays [5] and graphene [6], and three dimensional such as spherical [7] and cubical nanoparticles [8]. Carbonaceous nanofillers such as nanotubes and graphene display excellent properties due to their high mechanical strength and high aspect ratio. Graphene is a two-dimensional single atom thick sheet composed of sp^2carbon structure arranged in a honeycomb structure (Figure 1). It can be considered as a fundamental building block for all sp^2 hybridized carbon allotropes. It can be wrapped up into 0D fullerenes, rolled into 1D nanotubes or 3D graphite when a number of graphene layers stack up. The 2D geometry of graphene nanosheets is responsible for the maximum value of its surface to volume ratio. These are favorable for good reinforcement of polymers.

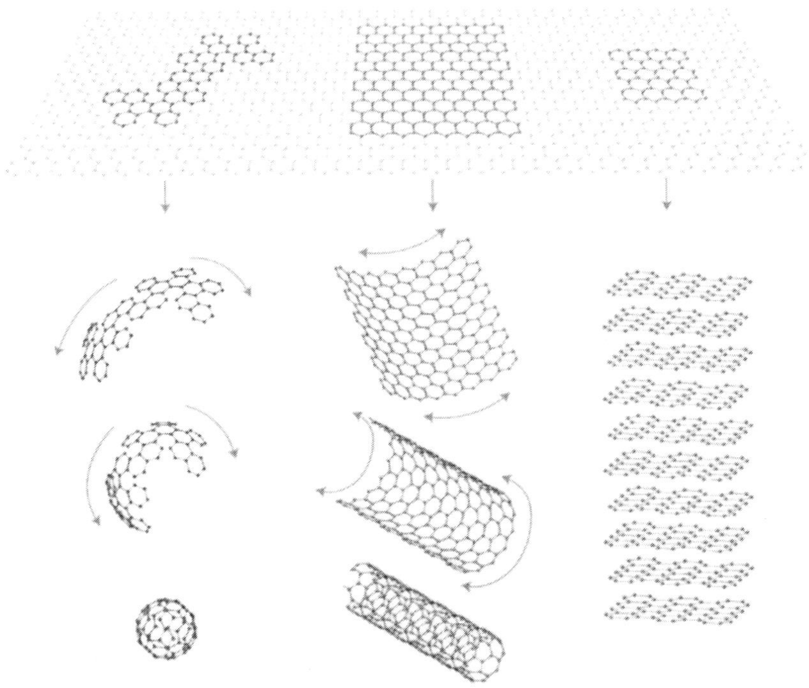

Figure 1. Graphene is a 2D building material for carbon materials of such as 0D buckyballs, 1D nanotubes or 3D graphite. Reproduced from Reference [13] with permission.

Carbon nano tubes have attracted attention because of their unusual structures and properties. High Young's modulus in the direction of the nanotube's axis, electrical conductivities that vary from insulating to metallic, and their hollow structures are attractive features. Carbon nanotubes are half as dense as aluminum and have tensile strengths 20 times that of high strength alloys [9]. Studies [10,11] have shown that nanotubes display extraordinary mechanical properties—tensile modulus of 1 TPa, tensile strength in the range of 50–150 GPa and a failure strain in excess of 5%.

A graphene nanosheet with a Young's modulus of 1 TPa and ultimate strength of 130 GPa is one of the strongest materials known [12]. It has a high specific area of 2600 m^2/g, very high electric conductivity (6000 S/cm) [13], thermal conductivity (~5000 W/mK) [14] and high gas impermeability [15].

Clays are naturally found as platelets, stacked from a few to as many as one thousand sheets. A single sheet of MMT was reported to have an in plane Young's modulus ranging between 178 and 265 GPa [16,17]. These excellent properties of nanofillers make them suitable candidates for reinforcing polymer matrix. Montmorillonite (MMT), is the most widely used clay nanofiller, sandwiched between two silicate layers of an octahedral sheet of alumina. The nanometer-scale sheets of aluminosilicates have dimensions of 1–5 nm thickness and 100–500 nm in diameter (Figure 2). These dimensions lead to platelets of high (>50) aspect ratio. Hence, when blended with polymer, it enables stress transfer from the polymer to the mineral. The stiffness of the clay minerals results in increased mechanical properties of the blend.

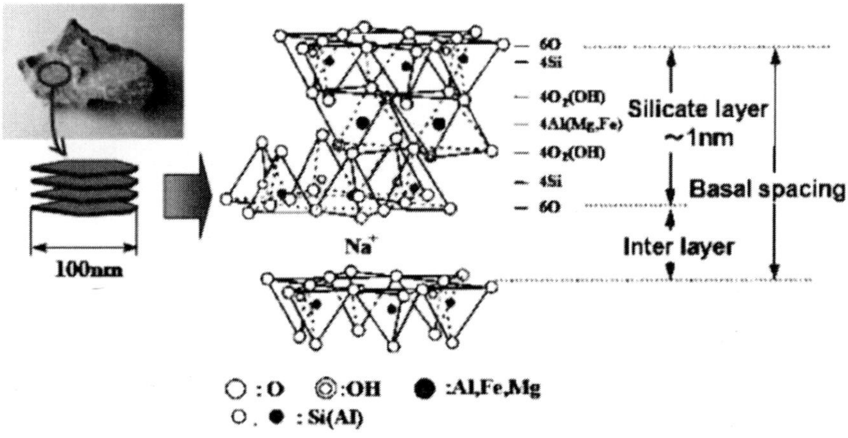

Figure 2. Structure of montmorillonite. Reproduced from reference [5] with permission.

This article reviews the processing techniques for developing nanocomposites and summarizes their properties. The focus will be between 1D *versus* 2D carbonaceous nanofillers (nanotubes *versus* graphene), and between two different 2D nanofillers (graphene *versus* clay). We limit the discussion to 1D and 2D fillers as it has been reported that 2D fillers provide higher degree of reinforcement than spherical shaped fillers [18].

2. CONSIDERATIONS FOR DEVELOPING NANOCOMPOSITES

Because of the contrast in composition, interaction, and properties between dissimilar components in nanocomposites, several key factors affect the role that nanoparticles play as reinforcing fillers in a polymer-matrix. For example: (i) fillers should have excellent mechanical properties such as strength and Young's modulus; (ii) they should have high aspect ratio and high surface area to enable interaction with the polymer; and (iii) they should be well dispersed and avoid agglomeration.

The dispersion of nanoparticle in the matrix to the point where individual particles are coated by the polymer is extremely critical. Increased dispersion help achieve good load transfer to the nanofiller network, resulting in more uniform stress distribution. The mismatch between the properties of nanoparticle and polymer is mitigated by the increased interfacial area between the filler and matrix, leading to improved strength. Alignment of the filler in the matrix, while important, is not critical. While alignment maximizes modulus and strength, it makes the composite anisotropic. Bonding between the filler and the polymer is essential to allow the external stress applied to the composite to be transferred to the nanofillers, enabling them to bear most of the applied load. Hence, nanoparticles are often functionalized to improve their dispersibility, and enable their interactions with polymers. In addition to criterion listed above, the cost of nanomaterial is also an issue.

Several mathematical models such as, Kerner, Nielsen, Halpin-Tsai, and Mori-Tanaka [19,20], developed for conventional composites, can be used to obtain a qualitative measure of Young's modulus of the nanocomposites based on filler geometry, dispersion, and interfacial properties. Results from these models indicate that as the filler content increases, the Young's modulus increases linearly though the differences in the predicted values among various models can easily reach 200%. Spherical particles have a much weaker reinforcing effect on the nanocomposites properties. Randomly distributed platelets result in the strongest isotropic nanocomposites, while aligned fibers or platelet exhibit direction dependent reinforcement [21]. Dramatic changes in mechanical properties are reported to be obtained at low loadings (~2% or less) using exfoliated nanoclays [22,23], graphite nanoplatelets [24], and carbon nanotubes [25].

In addition to strong mechanical performance, nanocomposites also provide improved optical, thermal, and barrier properties. For example, clay nanoparticles possess excellent optical transparency [26,27], barrier properties, and fire retardency [28,29]. Carbon and metal nanoparticles in addition to reinforcing the polymer, also add electrical conductivity, catalytic activity, and plasmonic properties [30].

3. INTERFACIAL INTERACTIONS BETWEEN FILLER AND POLYMER

Mechanical performance of polymer nanocomposites is dependent upon the interfacial interaction between the nanofiller and the polymer matrix. TEM micrographs on fractured surfaces showed nanotube pull-out and nanotube fracture as well as bridging by nanotubes [31]. This implies that to utilize the reinforcing capability of carbon nanotube (CNT) and maximize the mechanical properties of the composites, strong interfacial bonding is necessary. If the interfacial region is stronger than the matrix, the matrix will yield. However, if the interfacial region is weaker than the matrix, de-bonding may initiate along the interface. The extent of interaction would depend on how well the filler is dispersed in the matrix. Because of the strong van der Waals forces and electrostatic interactions, nanofillers tend to aggregate in solvents. Although van der Waals forces are considered to be weak intermolecular forces, they become significant at the nanoscale due to the large surface area per unit mass of the material. Nanotubes tend to self-assemble along the length axis. Agglomerated nanotubes form ropes that slip when stressed due to their poor adhesion to the polymer matrix, affecting their elastic properties [3,32,33]. Nanotube bundles act as stress concentration points within the polymer matrix and can in some cases, reduce the mechanical properties of the original polymer. The reduced aspect ratio also leads to a reduction in reinforcement. Graphite is composed of many graphene sheets held together by van der Waals forces to form rigid platelets several hundreds of nanometers thick. Since most of the graphene sheets in the stacks are unable to interact with the polymeric matrix, its performance is limited. Similar issues arise in the case of clay. In its pristine state, clay exists as stacks of many platelets. In aqueous solution, the hydration of the sodium ions causes the clay to swell and disperse the platelets which allows it to achieve increased interfacial or surface area to interact with polymer. The interaction between the clay and polymer occurs through ion-dipole bonding to the exchangeable cation on the surface of the clay.

Extensive work has been conducted in dispersing nanotubes in solvents. The dispersion procedure involves sonication in a bath for a period of time. The mechanical energy generated during sonication overcomes the van der Waals forces between the nanotube bundles or graphene/clay platelets leading to exfoliation of the filler. Surfactants adsorb on the surface of the filler, and stabilize colloidal particles by the mechanism described by the DLVO (Derjaguin-Landau-Verwey-Overbeek) theory. The surfactant concentration for CNT dispersion generally needs to exceed a critical micelle concentration [34] and also needs to exceed the CNT concentration [34,35,36] in the solution.

Surfactants are most commonly used to disperse nanotubes in water. Common surfactants include sodium dodecyl benzenesulfonate (SDBS), sodium dodecyl sulfate (SDS), cetyltrimethylammonium bromide (CATB), sodium *n*-lauroylsarcosinate, nonyphenol ethoxylate. Other surfactants that have been successfully used to disperse nanotubes are cholate, oxycholate, and oligonucleotides. These are summarized in several reviews [37,38,39]. Attaching functional groups to the nanotubes such as hydroxyl [40] and carboxylic acids [41] improves dispersion in water. Other groups that enhance water dispersion include glucosamine [42], proteins [43], peptides [44], starch [45], amine-containing dendrimers [35], poly(ethylene glycol) [46], poly(styrene sulfonate) [47], and poly(vinyl alcohol) [48,49]. Certain organic solvents such as *N*-methyl pyrrolidone (NMP), α-caprolactone , and dimethyl formamide (DMF) have been shown to disperse carbon nanotubes [50]. Functionalization can also improve dispersion in solvents. Strong acids promote solubilization via protonation of the nanotubes [51]. However, during the functionalization process the use of an ultrasonicator, acids and strong oxidants can damage the properties of nanotubes such as shortening of the tube length or unzipping of the CNTs [52,53]. Below a critical length, nanotubes cannot transfer their load bearing ability (reduced aspect ratio) to the polymer matrix.

To increase interaction between the polymer and the filler, polymers can also be grafted to CNTs. There are two methods: the first, commonly referred to as "grafting to" approach, involves preformed polymer chains reacting with the surface of nanotubes. Examples include grafting of PS to oxidized single-walled CNT and multi-walled CNT, as well as PVA grafted by carbodiimide activated esterification reaction of oxidized nanotubes. Other techniques include reacting to oxidized nanotubes via esterification or amidation, nucleophilic addition, cycloaddition, condensation reactions, and sonochemical reactions. The main drawback of this first procedure is the low graft density caused by steric hindrance of the initially bound polymer chains on the nanotube surface to additional macromolecules. The second method, referred to as "grafting from" approach involves the polymerization of monomers on nanotube surfaces. In this process steric hindrance is not an issue and a high molecular weight polymer can be achieved. This process, however, requires accurate process control for the polymerization. "Grafting from" techniques include atom transfer radical polymerization (ATRP), reversible addition-fragmentation chain transfer, ring opening polymerization, free radical polymerization, cationic/anionic polymerization, condensation polymerization, reduction/oxidation polymerization, metallocene catalysis polymerization, electrochemical grafting, and nitroxide-mediated radical polymerization . These grafting techniques are summarized in several excellent reviews [54,55,56,57].

Depending on the exfoliation procedure, graphene can be dispersed with the aid of surfactant. Graphene can be dispersed in water with the aid of surfactants such as SDS, SDBS, CATB, and tetradecyl trimethyl ammonium bromide. Organic solvents such as DMF, NMP, and cyclohexanone have successfully exfoliated graphene [38]. Graphene exfoliation and dispersion in low boiling solvents such as chloroform, acetone, and isopropanol have also been reported [58]. Unlike in the case for CNT, the surfactant concentration does not have to exceed that of the graphene for dispersion. Surfactants such as sodium cholate perform better at concentrations below the critical micelle concentration [59] for dispersion. The dispersion quality of both CNT and graphene improves as the filler concentration decreases [36,60].

Graphene without functionalities interacts with polymer through van der Waals force, π-π stacking, and hydrophobic interactions [21]. These interactions are, however, weak. Functional groups can be inserted by oxidizing graphite using the Hummers method [61] using a mixture of sulfuric acid, sodium nitrate, and potassium permanganate. The resulting graphene oxide (GO), which generates homogeneous colloidal dispersions in water, alcohol, and organic solvents, possesses oxygen-containing polar functionalities such as carbonyl, hydroxyl, epoxides, and carboxyl groups [62,63]. The epoxide and hydroxyl groups are located in the basal plane of the graphene sheets while the carbonyl and carboxyl groups are located at the edges [64]. The oxygen groups in the graphene oxide reduce the van der Waals forces enabling water molecules to penetrate into the interlayer space of the platelets. Under the influence of the sonication and reduction in van der Waals force, exfoliation of GO occurs. The oxygen functionalities create electrostatic repulsion that prevents the re-aggregation of the exfoliated graphene oxide in water. GO has also been found to disperse in certain organic solvents such as *N,N*-dimethylformamide, *N*-methyl-2-pyrrolidine, tetrahydrofuran, and ethylene glycol [65]. Methods used in reacting polymers to nanotubes are also applicable in the case of graphene. Oxygen functionalities can form amide groups by reacting the amine group in *N*-ethyl-*N′*-(3 dimethyl aminopropyl) carbodiimide methiodide (EDC) with the carboxyl group in GO [62]. ATRP techniques have been used to graft poly (2-(dimethylamino)ethyl methylacrylate) to GO [66].

Graphene oxide is electrically insulating due to the disruption of the graphite structure during the oxidation process. This can be reversed by reducing GO by chemical, thermal, in alkaline environment, in supercritical water and by hydrogen plasma treatment [67]. However, the reduction of GO dispersion without stabilizer causes precipitation of graphite particles due to the rapid and irreversible aggregation of graphene sheets. Hence, prior to reduction, the surface of GO sheets needs to be modified by covalent modifications or non-covalent functionalizations [68]. Reducing agents that are generally used for the reduction of organic ketones, carboxylic acids, and epoxy functional groups can be used for the reduction of pure GO or functionalized GO [69]. Chemical methods include using reductants such as hydrazine [69,70], dimethyl hydrazine [70,71], hydroquinone [72], and sodium borohydride [73]. Thermal heating to above 2000 °C under various atmospheres (vacuum, Ar, H2, NH3) reduces graphite oxide as the oxygenated functional groups are removed during the release of gas molecules such as steam, carbon dioxide, and carbon monoxide [74,75]. Reduction of GO can also be achieved by electrochemical [76,77] or photocatalytic means [78]. The different techniques used to functionalize graphene are summarized in several review articles [67,79,80,81]. However, one drawback of reduced GO is that during the process lattice defects are incurred and this drops the modulus to approximately 220 GPa [82,83].

Non-covalent functionalization of graphene has been achieved via π-π stacking or van der Waals interactions. Such methods prevent any structural damage that affects the electronic properties. Graft and block copolymers have been used as compatibilizers for polymer/CNT composites. Here one chain of the copolymers interacts with the carbon nanofillers by π-π interactions while the other chain is miscible with the matrix polymer. This results in well dispersed CNTs in the polymer matrix. Graphene can be wrapped in poly(sodium 4-styrenesulfonate) that renders it soluble in water [71]. Aqueous dispersion of graphene was prepared using tetracyanoquinodimethane anion, pyrene derivatives, and sulfonated polyaniline. The tetracyanoquinodimethane stabilized graphene sheets were also found to disperse in DMF and DMSO [67,84].

The functional groups on graphene oxide open the possibility of interaction with polymer during processing. Hydroxyl terminated polymers can react with the carboxyl groups on the graphene oxide through esterification. GO can form nanocomposites with hydrophilic polymers such as poly(vinyl alcohol) (PVOH) [85], and poly(ethylene oxide) (PEO) [86]. The carboxyl and hydroxyl group of the GO reacts with organic isocyanates. These isocyanate modified graphene oxides have been known to form composites with polystyrene (PS), acrylonitrile-butadiene-styrene (ABS), and styrene-butadiene rubbers (SBR) [87]. Other polymers that have been incorporated with functionalized graphene include poly (methyl methacrylate) (PMMA), poly vinvyl acetate (PVA), PS, polyaniline (PANI), polyurethane (PU), nylon as well as polyesters such polycaprolactone (PCL), polyethylene terephthalate (PET) and polylactic acid (PLA) [21,84,87,88,89]. The effect of the degree of functionalization, the molecular structure, and the molecular weight of the functional groups on the mechanical properties of functionalized graphene were investigated. While the Young's modulus was found to be insensitive to the molecular weight of the functional group, a decrease in the Young's modulus with increasing levels of functionalization was observed [90].

Polymer chains can be grafted to the surface of modified GO—using the "grafting to" and "grafting from" techniques used to graft polymers to CNT surfaces. This technique increases interfacial interaction between the polymer and the GO through covalent bonding. Grafting of hydroxyl terminate polyvinylcarbazole to GO using diisocyanate [91], or PVA grafted to GO sheets by esterification in DMSO [85] are examples of "grafting to" techniques. "Grafting from" occurs during *in situ*polymerization. ATRP is an effective method used for grafting polymer chains [68,92].

For polymer clay nanocomposites, either the polymer or the clay needs to be modified. Natural clay is hydrophilic and can readily interact with polar polymers like PEO, and PEG. The interaction happens due to the formation of hydrogen bonds between the hydroxyl groups in PVA and

PEG, or oxyethylene groups in PEO [93,94,95,96]. The extent of modification of these clay platelets affects the performance of the composite. Modifications to make MMT organophillic include surfactants such as alkyl and quaternary ammonium halides. Surfactants with a single long alkyl tail gave the highest level of exfoliation when combined with nylon, which decreased as longer alkyl chain surfactants are added [97]. Molecular simulations [98] indicate that the presence of in the alkyl chain of ammonium ions (-OH and -COOH) increased binding energies with nylon. A research group in the Toyota corporation used a solution of 12-aminolauric acid in concentrated hydrochloric acid to modify the surface of MMT with subsequent polymerization with ε-caprolactam to reinforce nylon [99]. However, nonpolar polyolefins have minimal attraction to polar silicate surfaces, and an increase in the number of alkyls on the surfactant improves dispersion [100].

Polymer clay composites can be classified into three types (Figure 3)—(i) immiscible or conventional composites; (ii) intercalated nanocomposites; and (iii) miscible or exfoliated nanocomposites. Complete exfoliation is seldom achieved in practice. Modification of MMT by quaternary ammonium compounds changes the morphology of conventional composite to intercalated [101]. In conventional composites, the clay platelets exist as aggregates or tactoids as in the original clay powder. The wide angle X-ray diffraction patterns for the composite are the same as that for the organo-clay powder. For a completely exfoliated clay, on the other hand, the X-ray peak is absent [22].

Clays are capable of modification through cation exchange, silane grafting, and adsorption of polar polymers [102]. Cation exchange depends on crystal size, the pH, and the type of exchangeable ions [103]. The amount of cations that can be exchanged depends on the amount of exchangeable sites and the structure of the silicate. The exchangeable cation affects the interaction between organic molecules and clay, particularly the likelihood of intercalation or exfoliation [104]. Organosilanes contains organic moiety where a covalent bond can be created between the silicate filler and the polymer matrix.

The ion-exchange reactions with cationic surfactants include primary, secondary, tertiary, and quaternary alkylammonium or alkylphosphonium cations. These surfactants can provide functional groups on the clay that react with the polymer matrix, or in some cases initiate the polymerization of monomers to improve the strength of the interface between the inorganic and the polymer matrix. The increased modulus and impact properties observed during the processing of nylon and clay are due to the formation of ionic bonds between the amine groups on nylon and clay sheet. These exchangeable cations affect interaction with organic molecules, which control the morphology (intercalation or exfoliation). Wilke [105,106,107] developed clays treated with trpylium or triphenyl hexadecylstibonium trifluromethylsulfonate or that containing oligomeric styrene, methyl methacrylate, or ε-caprolatone for reinforcing PS, PE, polypropylene (PP), PMMA or ABS as they are more stable than quaternary ammonium treated clay.

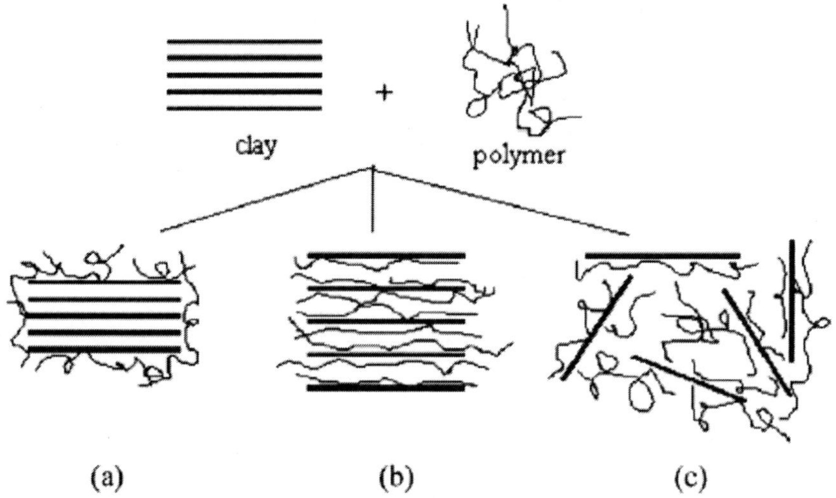

Figure 3. Schematic representation of different types of composite; (**a**) conventional composite; (**b**) intercalated composite; and (**c**) exfoliated composite. (Reproduced from reference [101], with permission).

Unlike polyamides, more commonly used commodity polymers such as polyolefins do not have any polar groups capable of interacting with the aluminosilicate surface of the clay. One way to address the shortcoming is to add maleic anhydride grafted polyolefin to the mixture of clay and polyolefin. The maleated polyolefin fulfils the role of a compatibilizer between two immiscible polymers. This approach has been found useful for developing PP-clay [108,109] and PE-clay [100] systems. The use of maleic anhydride functionalized polymers gave better mechanical properties than when the compatibilizer was absent.

4. NANOCOMPOSITE PROCESSING

The ultimate properties of the nanocomposite are dependent on the processing methods and processing conditions. Most composites can be processed using one or more of the following methods: (a) melt processing; (b) solvent processing; (c) *in situ* polymerization; (d) electrospinning; and (e) layer by layer (LBL) assembly.

Melt blending is one of the most economical and environmentally friendly methods of fabricating composites. This is the processing method of choice for most industries. The compounding is generally achieved in a single or twin-screw extruder where the polymer and the nanoparticle mixture are heated to form a melt. The mixer imparts shear and elongational stress to the process helping to break apart the filler agglomerates and dispersing them uniformly in the polymer matrix. Better dispersion is achieved with MWCNT than SWCNT [110]. The extruder is a versatile device whereby simply changing the screw configuration better control of shear and mixing is obtained. Higher shear rates tends to provide better dispersion. Production rates and material through-puts in a continuous extrusion process can be high. Another advantage of melt processing is that it does not require the use of organic solvents during processing. The compounded nanoparticle-polymer composite can be further processed using other polymer-processing techniques such as injection molding, profile extrusion, blow molding *etc*. Because of the large number of variables involved (temperature, screw-speed, residence time, and shear stress) the mixing process needs to be fine-tuned for optimal properties. Differing characteristics such as agglomerate structure, packing density, length to diameter ratio and purity affect the dispersibility of the nanotubes MNWT in polymeric matrix. In addition, the polymer matrix (specifically the melt viscosity) also affects the degree of dispersion. Unfortunately, the shear forces generated in most mixing equipment are not large enough to break and disperse the CNT in the polymer matrix efficiently. Special mixers where shear rates are an order of magnitude higher than obtained in a typical screw-extruder are often used leading to better dispersion and improved properties. However, the high shear also has the potential for degrading both the polymer and the CNT.

Most of the work reported in the literature has involved polymers such as low density PE (LDPE), HDPE, PP, PS, PMMA, polyamide, polyesters, and polycarbonate (PC). There are several reviews that detail the important findings [3,25,32,33,111,112,113]. Melt processing has shown modest improvement in mechanical properties. It has been reported [114] that intensity in the mixing section improves dispersibility over the kneading section. Polymers containing functional groups capable of reacting with functional groups on nanotubes improves dispersions [115]. Extensional flow gave better dispersion of nanotubes than shear flows [116].

There are limited studies on the melt blending of graphene and polymers. The low thermal stability of most chemically modified graphene and the low bulk density of graphene makes the use of melt processing difficult. Use of high shear melt mixing has been used to fabricate graphene based nanocomposite with PLA [117], PET [118], PP [119], nylon 6 [120], PC [121], PS [122], and elastomers [123]. However, high shear forces can cause buckling, rolling or shortening of graphene sheets [82], thus reducing its aspect ratio.

Melt mixing of clay with polymer has met with great success. There is a body of literature on melt compounding of clay and polymer. A wide range of polymers such as PS [124,125], polyolefins [126], polycarbonate [127,128], PCL [129,130], lactides [131,132,133], PMMA [107,134,135], polyamide [97,99,136,137,138], ABS [139,140], and PEO [141,142] have been successfully melt compounded with clay to obtain various degrees of exfoliation. The intercalation during melt processing of clay-PEO can be further improved by using microwave irradiation [143]. Liu *et al.* [144] obtained an exfoliated composite by compounding organic clay and nylon 6 using a twin-screw extruder. In general, the degree of exfoliation between nylon 6 and clay is quite high because of its excellent affinity for the silicate surface [22]. Stresses generated during melt blending can break up the clay aggregates. The greater the affinity between the clay and the polymer, the greater is the dispersion of the individual platelets in the polymer matrix. Polyolefins which have no polar groups to interact with clay often have to be modified. Polyolefins with maleic anhydride have better properties than unmodified polyolefins when compounded with clay platelets [100,108,145]. Longer residence times in the extruder leads to better dispersion [146]. Similarly, higher melt viscosity leads to higher stresses and better dispersion of the clay particles [22,146].

Solution mixing or solvent casting is another method of producing composites containing graphene or nanotubes and polymer. As the name suggests, solvent casting involves the agitation of the nanoparticle in a polymer that is dissolved in a solvent before casting in a mold and evaporating the solvent. Both thermoplastic and thermoset materials have been produced. Polymers such as PMMA [51], poly(vinyl alcohol) [49,147], polyhydroxyaminoether (PHAE) [148], PS [149], PE [150], PEO [151] and epoxy [152,153,154] with CNTs have been processed.

The lower viscosity of the polymer in solution (as opposed to a melt) coupled with agitation by mechanical stirrer or ultrasonication aids in better dispersion of nanoparticles in the polymeric matrix. Different solvents, from aqueous to organic can be used. However, the removal of organic solvent after casting has environmental implication. Composites involving modified graphene and polymers such as PVA, PMMA, PP, PS, LLDPE, nylon, epoxy, PANI, and PU, have been prepared using the solution mixing and solvent casting method [84]. Solution casting involving clay often uses water soluble polymers such as PEO [96,155,156], lactide [157], and PVA [158,159] though casting using organic solvent [160] has also been attempted. The clay particles are swollen in the solvent. When the polymer solution is added to suspended clay during heating and dispersion, the polymer chains intercalate and displace the solvent within the interlayers of the clay stack. The polarity of the solvent is critical in facilitating the intercalation of the polymer into the space between the clay platelets [101].

A variety of CNT-polymer composite has been prepared using *in situ* polymerization. This technique can be used to produce both thermoset and thermoplastic materials. Here the nanotubes are dispersed in the monomer which is then polymerized. Dispersants may be added to assist in the de-agglomeration of the nanotubes [161]. Alternately functionalization [116,162] or polymer adsorption [163] techniques have been used to aid in dispersion. Polymerization is initiated by increasing the temperature, adding a chemical that initiates the reaction or by mixing two monomers. Since nanotubes are microwave absorbing causing an increase in temperature, microwaves have been used to induce polymerization [164]. One of the advantages of this technique is that it allows the grafting of polymer molecules on to the walls of the tube. The technique is useful in making CNT composites with polymers that are insoluble in most common solvents or are thermally unstable (thereby making melt processing difficult). Some of the composites developed include PE [165], PP [166], PMMA [167], PU [168,169], PCL [170], and PLA [171]. The *in situ* polymerization technique has been used to form composites of poly (vinyl acetate) [172], nylon 6 composite by bulk condensation polymerization of caprolactam in its mixture of GO [173], *in situ* synthesis of PU in DMF [123], emulsion polymerization of styrene in water with GO [174], or PMMA with GO [175].

The solvent casting method of making clay polymer composites has been attempted for water soluble polymers such as PEO, PVA, and poly(vinyl pyrolidone) (PVP), though solvents such as toluene [176], chloroform [156] and, acetonitrile [96,155] have been used. There have been problems with the quality of clay dispersion [158,177] and the volume fraction is fairly low and has not been extensively explored. *In situ* polymerization, on the other hand, is a versatile technique that has received increased attention. Heterophase polymerization in aqueous media is an attractive option [102], as it results in product with low viscosity compared to bulk polymerization and is environmentally friendly. Composite synthesis through emulsion polymerization involving PMMA [178], poly (styrene-acrylonitrite) [179], poly (styrene-co-butyl acrylate) [180], and polystyrene [181,182] have been reported. This route enables the synthesis of composite containing clay platelets located either at the surface or embedded inside the polymer particle. Compared to bulk polymerization and solution polymerization, emulsion polymerization leads to better dispersion of clay platelets in the polymer matrix after the removal of water [102].

An alternate technique to fabricate polymer/nanofiller composite fibers is electrospinning. This technique allows the alignment of the CNTs along the fiber axis. The diameter of electrospun polymeric fibers ranges from tens of nanometers to several microns. The elements of a basic electrospinning unit include an electrode connected to a high voltage power supply that is inserted into a syringe-like container containing the polymeric solution. A schematic of the set-up is shown in Figure 4. Connected to the syringe is a capillary. The syringe-capillary set up can be mounted vertically [183], horizontally [184] or tilted at a defined angle [185]. A grounded collector plate, which is connected to the other end of the electrode, is placed at a distance of 10–30 cm from the tip of the capillary (Figure 4). The polymer solution at the end of the capillary upon the application of high voltage becomes charged. As the voltage is increased, a charge is induced on the surface of the liquid. Mutual charge repulsion leads to development of a force directly opposite to the surface tension. A jet is ejected from the suspended liquid meniscus at the end of the capillary when the applied electric field overcomes the surface tension of the liquid. Further increase in the electric field causes the hemispherical surface of the droplet at the tip of the capillary tube to elongate and form a conical shape known as the Taylor cone. When the repulsive electrostatic force overcomes the surface tension of the fluid, the charged jet is ejected from the tip of the Taylor cone. Within a few centimeters of travel from the tip, the discharged jet undergoes bending instability (Raleigh instability) and begins to whip and splits into bundles of smaller fibers. In addition to bending instability, the jet undergoes elongation (strain $\sim 10^5$ and rate of strain $\sim 10^3$ s^{-1}) which causes it to become very long and thin (diameter in the range of nanometers to micrometers). The solvent evaporates, leading to the formation of a skin and solidification of the fluid jet followed by the collection of solid charged polymer fibers on the collector, usually in the form of non-woven fabric.

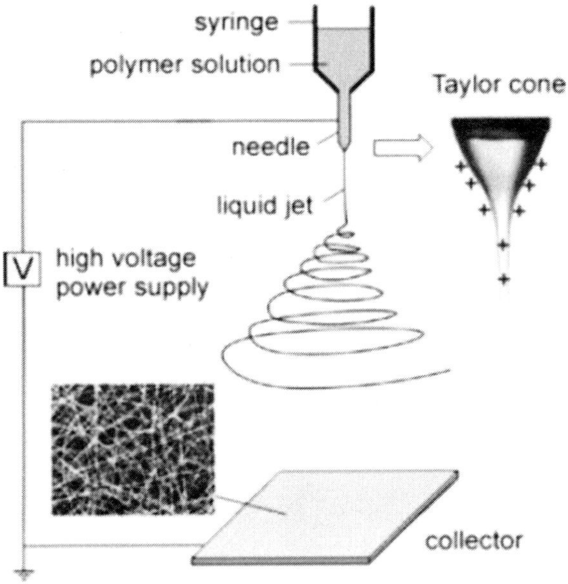

Figure 4. Schematic of basic electrospinning set-up with SEM image of nonwoven mat of poly(vinylpyyrolidone) nanofiber deposited on the collector (From Reference [187] with permission).

Parameters that affect the formation of nanofibers during the electrospinning process include (1) solution properties—viscosity, elasticity, conductivity, and surface tension; (2) system properties—hydrostatic pressure in the capillary, applied voltage, distance between tip and collecting screen, and (3) ambient parameters—solution temperature, humidity, and air velocity [186]. Comprehensive reviews on this topic can be found in several monologues [186,187]. Parameters that control fiber diameter are concentration of the spinning solution, electrical conductivity of the solution, and the feeding rate of the solution through the nozzle.

The potential of incorporating nanometer-sized particulates into fibers has made this process even more attractive for the production of composite fibers. The possibility of orientating the filler particles within the fibers during processing enables the possibility for manipulation of nanoparticles in which they are embedded and oriented. The critical material parameters for manufacturing such composite nanofibers include the geometry of the fillers, and the extent of homogeneous dispersion of filler within the polymer solution.

Graphene oxide has been electrospun with thermoplastic PU [188], polyimide [189], gelatin [190], PVA [191,192], nylon 6 [193], PANI [194], PVP [194], and PVA [195]. A small addition of GO increased both the strength and modulus of the composite [189,191]. Clay has been electrospun with PMMA [196], PVA [197], PEO [198], PVF [199], PLA [200], nylon 6 [201], and poly (ethylene-co-vinyl acetate) [202]. A number of CNT/polymer composites (mostly consisting of MWNT) have been successfully electrospun making it a versatile fiber processing technique. The alignment of the CNT in the polymer enhances the aspect ratio for reinforcing and increases the area for interfacial bonding [203]. Polymers that have been electrospun with CNT's include polyacrylonitrile [204,205], PVA [206,207], PEO [206,208], PMMA [209], PU [210], PCL [211], PLA [184], PS [210,212], nylon 6,6 [213], and silk fibrion [214]. Significant increases in fiber modulus were reported upon the incorporation of nanotubes [184,205,210].

Another technique, often used to assemble multilayer and multi-material thin films is the layer by layer (LBL) assembly. The technique involves immersing a negatively (or positively) charged substrate in an oppositely charged polyelectrolyte (PE) which is adsorbed onto the substrate. After equilibrium is reached, the substrate is removed, rinsed, dried, and immersed in a negatively charged polyelectrolyte solution (Figure 5). This process is repeated until the desired thickness is achieved. The thickness of the film depends on the concentration of polymer in the solution, ionic strength, molecular weight of polymer, temperature during assembly, ionic strength, and pH [215]. The absorption of the polyelectrolyte is irreversible and charge overcompensation leads to charge reversal at the surface [216]. Different materials can be inserted between layers as long as they have the opposite charge.

One of the advantage of LBL assembly is the high level of dispersion of nanoparticle into a composite which occurs as a result of direct absorption of nanoparticle from a solution to a solid state without phase segregation. The first report of LBL assembly of clay nanosheets with polyelectrolytes was in 1994 [217]. Graphene-based LBL assembly was reported in 1996 using non-exfoliated graphite oxide platelets and PE

[218]. GO sheets have been incorporated into LBL assembly through stable GO suspensions containing negatively charged GO-COO^- or positively charged GO-NH_3^+ prepared by introducing amine groups on the surface of negatively charged GO sheets [219].

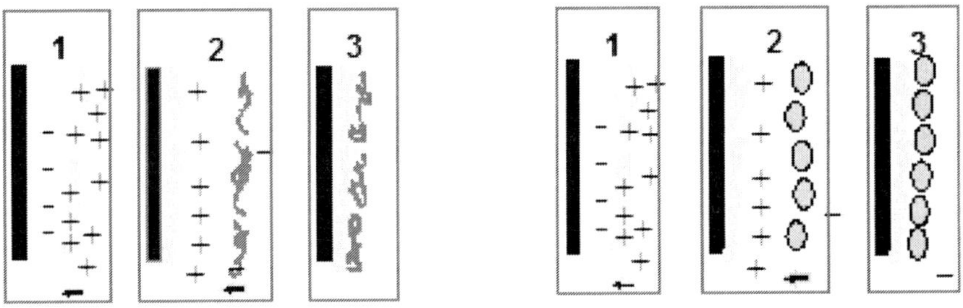

Figure 5. A negatively charged substrate is immersed in the solution of positively charged polyelectrolyte. The latter forms a submonolayer on the surface of the substrate, which switches the surface charge to positive. After rinsing with water, it is then immersed in the dispersion of negatively charged polyelectrolyte or nanoparticles. This results in the formation of a new layer, which switches the surface charge to negative. The whole cycle can be repeated as many times as is desired.

Initial results from multilayered film assembly showed linear growth of mass and film thickness for [220]. In these films each polyelectrolyte interpenetrates only its neighboring layers. However, films that experience exponential growth have also been reported [221]. This exponential growth pattern was attributed to the vertical diffusion of polyelectrolyte into the film. Diffusion is controlled by the molecular weight (MW) of the polyelectrolyte, with higher MW diffusing much more slowly. Other factors that affect diffusion include polymer charge density, and the nature of chemical groups present on the polymer.

LBL assembly initially focused on construction films based on electrostatic interaction, subsequent works have focused on developing LBL composites based on hydrogen bonding [222,223], charge-transfer interactions [224,225], coordination bonding and covalent bonding [226,227,228]. Through hydrogen bonding, a number of additional materials can be incorporated into multilayered composites in a water solution or organic phase. A number of polymers can act as donors and acceptors for hydrogen bonding. Hydrogen-bonding multilayer film assembly is based on the alternate deposition of polymers containing a hydrogen bond acceptor and a hydrogen bond donor, respectively. Compared to films assembled using electrostatic attractions, the pH range where hydrogen bonded multilayers form stable films are limited. Disassembly can be achieved through fine tuning the pH by varying the hydrogen-bonding pairs or the conditions under which the layers are assembled. However, the films can also be made stable by cross-linking using chemical, thermal, and photochemical techniques.

Covalently bonded multilayered films have also been assembled using the LBL techniques. The presence of covalent bonds imparts stability to the films. The strength of the composites depends on the strong adhesion between the two polymers. The films can be assembled in organic solvents, where electrostatic interaction is impossible because a polycation and polyanion will form a salt and precipitate in solution. LBL composites containing carbon nanotubes have been assembled using electrostatic interactions [229,230] or hydrogen bonding [231,232]. Covalent cross-linking can increase the modulus [233,234] and stability of the films. This is achieved by using polymers that have functional groups that are capable of reacting with one another or can react with a bi-functional agent (using diamines, diimides or dialdehydes).

The general concept of LBL assembly makes this a versatile method for combining different kinds of materials (CNTs, clays, nanoparticles (NPs), polymers, proteins, *etc.*) in making nanocomposites. This technique has the capability of controlling the morphology at nan oscale level, to enable a more compact and ordered structures compared to other solution-based methods. LBL assembled thin films can find applications in transparent conducting films, field effect transistors, and supercapacitors. Slow deposition speeds and cumbersome assembly process remain the main drawback of this process.

5. PROPERTIES OF NANOCOMPOSITES

Reinforcement: While it is well know that nanofillers increase the mechanical properties of composites, it is widely recognized that the excellent properties of nanofillers have yet to be realized. This is particularly true at higher volume fraction. Using the Halpin-Tsai [235] equation, for an aspect ratio of 1000 for 1D nanotubes or 2D platelets, a modulus enhancement by a factor of six is predicted at 0.01 filler volume fraction. Molecular simulations predict similar reinforcing potentials for aspect ratio of approximately 100 [236]. Relative improvements in modulus are expected to be much higher than the increase in tensile strength. Strain at break generally decreases with nanofiller loading. For nanotubes, properties generally increase with increasing nanotube content at low volume fractions but then decrease at higher fractions due to issues related to dispersions and agglomeration. It has also been reported that higher surface area leads to better reinforcement [237], except for single walled nanotubes which the authors attributed to poor dispersion. Even with improved adhesion and dispersion in the polymer matrix, the nanotubes remain randomly dispersed. Attempts have been made to align nanotubes to increase reinforcement. Alignment techniques include melt drawing [9], polymer stretching [148,238,239], alternating-current electric field [240,241,242,243], surface acoustic waves [244], direct-current electric field [241,242,243,245] and magnetic fields [246,247,248]. Studies [9,238,239,249,250,251] have shown that in composites where the nanotubes were aligned, a significant increase in the modulus was obtained over non-aligned composites. However, alignment of nanotubes in the composite also caused anisotropy—with improvement in the perpendicular direction being significantly less [252]. The use of magnetic field as a technique to align nanotubes gave conflicting results on modulus enhancement [246]. There is a great body of literature that covers various aspects of mechanical property enhancements of different polymer systems for various types of nanotubes. These have been summarized in several excellent reviews [54,111] and monographs [25]. The mechanical properties of selected nanotube polymer compositions are shown in Table 1. It is difficult to compare the results as the type of nanotube, the method of processing, the type of functional groups, the aspect ratio, and the type of polymer all affect the properties.

Table 1. Mechanical properties of selected carbon nanotube (CNT) based polymers.

Polymer	CNT Type	wt (%)	Processing Method	σ/σ_0	E/E_0	Reference
PS	MWCNT	1.00	Solution Casting	1.25	1.42	[31]
PS	MWCNT	5.00	Solution Casting	1.5	2.2	[149]
PS	MWCNT	40.00	Melt Mixing	1.1	2.0	[262]
LDPE	MWCNT	10.00	Melt Mixing	1.56	1.89	[263]
LDPE	MWCNT	3.00	Ball Milling	3.50	1.3	[264]
PP	SWCNT	1.00	Solution mixing-fiber spinning	1.5	2.0	[265]
PP	SWCNT	0.75	Shear Mixing	1.15	1.4	[266]
PP	MWCNT	0.25	Melt fiber spinning	1.1	2.3	[267]
PP	MWCNT	1.00	Melt fiber spinning	5.0	3.7	[268]
PMMA	Oxidized MWCNT	5.00	*In situ* bulk polymerization	1.30	–	[269]
PMMA	Oxidized MWCNT	1.50	*In situ* bulk polymerization	1.75	–	[270]
PMMA	MWCNT	1.00	Melt Extrusion	1.0	1.0	[271]
PMMA	SWCNT	2.00	Solution casting	1.9	–	[51]
PMMA	PMMA-*g*-MWCNT	0.15	Solution casting	3.6	1.9	[272]
PVA	Hydroxy-modified SWCNT	0.80	Solution casting	1.47	1.79	[49]
PVA	Oxidized MWCNT	0.20	Solution casting	1.52	1.46	[273]
PVA	Oxidized SWCNT	0.20	Solution casting	1.42	1.29	[273]
SBR	MWCNT	10.00	Solution casting	4.0	5.0	[274]
Nylon 6	MWCNT	1.00	Melt Blending	1.2	1.1	[275]
Nylon 6	MWCNT	2.00	Melt Blending	1.62	2.14	[276]
Nylon 6	Oxidized SWCNT	0.5	*In situ* polymerization	2.5	3.5	[277]
PI	Oxidized MWCNT	0.38	Solution casting	1.6	1.5	[278]
PI	Oxidized MWCNT	1.00	Solution casting	1.23	1.40	[279]
PI	SWCNT	1.00	Melt Extrusion	1.0	1.5	[280]
PU	Oxidized MWCNT	1.50	*In situ* condensation	1.3	2.4	[169]
PU	Oxidized MWCNT	2.00	*In situ* condensation	1.15	1.4	[281]
PU	Oxidized MWCNT	4.00	*In situ* condensation	1.4	2.3	[282]
Epoxy	SWCNT	5.00	Solution casting	1.07	1.0	[283]
Epoxy	MWCNT	0.5	Solution casting	1.62	1.54	[154]

Notes: MWCNT—Purified MWCNT, SWCNT—Purified SWCNT; σ_0—tensile strength of pure polymer; E_0—modulus of pure polymer.

As in the case of nanotubes, modulus of graphene filled nanocomposites increased with loading fraction (Table 2). Results indicate that the strength of the interface is critical to the enhancement of the mechanical properties. Modulus increase is more pronounced for elastomeric matrices due to their lower intrinsic modulus. Studies have shown that the mechanical reinforcement of graphene is superior over fillers such as carbon black or single wall nanotube [88]. For epoxy polymers, functionalized graphene sheets have better fracture toughness, fracture energy stiffness, strength, and fatigue resistance at lower loading fractions as compared to nanotubes. [24,253]. Tensile strength has generally also been shown to increase with increase in graphene content, though there are instances when tensile strength decreased [254]. Elongation generally decreased or remained the same. As with nanotube based composites, the improvement in mechanical properties observed falls well short of that predicted theoretically.

The mechanical properties of nanoclay mirrortrends are seen in nanotube polymer composites or graphene polymer. Addition increases modulus and tensile strength but decreases elongation at break. However, with clay, the relative reinforcement [2] for a given volume percent of filler in the composite is significantly lower than for graphene composite [88] or carbon nanotube based composite [54]. The tensile modulus of a polymeric material has been shown to be significantly improved when nanocomposites are formed with either pristine or organically modified clays. For nylon 6, tensile strength increased by 42% and modulus by 90% [255,256]. Interestingly, the stiffness increases with the increasing molecular weight of the matrix at any given loading, even though all the moduli of the neat nylon 6 are quite similar [257]. Similarly, the increase in strength relative to the neat matrix for the high molecular weight composite is nearly double compared to that of the low molecular weight composite. While nylon 6 interacts with silicate surface through hydrogen bonding, nancomposites with polyolefins require modification of both polymer and clay. For PP-clay composition, the modulus increases with clay content untill 3 wt% of clay, after which with an increase in clay content the modulus shows minimal improvement [109]. When MA functional groups are incorporated in the PP the stresses are much more effectively transferred from the polymer matrix to the inorganic filler, and thus a higher increase in Young's modulus was observed. For maleic anhydride functionalized LLDPE, modulus increase is higher with MMT with two alkyl tails [100]. The addition of LLDPE-g-MA is advantageous with clay content of greater than 2.5 wt%.

Mechanical properties of thin films developed using LBL have shown promise in terms of mechanical reinforcement. Mamedov [233] used poly(ethyleneamine) (PEI) as the polycation and acid modified SWNT or poly(acrylic acid) (PAA) as the polyanion. The film was heated and cross-linked using glutaraldehyde. The tensile strength of a 40-layer PEI/SWNT/PAA film was reported to be 220 ± 40 MPa while that of PEI/PAA (without CNT) film of the same number of layers was 9 MPa. The modulus of the composite containing SWNT was 35 GPa. These values are better than any known engineering plastics or are obtained by blending with carbon fillers and are similar to those of ceramics. Hu [258] reported linear increase in modulus of silk fibrion graphene thin film with increasing graphene oxide. Kulkarni *et al.* [259] observed 500% increase in break energy, eight fold increase in modulus, 120% increase in modulus, and increase in ultimate strain in LBL of negatively charged graphene oxide in a polyelectrolyte multilayer. The measured modulus exceeds that predicted by mathematical models. PDDA/MTM multilayers have high strength, flexibility, and resistance to crack propagation [260]. Free standing films of 50, 100, 200 bilayers of PDDA/MMT displayed tensile strength of ~100 MPa and $E \sim 11$ GPa—the increase was 10× and 50× over virgin polymer [261]. PVA/MMT films covalently cross-linked using glutaraldehyde [234], displayed tensile strength of 400 MPa and $E = 106$ GPa.

Electrical Conductivity: The excellent electrical conductivity of CNT's and graphene can be exploited to make traditionally insulating polymer matrices into electrically conductive materials for various applications in conductive adhesives, antistatic coatings, and films. There is a critical loading (percolation threshold) when the composite transitions from an insulator to a conductor due to the formation of a continuous conducting network. Below the percolating threshold, the electrical properties are dominated by the dielectric matrix and hence the composite is non-conductive as the fillers do not form a continuous network for electrons to flow. Beyond the percolation threshold, a small increase in loading will result in a significant increase in conductivity. As filler content increases, the fillers begin to form a contact with each other. At percolation threshold, conduction paths are created in the insulating matrix causing an increase in conduction. However, beyond a certain level of filler concentration a plateau in conductivity is reached. Factors affecting percolation threshold include aspect ratio, functionalization, processing, polymer type, dispersion *etc.* [53]. For example, it has been reported [292] that nanocomposites prepared by *in situ* polymerization showed significant increase in electrical conductivity compared to melt blending. There are no clear trends regarding the type of polymer and its effect on conductivity, although electrically conducting polymers have higher conductivities in the case of CNTs. According to theoretical prediction [293], rod-like structures percolate at one half the volume fraction of disk-like structures. However, there are examples that indicate that graphene has a lower electrical percolation threshold than CNTs [294]. Contradicting results have been published concerning the effect of aspect ratio on the percolation threshold [295,296]. Percolation threshold becomes greater as particles are aligned parallel. In general, SWNT composites have lower conductivities than MWNT composites due to the high contact resistance in SWNT composites because of smaller diameter. Using a similar processing method it was observed that the conductivity of graphene PS films had conductivity several orders of magnitude lower than films with CNTs [297]. For aligned CNT/epoxy composites, the electrical percolation threshold was 0.0025 wt% [298,299]. In comparison, the lowest percolation threshold for graphene based composite was 0.19 wt% for PS solvent blended with isocyanate-treated GO [69,87]. Electric conductivity of CNT and graphene based composites are summarized in several review [54,84,88,300].

Thermal Conductivity—The thermal conductivity of carbon nanotubes has been estimated to be in the range between 650 and 10,000 W/mK. The thermal conductivity of a typical polymer ranges between 0.3 and 0.4 W/mK. It was anticipated that nanotube based composites would experience a significant increase in thermal conductivity (similar to electrical conductivity enhancements). In reality the increase has been rather modest (typically less than 1 W/mK). The reason can be attributed to large resistance to heat transfer at the nanotube-polymer interface. Functionalized nanotubes gave higher thermal conductivity than unfunctionalized tubes indicating that higher dispersion aids conductivity [301]. There have been few studies relating to the thermal conductivity of graphene/polymer composites. The most improvement in the thermal conductivity of graphene based nanocomposites were those where the nanocomposites were produced via *in situ* polymerization using chemically modified graphene [92,302]. Since thermal conductivity of composites increases linearly with filler content, a 20 fold increase was obtained by loading the composite with nanofiller [303].

Thermal Stability: Polymers have a high thermal expansion coefficient when compared to metals. The addition of fillers like clay or nanotubes of graphene reduces the thermal expansion of polymers by constraining the movement of a significant volume of polymer chains because of their interaction with the filler. Graphite has a positive thermal expansion coefficient and when incorporated into polymers does not reduce the expansion of polymers [6]. However, incorporating reduced GO or

SWNTs into resins displayed the effect of reduced thermal expansion [304]. Graphene oxide or single walled carbon nanotube have a negative thermal expansion coefficient and hence, composites containing GO or SWNT increase the thermal stability by decreasing the coefficient of expansion. Liu *et al.* [305] reported that the degradation temperature of PS increased from 400 to 450 °C when impregnated with graphene sheets. Thermal stabilities of graphene/PMMA were reported to be higher than that of PMMA [285]. Clay nanocomposite samples prepared using injection molding displayed anisotropy; the expansion coefficient in the flow direction was lower than in the perpendicular direction [127,306] and this difference was attributed to the orientation of platelets in the respective direction. Similar results were observed when MMT was aligned using magnetic or electric field [307,308].

Glass Transition Temperature: Fillers can be an impediment to the motion of polymer chains due to the interfacial interaction between the polymer and the filler. Studies have confirmed that both T_g and the breadth of the transition can be affected by nanofillers [309,310]. Factors that influence T_g include sample thickness [311,312], sample preparation and measurement [313], nanoparticle dimension [309,310,314], and chemical structure of the polymers [315]. The interaction of the filler with the surface will determine the degree of change of T_g. Surfaces that interact strongly with the polymer causes an increase in T_g [316,317] relative to the bulk. The unadsorbed material can have the same or lower T_g than the bulk depending on the nature of the adsorbed layer. Even hydrogen bonds at the polymer-substrate interface can increase T_g relative to bulk values [313,317]. Graphene platelets with higher aspect ratio, higher surface roughness, and which are well dispersed in the polymer lead to a composite with higher T_g. Liao [313] observed that solvent and melt blending processes lead to insignificant changes in the T_g of polymer-graphene or polymer-GO composites, while *in situ* polymerization with unmodified graphene or solvent blending with chemically modified graphene or GO causes an increase in T_g. The authors attributed this to the covalent bonding between the graphene and the polymer. The type of polymer also affects how and whether the T_g increases, decreases or remains unchanged. For example, nanospherical silica showed an increase in T_g with PVP, decrease in T_g with PMMA, while $T_g$was unchanged with PS [318]. Mixing a polymer and nanotube can increase or decrease T_g depending on the surface functionalization of nanotubes. However, polyimide mixed using a dispersion-reaction scheme with non-functionalized depressed T_g, with acid-functionalized, and amine-functionalized tubes, elevated T_g [319]. Another important effect due to nanotube polymer interaction is the amount of material participating in the glass transition

[320]. Also, nanotube dimension does not affect polymer/nanotube interaction. However, nanotubes can also affect the growth rate of crystals as they alter chain mobility and provide impediment to growth [110]. Higher T_g in some exfoliated and intercalated polymer clay nanocomposites has been attributed to the large interlayer distance between the clay platelets and the strong polymer—filler interactions that exist in the system [321,322,323]. When the interlayer distance is less than the characteristic length of polymer chains for relaxation, T_g is either depressed or absent [125,324].

Barrier and Membrane Separation Properties—Composites containing fillers with large aspect ratio can impede and alter the diffusion path of penetrating molecules. Well dispersed fillers create a tortuous path for permeants to travel. A decrease in gas permeability, that is independent of the type of gas [99] was observed for clay reinforced composites. A 1%-loading of clay in PET showed a two fold reduction in O_2 permeability. Messersmith and Giannelis [325] reported that the water vapor permeability of PCL clay nanocomposites decreased over neat PCL. Permeability was also observed to decrease as the aspect ratio of the platelets increased. Defect-free graphene sheets are impermeable to gas molecules [15], and hence, graphene polymer composites films can be used as protective elements in electronics and fuel cells that are sensitive to the presence of gases such as oxygen and moisture [123,326]. GO conjugated polymer nanocomposite films have been shown to significantly reduce oxygen and carbon dioxide permeation [326,327]. Graphene polyimide composite films have been reported to display high moisture barrier properties [328]. It has been reported that modified GO reduced the permeability of thermoplastic polyurethane more than modified MMT platelet layers at similar loadings.

Flammability Resistance: Polymers will burn easily compared to metals or ceramics. Above a certain temperature a polymer decomposes releasing gaseous products that react with the oxygen in the air and burn. Well dispersed nanoplatelet/nanotube in the polymeric matrix capable of forming a continuous network, form a protective layer on the surface. This protective layer which acts as a heat shield, in turn, prevents the gaseous degradation products from diffusing through it and reacting with the oxygen in the air. Nanofiller impregnated polymer shows a significant reduction in the maximum heat release [329] compared to neat polymer, though the total heat release remained unchanged. Aspect ratio of dispersed silicate layers and the processing method have a strong effect on the fire-retardant properties [330,331]. Both clays and nanotubes have been investigated as flame retardants. Nanotubes have been reported to be more effective retardants [29,332] over clay. However, because of the heat localization due to the high thermal conductivity and low specific heat, time to ignition is lower with nanotube addition. Poorly dispersed nanofillers or low concentration of fillers result in the formation of a discontinuous network leading to much poorer flame resistance [29].

Table 2. Mechanical properties of graphene.

Polymer Matrix	Graphene Type	wt (%)	Processing Method	σ/σ_0	E/E_0	Reference
Epoxy	Expanded	1	Sonication	0.8	1.08	[254]
	Expanded	1	Shear	0.93	1.11	
	Expanded	1	Sonication and Shear	0.94	1.15	
PMMA	Expanded	21	Solution casting	-	1.21	[284]
	GNP	5	Solution casting	-	2.33	[285]
	GO	-	*In situ* Polymerization	-	1.54	[286]
	TRG	1.2	Solution mixing	1.80	-	[285]
HDPE	Expanded	3	Melt processing	1.04	2.0	[287]
PVA	GO	0.7	Solution casting	1.76	-	[288]
	Graphene	4.5	Solution casting	2.5	-	[289]
	GO	6.3	Solution casting	1.7	2.23	[290]
TPU	Graphene	13.3	Solution casting	-	3.0	[291]
	TRG	4.0	Melt	-	3.5	[123]
		4.0	Solution casting	-	7.8	
		3.78	*In situ* polymerization	-	3.1	
PC	TGO	2.5	Melt Processing	-	1.21	[121]
PC	Graphite	15	Melt Processing	-	2.48	

Notes: TRG—Thermally Reduced Graphene; GNP—Graphite Nanoplatelets; TGO—Thermally Exfoliated Graphene Oxide. σ/σ_0 is the normalized tensile strength (tensile strength of composite/tensile strength of polymer matrix); E/E_0 is the normalized modulus (modulus of composite/modulus of polymer).

6. FUTURE OUTLOOK

Carbon based nanofillers are relatively at an early stage of development. Nanoclay based nanocomposites have met with success. However, there are several fundamental challenges that need to be addressed before complete understanding of the nanofiller in polymer composites is attained.

1) It has been well documented that dispersion of nanofillers is critical in achieving properties of nanocomposites. However, many of the processing techniques used to manufacture these composites are not economically viable. Solvent processing, LBL assembly, and electrospinning while resulting in better dispersions of nanofiller in the polymer are not cost-effective. Melt processing, the only economically viable processing technique, generally leads to poor dispersion and less than optimal properties.

2) The problems associated with mechanical reinforcement of melt processed composites need urgent attention. At higher volume fractions, SWNT remains agglomerated. MWNTs are easy to disperse at much higher loading. There is an increase in interfacial area with decrease in tube diameter. For example, at 0.5 wt% of nanotubes MWNT has 70% of the interfacial area of SWNT [29]. Hence, SWNT lose their intrinsic advantage of higher aspect ratio. Similarly, composites at higher loading do not perform as well as at lower loadings. This puts a ceiling on the magnitude of reinforcement.

3) Development and quality of polymer composites with CNT, graphene or clay depend upon a number of factors such as types of CNTs (MWCNT or SWCNT), layers of graphene or clay, purity, length of CNTs, diameter and length of CNT (aspect ratio), loading of nanofillers, dispersion in the matrix, alignment (tough to align graphene or nanoclay), and interaction between the polymer and the nanofiller. However, there are no systematic studies that compare the effect of aspect ratio, nanofiller purity, degree of functionalization, and type of functional group on the properties of the composite. For example, minimal reinforcement is obtained from graphene flakes with an aspect ratio of 1000, while both modulus and strength doubled for graphene with an aspect ratio of 2000 [333].

4) Most work has been conducted using a single type of nanofiller in a polymer matrix. Simultaneous incorporation of different nanofillers

may significantly enhance the properties of composites. The synergy in properties between multiple nanofillers needs to be investigated. For example, 1D fillers might interfere with the stacking of 2D platelets. Incorporating CNTs into glass fiber composites inhibits crack formations due to the large density of nucleation sites provided by CNT [334].

5) Load transfer between nanofiller and polymer has been achieved by both non covalent and covalent modifications with functional groups. Introduction of covalently functional bonds disrupts the π conjugation of CNTs and graphene, leading to a negative effect on the electrical properties of the resulting composites. It has been reported that a combination of non-covalent and covalently fuctionalization on CNT can enhance compatibilizer-polymer interaction leading to better mechanical and electrical properties [335,336]. Hence, it is critical to develop an understanding of the interface between the non-covalently functionalized CNT/graphene and the polymeric matrix to enable simultaneous enhancement of both mechanical and electrical properties of the composites.

6) It has been reported that nanofiller can act as nucleating agent and affect polymer crystallinity [113,337,338]. There should be more attempts to correlate the extent of change crystallinity with the mechanical properties of composite. Similarly, the effect of surfactants on T_g of nanoparticles should be careful investigated [313].

7) Modification of organoclays that are stable at higher temperatures are also desirable. Organically modified clay starts to decompose at approximately 180 °C [339,340] This would enable the use of high temperature polymers like PEEK

7. SUMMARY AND CONCLUSIONS

The current progress on the state of nanoparticle filled polymer nanocomposites and their potential application has been reviewed. The development of nanofiller reinforced polymer matrix has been an active area of research for well over two decades. Amongst the various nanofillers used for reinforcing, clay (montmorillonite MMT), carbon nanotube and graphene have been the most studied. The research field initially started with polymer-clay nanocomposites in the late 80s, followed by polymer-carbon nanotube nanocomposites in the late 90s. Graphene based composite research papers started around 2006 and have experienced exponential growth since 2010.

The reinforcement of polymeric systems by nanoscale sized fillers has opened up the possibility of improving modulus and strength of composites using a much lower filler content. The extent of reinforcement depends on the filler type, the type of functional group on the filler, the aspect ratio of the filler, the amount of filler, the type of polymer, and the method of processing. Homogeneous dispersion of the nanofiller in the polymeric matrix and strong interaction between the filler and the polymer is absolutely necessary for good reinforcement. At higher nanofiller content, composite properties decrease indicating difficulties in dispersing the fillers.

Alignment of nanofiller in the matrix increases mechanical properties but causes anisotropy. The alignment is much easier in the case of nanotubes because of their 1D structure. Composites films made with alternate layers of nanofillers and polyelectrolyte have shown great promise in terms of mechanical properties. These fillers can have high filler loading. The LBL assembly technique also minimizes damage to the nanofiller during assembly. Solution based techniques provide better mechanical reinforcements than melt based techniques. Large scale use of polymer/nanofillers can only be possible if the nanofillers are easily dispersed in the melt using conventional polymer processing such as extrusion and injection molding or fiber spinning in both melt and solution phases.

Covalently functionalized nanotubes provide better polymer reinforcement, allowing for better stress transfer between the polymer matrix and the nanofiller. However, covalent functionalization can destroy intrinsic electrical properties of CNT [111], requiring higher percolation threshold. CNTs can improve the thermal conductivity, flame retardancy, and the thermal stability of the composites.

Graphene based composite is attractive as it leads to improved mechanical properties, electrical conductivity, thermal conductivity, and thermal stability over neat polymers (similar to CNTs). Graphene filled polymers showed as much as 50 °C or larger increase in thermal stability. The percolation threshold of graphene was comparable to that of carbon nanotubes [87]. The conductivity depends on the method of synthesis and on the surface modification, particularly on the number of defects (reactive sites) generated during the oxidation-reduction process on the surface [88]. Homogenously dispersed graphene also reduced barrier properties. One barrier to increased use is the low bulk density of graphene sheets that make handling difficult.

Like the other nanofillers, the mechanical and barrier properties of nanoclay-polymer nanocomposites depend on the extent of dispersion of the silicate layers in the polymer matrix. Nanoclay can be modified by a wide range of organo-modifiers to enhance compatibilization with different polymers and help achieving the intercalation and exfoliation morphologies necessary for achieving the best mechanical properties. Processing can be performed using three primary techniques: solution blending, *in situ* polymerization, and melt blending or extrusion. Melt intercalation is considered the most promising approach for the fabrication because of its versatility, compatibility with current polymer processing equipment, and environmental friendliness due to the absence of solvents. Addition of pristine and organically modified clay increases the thermal stability of polymers. One drawback of clay concerns its stability at high temperatures.

LBL assembly offers promise as it offers the ability to control the morphology at the nanometer scale and allow a much higher loading of nanofillers compared to other processing techniques. Such morphological control has enabled composites to attain record properties. Furthermore, LBL technique can be used to incorporate inorganic nanoparticles, and nanotubes into organic polymers. Processing techniques such as elctrospinning or LBL assembly are unlike to find widespread usage in industrial settings due to their low production rate.

While clay based nanocomposites have found application in the automotive industry, there has been an unfulfilled expectation in the area of CNT/polymer nanocomposites, primarily due to high material costs and difficulties in processing. Cost remains an issue with both nanotubes and graphene. Pricing for research grade nanotubes range from around $5/g for MWCNT to $75/g for SWCNT [341,342]. Graphene oxide costs $100/g [341,342]. Nanoclay is relatively less expensive and melt-compounded composites containing nanoclay have found usage in industrial applications such as in automobiles [22]. Carbon nanotube and graphene based composites might find some applications in advanced composites, replacing carbon fiber, or even in the electronics area.

ACKNOWLEDGMENTS

The author thanks National Science Foundation division of Civil, Mechanical and Manufacturing Innovation for partial support of this work through award # 1462080.

CONFLICTS OF INTEREST

The authors declare no conflict of interest.

REFERENCES

1. Edwards, D.C. Polymer-filler interactions in rubber reinforcement. J. Mater. Sci. 1990, 25, 4175–4185.

2. Fornes, T.D.; Paul, D.R. Modeling properties of nylon 6/clay nanocomposites using composite theories. Polymer 2003,44, 4993–5013.

3. Coleman, J.N.; Khan, U.; Blau, W.J.; Gun'ko, Y.K. Small but strong: A review of the mechani-cal properties of carbon nanotube-polymer composites. Carbon 2006, 44, 1624–1652.

4. Lin, B.; Gelves, G.A.; Haber, J.A.; Sundararaj, U. Electrical, Rheological, and Mechanical Properties of Polystyrene/Copper Nanowire Nanocomposites. Ind. Eng. Chem. Res. 2007, 46, 2481–2487.

5. Okada, A.; Usuki, A. Twenty Years of Polymer-Clay Nanocomposites. Macromol. Mater. Eng. 2006, 291, 1449–1476.

6. Fasolino, A.; Los, J.H.; Katsnelson, M.I. Intrinsic ripples in graphene. Nat. Mater. 2007, 6, 858–861.

7. Liu, Y.; Wang, A.; Claus, R. Molecular Self-Assembly of TiO2/Polymer Nanocomposite Films. J. Phys. Chem. B 1997,101, 1385–1388.

8. Huang, J.; He, C.; Xiao, Y.; Mya, K.Y.; Dai, J.; Siow, Y.P. Polyimide/POSS nanocomposites: Interfacial interaction, thermal properties and mechanical properties. Polymer 2003, 44, 4491–4499.

9. Thostenson, E.K.; Chou, T.W. Aligned multi-walled carbon nanotube-reinforced composites: Processing and mechanical characterization. J. Phys. D Appl. Phys. 2002, 35, L77–L80.

10. Wong, E.W.; Sheehan, P.E.; Lieber, C.M. Nanobeam mechanics: Elasticity, strength, and toughness of nanorods and nanotubes. Science 1997, 277, 1971–1975.

11. Yu, M.; Dyer, M.J.; Kelly, T.F.; Ruoff, R.S. Strength and breaking mechanism of multiwalled carbon nanotubes under tensile load. Science 2000, 287, 637–640.

12. Lee, C.; Wei, X.; Kysar, J.W.; Hone, J. Measurement of the Elastic Properties and Intrinsic Strength of Monolayer Graphene. Science 2008, 321, 385–388.

13. Geim, A.K.; Novoselov, K.S. The rise of graphene. Nat. Mater. 2007, 6, 183–191.

14. Balandin, A.A.; Ghosh, S.; Bao, W.; Calizo, I.; Teweldebrhan, D.; Miao, F.; Lau, C.N. Superior Thermal Conductivity of Single-Layer Graphene. Nano Lett. 2008, 8, 902–907. [Google Schol-ar] [CrossRef] [PubMed]

15. Bunch, J.S.; Alden, J.S.; Zande, A.M.; Parpia, J.M.; Craighead, H.G.; McEuen, P.L. Imperme-able Atomic Membranes from Graphene Sheets. Nano Lett. 2008, 8, 2458–2462.

16. Chen, B.; Evans, J.R.G. Elastic moduli of clay platelets. Scripta Mater. 2006, 54, 1581–1585.

17. Suter, J.L.; Coveney, P.V.; Greenwell, H.C.; Thyveetil, M.-A. Large-scale molecular dynamics study of montmorillonite clay: Emergence of undulatory fluctuations and determination of material properties. J. Phys. Chem. C 2007, 111, 8248–8259.

18. Brune, D.A.; Bicerano, J. Micromechanics of nanocomposites: Comparison of tensile and compressive elastic moduli, and prediction of effects of incomplete exfoliation and imperfect alignment on modulus. Polymer 2002, 43, 369–387.

19. Ward, I.M.; Sweeney, J. An Introduction to the Mechanical Properties of Solid Polymers; John Wiley: Sussex, UK, 2004.

20. Nielsen, L.E.; Landel, R.F. Mechanical Properties of Polymers and Composites; Marcel Dekker: New York, NY, USA, 1994.

21. Hu, K.; Kulkarni, D.D.; Choi, I.; Tsukruk, V.V. Graphene-polymer nanocomposites for structuraland functional applications. Prog. Polym. Sci. 2014, 39, 1934–1972.

22. Paul, D.R.; Robeson, L.M. Polymer Nanotechnology: Nanocomposites. Polymer 2008, 49, 3187–3204.

23. Biswas, M.; Ray, S.S. Recent progress in synthesis and evaluation of polymer-montmorillonite nanocomposites. Adv. Polym. Sci. 2001, 155, 167–221.

24. Rafiee, M.A.; Rafiee, J.; Wang, Z.; Song, H.; Yu, Z.Z.; Koratkar, N. Enhanced Mechanical Properties of Nanocomposites at Low Graphene Content. ACS Nano 2009, 3, 3884–3890.

25. Grady, B.P. Carbon Nanotube-Polymer Composites Manufacture, Properties, and Applications; John Wiley and Sons: New York, NY, USA, 2011.

26. Das, P.; Jani-Markus, S.; Malho, B.Z.; Klemradt, U.; Walther, A. Facile Access to Large-Scale, Self-Assembled, Nacre-Inspired, High-Performance Materials with Tunable Nanoscale Perio-dicities. ACS Appl. Mater. Interfaces 2013, 5, 3738–3747.

27. Podsiadlo, P.; Tang, Z.; Shim, B.S.; Kotov, N.A. Counterintuitive Effect of Molecular Strength and Role of Molecular Rigidity on Mechanical Properties of Layer-by-Layer Assembled Nano-composites. Nano Lett. 2007, 7, 1224–1231.

28. Bitinis, N.; Hernandez, M.; Verdejo, R.; Kenny, J.M.; Lopez-Manchado, M.A. Recent Advances in Clay/Polymer Nanocomposites. Adv. Mater. 2011, 23, 5229–5236.

29. Kashiwagi, T.; Du, F.M.; Douglas, J.F.; Winey, K.I.; Harris, R.H.; Shields, J.R. Nanoparticle networks reduce the flammability of polymer nanocomposites. Nat. Mater. 2005, 4, 928–933.

30. Tu, W.; Zhou, Y.; Zou, Z. Versatile Graphene-Promoting Photocatalytic Performance of Semiconductors: Basic Principles, Synthesis, Solar Energy Conversion, and Environmental Applications. Adv. Funct. Mater. 2013, 23, 4996–5008.

31. Qian, D.; Dickey, E.C.; Andrews, R.; Rantell, T. Load transfer and deformation mechanisms in carbon nanotube-polystyrene composites. Appl. Phys. Lett. 2000, 76, 2868–2870.

32. Coleman, J.N.; Khan, U.; Gun'ko, Y.K. Mechanical Reinforcement of Polymers Using Carbon Nanotubes. Adv. Mater.2006, 18, 637–640.

33. Byrne, M.T.; Gun'ko, Y.K. Recent Advances in Research on Carbon Nanotube-Polymer Composites. Adv. Mater. 2010,22, 1672–1688.

34. Sun, Z.; Nicolosi, V.; Rickard, D.; Bergin, S.D.; Aherne, D.; Coleman, J.N. Quantitative Evaluation of Surfactant-stabilized Single-walled Carbon Nanotubes: Dispersion Quality and Its Correlation with Zeta Potential. J. Phys. Chem. C 2008, 112, 10692–10699.

35. Amiran, J.; Nicolosi, V.; Bergin, S.D.; Khan, U.; Lyons, P.E.; Coleman, J.N. High quality dispersions of functionalized single walled nanotubes at high concentration. J. Phys. Chem. C 2008, 112, 3519–3524.

36. Coleman, J.N. Liquid-Phase Exfoliation of Nanotubes and Graphene. Adv. Funct. Ma-ter. 2009, 19, 3680–3695.

37. Vaisman, L.; Wagner, H.D.; Marom, G. The role of surfactants in dispersion of carbon nanotubes. Adv. Colloid Interface Sci. 2006, 128–130, 37–46.

38. Tkalya, E.E.; Ghislandi, M.; de With, G.; Koning, C.E. The use of surfactants for dispersing carbon nanotubes and graphene to make conductive nanocomposites. Curr. Opin. Colloid In-terface Sci. 2012, 17, 225–232.

39. Constanze, L.; Huzil, J.T.; Marina, V.I.; Marianna, F. Non-Covalent Functionalization of Carbon Nanotubes with Surfactants for Pharmaceutical Applications—A Critical Mini-Review. Drug Deliv. Lett. 2011, 1, 45–57.

40. Zhang, L.; Kiny, V.U.; Peng, H.; Zhu, J.; Lobo, R.F.M.; Margrave, J.L.; Khabashesku, V.N. Sidewall Functionalization of Single-Walled Carbon Nanotubes with Hydroxyl Group-Terminated Moieties. Chem. Mater. 2004, 16, 2055–2061.

41. Choi, J.Y.; Han, S.W.; Huh, W.S.; Tan, L.S.; Baek, J.B. In situ grafting of carboxylic acid-terminated hyperbranched poly(ether-ketone) to the surface of carbon nano-tubes. Polymer 2007, 48, 4034–4040.

42. Pompeo, F.; Resasco, D.E. Water solubilization of single-walled carbon nanotubes by functionalization with glucosamine. Nano Lett. 2002, 2, 369–373.

43. Zhang, Y.; Li, J.; Shen, Y.; Wang, M.; Li, J. Poly-L-lysine Functionalization of Single-Walled Carbon Nanotubes. J. Phys. Chem. B 2004, 108, 15343–15346.

44. Li, X.; Zhan, Q.; Dai, L. Direct Measurements of Interactions between Polypeptides and Carbon Nanotubes. J. Phys. Chem. B 2006, 110, 12621–12625.

45. Star, A.; Steuerman, D.W.; Heath, J.R.; Stoddart, J.F. Starched Carbon Nanotubes. Angew. Chem. Int. Ed. 2002, 41, 2508–2512.

46. Huang, W.; Fernando, S.; Allard, L.F.; Sun, Y.-P. Solubilization of Single-Walled Carbon Nanotubes with Diamine-Terminated Oligomeric Poly(ethylene Glycol) in Different Func-tionalization Reactions. Nano Lett. 2003, 3, 565–568.

47. Qin, S.; Qin, D.; Ford, W.T.; Herrera, J.E.; Resasco, D.E.; Bachilo, S.M.; Weisman, R.B. Solubilization and Purification of Single-Wall Carbon Nanotubes in Water by in Situ Radical Polymerization of Sodium 4-Styrenesulfonate.Macromolecules 2004, 37, 3965–3967.

48. Fernando, K.A.S.; Lin, Y.; Zhou, B.; Grah, M.; Joseph, R.; Allard, L.F.; Sun, Y.-P. Poly(ethylene-co-vinyl alcohol) Functionalized Single-Walled Carbon Nanotubes and Related Nanocomposites. J. Nanosci. Nanotechnol. 2005, 5, 1050–1055.

49. Liu, L.; Barber, A.H.; Nuriel, S.; Wagner, H.D. Mechanical Properties of Functionalized Single-Walled Carbon-Nanotube/Poly(vinyl alcohol) Nanocomposites. Adv. Funct. Ma-ter. 2005, 15, 975–980.

50. Ausman, K.D.; Piner, R.; Lourie, O.; Ruoff, R.S. Organic Solvent Dispersions of Single-Walled Carbon Nanotubes: Toward Solutions of Pristine Nanotubes. J. Phys. Chem. B 2000, 104, 8911–8915.

51. Du, F.; Fischer, J.E.; Winey, K.I. Coagulation method for preparing single-walled carbon nanotube/poly(methyl methacrylate) composites and their modulus, electrical conductivity, and thermal stability. J. Polym. Sci. B Polym. Phys. 2003, 41, 3333–3338.

52. Chen, J.; Hamon, M.A.; Hu, H.; Chen, Y.; Rao, A.M.; Eklund, P.C.; Haddon, R.C. Solution Properties of Single-Walled Carbon Nanotubes. Science 1998, 282, 95–98.

53. Mittal, G.; Dhand, V.; Rhee, K.Y.; Park, S.J.; Lee, W.R. A review on carbon nanotubes and graphene as fillers in reinforced polymer nanocomposites. J. Ind. Eng. Chem. 2015, 21, 11–25.

54. Spitalsky, Z.; Tasis, D.; Papagelis, K.; Galiotis, C. Carbon nanotube-polymer composites: Chemistry, processing, mechanical and electrical properties. Prog. Polym. Sci 2010, 35, 357–401.

55. Bahr, J.L.; Tour, J.M. Covalent chemistry of single-wall carbon nanotubes. J. Mater. Chem. 2002, 12, 1952–1958.

56. Sinnott, S.B. Chemical Functionalization of Carbon Nanotubes. J. Nanosci. Nanotech-nol. 2002, 2, 113–123.

57. Sun, Y.P.; Fu, K.; Lin, Y.; Huang, W. Functionalized Carbon Nanotubes: Properties and Applications. Acc. Chem. Res.2002, 35, 1096–1104.

58. O'Neill, A.; Khan, U.; Nirmalraj, P.N.; Boland, J.; Coleman, J.N. Graphene Dispersion and Exfoliation in Low Boiling Point Solvents. J. Phys. Chem. C 2011, 115, 5422–5428.

59. Lotya, M.; King, P.J.; Khan, U.; De, S.; Coleman, J.N. High-Concentration, Surfactant-Stabilized Graphene Dispersions.ACS Nano 2010, 4, 3155–3162.

60. Bergin, S.D.; Nicolosi, V.; Cathcart, H.; Rickard, D.; Sun, Z.; Blau, W.J.; Coleman, J.N. Large Populations of Individual Nanotubes in Surfactant-Based Dispersions without the Need for Ultracentrifugation. J. Phys. Chem. C 2008, 112, 972–977.

61. Hummers, W.; Offeman, R. Preparation of Graphite Oxide. J. Am. Chem. Soc. 1958, 80, 1339–1339.

62. Dreyer, D.R.; Park, S.; Bielwaski, C.W.; Ruoff, R.S. The Chemistry of Graphene Oxide. Chem. Soc. Rev. 2010, 39, 228–240.

63. Pei, S.; Cheng, H. The reduction of graphene oxide. Carbon 2012, 50, 3210–3228.

64. Lerf, A.; He, H.; Forster, M.; Klinowski, J. Structure of Graphite Oxide Revisited. J. Phys. Chem. B 1998, 102, 4477–4482.

65. Paredes, J.I.; Villar-Rodil, S.; Martínez-Alonso, A.; Tascón, J.M.D. Graphene Oxide Disper-sions in Organic Solvents.Langmuir 2008, 24, 10560–10564.

66. Yang, Y.; Wang, J.; Zhang, J.; Liu, J.; Yang, X.; Zhao, H. Exfoliated Graphite Oxide Decorated by PDMAEMA Chains and Polymer Particles. Langmuir 2009, 25, 11808–11814.

67. Kuila, T.; Bose, S.; Mishra, A.K.; Khanra, P.; Kim, N.H.; Lee, J.H. Chemical functionalization of graphene and its applications. Prog. Mater. Sci. 2012, 57, 1061–1105.

68. Fang, M.; Wang, K.; Lu, H.; Yang, Y.; Nutt, S. Single-layer graphene nanosheets with controlled grafting of polymer chains. J. Mater. Chem. 2010, 20, 1982–1992.

69. Stankovich, S.; Dikin, D.A.; Piner, R.D.; Kohlhaas, K.A.; Kleinhammes, A.; Jia, Y.; Wu, Y.; Nguyen, S.T.; Ruoff, R.S. Synthesis of graphene-based nanosheets via chemical reduction of exfoliated graphite oxide. Carbon 2007, 45, 1558–1565.

70. Li, D.; Muller, M.B.; Gilje, S.; Kaner, R.B.; Wallace, G.G. Processable aqueous dispersions of graphene nanosheets. Nat. Nanotechnol. 2008, 3, 101–105.

71. Stankovich, S.; Piner, R.; Chen, X.; Wu, N.; Nguyen, S.; Ruoff, R. Stable aqueous dispersions of graphitic nanoplatelets via the reduction of exfoliated graphite oxide in the presence of poly(sodium 4-styrenesulfonate). J. Mater. Chem. 2006,16, 155–158.

72. Wang, G.; Yang, J.; Park, J.; Gou, X.; Wang, B.; Liu, H.; Yao, J. Facile Synthesis and Charac-terization of Graphene Nanosheets. J. Phys. Chem. C 2008, 112, 8192–8195.

73. Bourlinos, A.B.; Gournis, D.; Petridis, D.; Szabó, T.; Szeri, A.; Dékány, I. Graphite Oxide: Chemical Reduction to Graphite and Surface Modification with Primary Aliphatic Amines and Amino Acids. Langmuir 2003, 19, 6050–6055.

74. Wang, X.; Zhi, L.; Müllen, K. Transparent, Conductive Graphene Electrodes for Dye-Sensitized Solar Cells. Nano Lett.2008, 8, 323–327.

75. Becerril, H.A.; Mao, J.; Liu, Z.; Stoltenberg, R.M.; Bao, Z.; Chen, Y. Evaluation of Solution-Processed Reduced Graphene Oxide Films as Transparent Conductors. ACS Nano 2008, 8, 463–470.

76. Ramesha, G.K.; Sampath, S. Electrochemical Reduction of Oriented Graphene Oxide Films: An in Situ Raman Spectroelectrochemical Study. J. Phys. Chem. C 2009, 113, 7985–7989.

77. Wang, Z.; Zhou, X.; Zhang, J.; Boey, F.; Zhang, H. Direct Electrochemical Reduction of Single-Layer Graphene Oxide and Subsequent Functionalization with Glucose Oxidase. J. Phys. Chem. C 2009, 113, 14071–14075.

78. Williams, G.; Seger, B.; Kamat, P.V. TiO2-Graphene Nanocomposites. UV-Assisted Photocata-lytic Reduction of Graphene Oxide. ACS Nano 2208, 2, 1487–1491.

79. Quintana, M.; Vazquez, E.; Prato, M. Organic Functionalization of Graphene in Disper-sions. Acc. Chem. Res. 2013, 46, 138–148.

80. Criado, A.; Melchionna, M.; Marchesan, S.; Prato, M. The Covalent Functionalization of Graphene on Substrates.Angew. Chem. Int. Ed. 2015, 54, 10734–10750.

81. Chua, C.K.; Pumera, M. Covalent chemistry on graphene. Chem. Soc. Rev. 2013, 42, 3222–3233.

82. An, X.; Simmons, T.; Shah, R.; Wolfe, C.; Lewis, K.M.; Washington, M.; Nayak, S.K.; Talapatra, S.; Kar, S. Stable Aqueous Dispersions of Noncovalently Functionalized Graphene from Graphite and their Multifunctional High-Performance Applications. Nano Lett. 2010, 10, 4295–4301.

83. Robinson, J.T.; Zalautdinov, M.; Baldwin, J.W.; Snow, E.S.; Wei, Z.; Seehan, P.; Houston, B.H. Wafer-sacle reduced graphene oxide films for nanomechanical devices. Nano Lett. 2008, 8, 3441–3445.

84. Kuilla, T.; Bhadra, S.; Yao, D.; Kim, N.H.; Bose, S.; Lee, J.H. Recent advances in graphene polymer composites. Prog. Polym. Sci 2010, 35, 1350–1375.

85. Salavagione, H.J.; Gomez, M.A.; Martınez, G. Polymeric Modification of Graphene through Esterification of Graphite Oxide and Poly(vinyl alcohol). Macromolecules 2009, 42, 6331–6334.

86. Matsuo, Y.; Tahara, K.; Sugie, Y. Structure and Thermal Properties of Poly(ethylene oxide)-intercalated Graphite Oxide. Carbon 1997, 35, 113–120.

87. Stankovich, S.; Dikin, D.A.; Dommett, G.H.B.; Kohlhaas, K.M.; Zimney, E.J.; Stach, E.A.; Piner, R.D.; Nguyen, S.T.; Ruoff, R.S. Graphene-based composite materials. Nature 2006, 442, 282–286.

88. Kim, H.; Abdala, A.A.; Macosko, C.W. Graphene/Polymer Nanocompo-sites. Macromolecules 2010, 43, 6515–6530.

89. Potts, J.R.; Dreyer, D.R.; Bielawski, C.W.; Ruoff, R.S. Graphene-based polymer nanocompo-sites. Polymer 2011, 52, 5–25.

90. Zheng, Q.; Geng, Y.; Wang, S.; Li, Z.; Kim, J. Effects of functional groups on the mechanical and wrinkling properties of graphene sheets. Carbon 2010, 48, 4315–4322.

91. Liu, G.; Zhuang, X.; Chen, Y.; Zhang, B.; Zhu, J.; Zhu, C.; Neoh, K.; Kang, E. Bistable electrical switching and electronic memory effect in a

solution-processable graphene oxide-donor polymer complex. Appl. Phys. Lett. 2009, 95.

92. Fang, M.; Wang, K.; Lu, H.; Yang, Y.; Nutt, S. Covalent polymer functionalization of graphene nanosheets and mechanical properties of composites. J. Mater. Chem. 2009, 19, 7098–7105.

93. Vaia, R.A.; Sauer, B.B.; Tse, O.K.; Giannelis, E.P. Relaxations of confined chains in polymer nanocomposites: Glass transition properties of poly(ethylene oxide) intercalated in montmoril-lonite. J. Polym. Sci. B Polym. Phys. 1997, 35, 59–67.

94. Vaia, R.A.; Vasudevan, S.; Krawiec, W.; Scanlon, L.G.; Giannelis, E.P. New polymer electrolyte nanocomposites: Melt intercalation of poly(ethylene oxide) in mica-type sili-cates. Adv. Mater. 1995, 7, 154–156.

95. Wu, J.; Lerner, M.M. Structural, thermal, and electrical characterization of layered nanocom-posites derived from sodium-montmorillonite and polyethers. Chem. Mater. 1993, 5, 835–838.

96. Aranda, P.; Ruiz-Hitzky, E. Poly(ethylene oxide)-silicate intercalation materials. Chem. Mater. 1992, 4, 1395–1403.

97. Fornes, T.D.; Hunter, D.L.; Paul, D.R. Nylon-6 Nanocomposites from Alkylammonium-Modified Clay: The Role of Alkyl Tails on Exfoliation. Macromolecules 2004, 37, 1793–1798.

98. Tanaka, G.; Goettler, L.A. Predicting the binding energy for nylon 6,6/clay nanocomposites by molecular modeling.Polymer 2002, 43, 541–553.

99. Kojima, Y.; Usuki, A.; Kawasumi, M.; Okada, A.; Fukushima, Y.; Kurauchi, T.; Kamigiato, O. Mechanical properties of nylon 6-clay hybrid. J. Mater. Res. 1993, 8, 1185–1189.

100. Hotta, S.; Paul, D.R. Nanocomposites formed from linear low density polyethylene and organoclays. Polymer 2004, 45, 7639–7654.

101. Chen, B.; Evans, J.R.G.; Greenwell, H.C.; Boulet, P.; Coveney, P.V.; Bowdenf, A.A.; Whiting, A. A critical appraisal of polymer-clay nanocomposites. Chem. Soc. Rev. 2008, 37, 568–594.

102. Faucheu, J.; Gauthier, C.; Chazeau, L.; Cavaillé, J.Y.; Mellon, V.; Lami, E.B. Miniemulsion polymerization for synthesis of structured clay/polymer nanocomposites: Short review and recent advances. Polymer 2010, 51, 6–17.

103. Pinnavaia, T.J. Intercalated Clay Catalyst. Science 1983, 220, 365–371.

104. Vaia, R.A.; Jandt, K.D.; Kramer, E.J.; Giannelis, E.P. Microstructural Evolution of Melt Intercalated Polymer-Organically Modified Layered Silicates Nanocomposites. Chem. Ma-ter. 1996, 8, 2628–2635.

105. Wang, D.; Wilkie, C.A. A stibonium-modified clay and its polystyrene nanocomposite. Polym. Degrad. Stab. 2003, 82, 309–315.

106. Zhang, J.; Wilkie, C.A. A carbocation substituted clay and its styrene nanocomposite. Polym. Degrad. Stab. 2004, 83, 301–307.

107. Su, S.; Jiang, D.D.; Wilkie, C.A. Poly(methyl methacrylate), polypropylene and polyethylene nanocomposite formation by melt blending using novel polymerically-modified clays. Polym. Degrad. Stab. 2004, 84, 321–331.

108. Kim, D.H.; Fasulo, P.D.; Rodgers, W.R.; Paul, D.R. Structure and properties of polypropyl-ene-based nanocomposies: Effect of PP-g-MA to organoclay ratio. Polymer 2007, 48, 5308–5323.

109. Manias, E. Polypropylene/montmorillonite nanocomposites. Review of the synthetic routes and materials properties.Chem. Mater. 2001, 13, 3516–3523.

110. Grady, B.P. Carbon Nanotube-Polymer Composites Manufacture, Properties, and Applications; John Wiley and Sons: New York, NY, USA, 2011; p. 145.

111. McClory, C.; Chin, S.J.; McNally, T. Polymer/Carbon Nanotube Composites. Aust. J. Chem. 2009, 62, 762–785.

112. Andrews, R.; Weisenberger, M.C. Carbon nanotube polymer composites. Curr. Opin. Solid State Mater. Sci. 2004, 8, 31–37.

113. Coleman, J.N.; Cadek, M.; Blake, R.; Nicolosi, V.; Ryan, K.P.; Belton, C.; Fonseca, A.; Nagy, J.B.; Gun'ko, Y.K.; Blau, W.J. High Performance Nanotube-Reinforced Plastics: Understand-ing the Mechanism of Strength Increase. Adv. Funct. Mater. 2004, 14, 791–798. [Google Schol-ar] [CrossRef]

114. Villmow, T.; Potschke, P.; Pegel, S.; Haussler, L.; Kretzschmar, B. Influence of twin-screw extrusion conditions on the dispersion of multi-walled carbon nanotubes in poly(lactic acid) matrix. Polymer 2008, 49, 3500–3509.

115. Wu, D.; Sun, Y.; Zhang, M. Kinetics Study on Melt Compounding of Carbon Nano-tube/Polypropylene Nanocomposites. J. Polym. Sci. B Polym. Phys. 2009, 47, 608–618.

116. Hong, J.S.; Kim, C. Extension-induced dispersion of multi-walled carbon nanotubes in non-Newtonian fluid. J. Rheol.2007, 51, 833–850.

117. Kim, I.H.; Jeong, Y.G. Polylactide/Exfoliated Graphite Nanocomposites with Enhanced Thermal Stability, Mechanical Modulus, and Electrical Conductivity. J. Polym. Sci. B Polym. Phys. 2010, 48, 850–858.

118. Zhang, H.B.; Zheng, W.G.; Yan, Q.; Yang, Y.; Wang, J.W.; Lu, Z.H.; Ji, G.Y.; Yu, Z.Z. Electrically conductive polyethylene terephthalate/graphene nanocomposites prepared by melt compounding. Polymer 2010, 51, 1191–1196.

119. Kalaitzidou, K.; Fukushima, H.; Drzal, L.T. Multifunctional polypropylene composites produced by incorporation of exfoliated graphite nanoplatelets. Carbon 2007, 45, 1446–1452.

120. Weng, W.; Chen, G.; Wu, D. Transport properties of electrically conducting nylon 6/foliated graphite nanocomposites.Polymer 2005, 46, 6250–6257.

121. Kim, H.; Macosko, C.W. Processing-property relationship of polycarbonate/graphene composites. Polymer 2009, 50, 3797–3809.

122. Chen, G.; Wu, C.; Weng, W.; Wu, D.; Yan, W. Preparation of polystyrene/graphite nanosheet composites. Polymer 2003,44, 1781–1784.

123. Kim, H.; Miura, Y.; Macosko, C.W. Graphene/Polyurethane Nanocomposites for Improved Gas Barrier and Electrical Conductivity. Chem. Mater. 2010, 22, 3441–3450.

124. Vaia, R.A.; Giannelis, E.P. Polymer melt intercalation in organically-modified layered silicates: Model predictions and experiment. Macromolecules 1997, 30, 8000–8009.

125. Vaia, R.A.; Ishii, H.; Giannelis, E.P. Synthesis and properties of two-dimensional nanostruc-tures by direct intercalation of polymer melts in layered silicates. Chem. Mater. 1993, 5, 1694–1696.

126. Ray, S.S.; Okamoto, M. Polymer/layered silicate nanocomposites: A review from preparation to processing. Prog. Polym. Sci 2003, 28, 1539–1641.

127. Yoon, P.J.; Hunter, D.L.; Paul, D.R. Polycarbonate nanocomposites. Part 1. Effect of or-ganoclay structure on morphology and properties. Polymer 2003, 44, 5323–5339.

128. Yoon, P.J.; Hunter, D.L.; Paul, D.R. Polycarbonate nanocomposites: Part 2. Degradation and color formation. Polymer2003, 44, 5341–5354.

129. Chen, B.; Evans, J.R.G. Poly(epsilon-caprolactone)-Clay Nanocomposites: Structure and Mechanical Properties. Macromolecules 2006, 39, 747–754.

130. Lepoittevin, B.; Devalckenaere, M.; Pantoustier, N.; Alexandre, M.; Kubies, D.; Calberg, C.; Jérôme, R.; Dubois, P. Poly(ε-caprolactone)/clay nanocomposites prepared by melt intercalation: Mechanical, thermal and rheological properties. Polymer 2002, 43, 4017–4023.

131. Ray, S.S.; Yamada, K.; Okamoto, M.; Ogami, A.; Ueda, K. New polylactide/layered silicate nanocomposites. 3. High-performance biodegradable materials. Chem. Mater. 2003, 15, 1456–1465.

132. Ray, S.S.; Maiti, P.; Okamoto, M.; Yamada, K.; Ueda, K. New Polylactide/Layered Silicate Nanocomposites. 1. Preparation, Characterization, and Properties. Macromolecules 2002, 35, 3104–3110.

133. Maiti, P.; Yamada, K.; Okamoto, M.; Ueda, K.; Okamoto, K. New polylactide/layered silicate nanocomposites: Role of organoclays. Chem. Mater. 2002, 14, 4654–4661.

134. Wang, J.H.; Young, T.H.; Lin, D.J.; Sun, M.K.; Huag, H.S.; Cheng, L.P. Preparation of Clay/PMMA Nanocomposites with Intercalated or Exfoliated Structure for Bone Cement Syn-thesis. Macromol. Mater. Eng. 2006, 291, 661–669.

135. Wang, Y.; Chen, W.C. Effect of clay modification on the dynamic mechanical and dielectric properties of PMMA nanocomposites via melt blending. Polymer 2013, 12, 128–144.

136. Shen, L.; Phang, I.Y.; Chen, L.; Liu, T.; Zeng, K. Nanoindentation and morphological studies on nylon 66 nanocomposites. I. Effect of clay loading. Polymer 2004, 45, 3341–3349.

137. Masenelli-Varlot, K.; Reynaud, E.; Vigier, G.; Varlet, J. Mechanical Properties of Clay-Reinforced Polyamide. J. Polym. Sci. B Polym. Phys. 2002, 40, 272–283.

138. Cho, J.W.; Paul, D.R. Nylon 6 Nanocomposites by Melt Compounding. Polymer 2001, 42, 1083–1094.

139. Stretz, H.A.; Paul, D.R.; Cassidy, P.E. Poly(styrene-co-acrylonitrile)/montmorillonite or-ganoclay mixtures: A model systems for ABS nanocomposites. Polymer 2005, 46, 3818–3830.

140. Ma, H.; Tong, L.; Xu, Z.; Fang, Z. Clay network in ABS-graft-MAH nanocomposites: Rheology and flammability.Polym. Degrad. Stab. 2007, 92, 1439–1445.

141. Abraham, T.N.; Ratna, D.; Siengchin, S.; Karger-Kocsis, J. Structure and properties of polyethylene oxideorgano clay nanocomposite prepared via melt mixing. Polym. Eng. Sci. 2009, 49, 379–390.

142. Choudhary, S.; Sengwa, R.J. Dielectric properties and structures of melt-compounded poly(ethylene oxide)-montmorillonite nanocomposites. J. Appl. Polym. Sci. 2012, 124, 4847–4853.

143. Aranda, P.; Mosqueda, E.; Pérez-Cappe, E.; Ruiz-Hitzky, E. Electrical characterization of poly(ethylene oxide)-clay nanocomposites prepared by microwave irradiation. J. Polym. Sci. B Polym. Phys. 2003, 41, 3249–3263.

144. Liu, L.; Qi, Z.; Zhu, X. Studies on Nylon 6/Clay Nanocomposites by Melt-Intercalation Process. J. Appl. Polym. Sci.1999, 71, 1133–1138.

145. Kawasumi, M.; Hasegawa, N.; Kato, M.; Usuki, A.; Okada, A. Preparation and Mechanical Properties of Polypropylene-Clay Hybrids. Macromolecules 1997, 30, 6333–6338.

146. Dennis, H.R.; Hunter, D.L.; Chang, D.; Kim, S.; White, J.L.; Cho, J.W.; Paul, D.R. Effect of melt processing conditions on the extent of the exfoliation in organoclay-based compo-sites. Polymer 2001, 42, 9513–9522.

147. Shaffer, M.S.P.; Windle, A.H. Fabrication and characterization of CNT-PVA compo-sites. Adv. Mater. 1999, 11, 937–941.

148. Jin, L.; Bower, C.; Zhou, O. Alignment of carbon nanotubes in a polymer matrix by mechani-cal stretching. Appl. Phys. Lett. 1998, 73, 1197–1199.

149. Safadi, B.; Andrews, R.; Grulke, E.A. Multiwalled carbon nanotube polymer composites: Synthesis and characterization of thin films. J. Appl. Polym. Sci. 2002, 84, 2660–2669.

150. Haggenmueller, R.; Fischer, J.E.; Winey, K.I. Single wall carbon nanotube/ polyethylene nanocomposites: Nucleating and templating polyethylene crystal-lites. Macromolecules 2006, 39, 2964–2971.

151. Geng, H.Z.; Rosen, R.; Zheng, B.; Shimoda, H.; Fleming, L.; Liu, J.; Zhou, O. Fabrication and Properties of Composites of Poly(ethylene oxide) and Functionalized Carbon Nanotubes. Adv. Mater. 2002, 14, 1387–1390.

152. Jang, J.; Baea, J.; Yoon, S.H. A study on the effect of surface treatment of carbon nanotubes for liquid crystalline epoxide-carbon nanotube composites. J. Mater. Chem. 2003, 13, 676–681.

153. Ajayan, P.M.; Stephan, O.; Colliex, C.; Trauth, D. Aligned carbon nanotube arrays formed by cutting a polymer resin-nanotube composite. Science 1994, 265, 1212–1214.

154. Špitalský, Z.; Matějka, L.; Šlouf, M.; Konyushenko, E.N.; Kovářová, J.; Zemek, J.; Kotek, J. Modification of carbon nanotubes and its effect on properties of carbon nanotube/epoxy nanocomposites. Polym. Compos. 2009, 30, 1378–1387.

155. Ruiz-Hitzky, E.; Aranda, P. Polymer-salt intercalation complexes in layer silicates. Adv. Mater. 1990, 2, 545–547.

156. Shen, Z.; Simon, G.P.; Cheng, Y.B. Comparison of solution intercalation and melt intercala-tion of polymer-clay nanocomposites. Polymer 2002, 43, 4251–4260.

157. Ogata, N.; Jimenez, G.; Kawai, H.; Ogihara, T. Structure and Thermal/Mechanical Properties of Poly(L-lactide)-Clay Blend. J. Polym. Sci. B Polym. Phys. 1997, 35, 389–396.

158. Sur, G.S.; Sun, H.L.; Lee, T.J.; Lyu, S.G.; Mark, J.E. Composites prepared by penetrating poly(ethylene oxide) chains into mesoporous silica. Colloid Polym. Sci. 2003, 281, 1040–1045.

159. Carrado, K.A.; Thiyagarajan, P.; Elde, D.L. Polyvinyl alcohol-clay complexes formed by direct synthesis. Clay Clay Miner. 1996, 44, 506–514.

160. Chang, J.H.; Park, D.K.; Ihn, K.J. Montmorillonite-Based Nanocomposites of Polybenzoxa-zole: Synthesis and Characterization (I). J. Polym. Sci. B Polym. Phys. 2001, 39, 471–476.

161. Geng, Y.; Liu, M.Y.; Li, J.; Shi, X.M.; Kim, J.K. Effects of surfactant treatment on the mechani-cal and electrical properties of CNT/epoxy nanocomposites. Compos. A 2008, 39, 1876–1883.

162. Ma, P.C.; Kim, J.K.; Tang, B.Z. Effects of silane functionalization on the properties of carbon nanotube/epoxy nanocomposites. Compos. Sci. Technol. 2007, 67, 2965–2972.

163. Liu, X.Q.; Chan-Park, M.B. Facile way to disperse single-walled carbon nanotubes using a noncovalent method and their reinforcing effect in poly(methyl methyacrylate) composites. J. Appl. Polym. Sci. 2009, 114, 3414–3419.

164. Chowdhury, S.R.; Chen, Y.; Wang, Y.; Mitra, S. Microwave-induced rapid nanocomposite using dispersed single-wall carbon nanotubes as the nuclei. J. Mater. Sci. 2009, 44, 1245–1250.

165. Trujillo, M.; Arnal, M.L.; Müller, A.J.; Laredo, E.; Bredeau, S.; Bonduel, D.; Dubois, P. Thermal and Morphological Characterization of Nanocomposites Prepared by in-Situ Polymerization of High-Density Polyethylene on Carbon Nanotubes. Macromolecules 2007, 40, 6268–6276.

166. Kaminsky, W.; Funck, A. In situ polymerization of olefins with nanoparticles by metallocene-catalysis. Macromol. Symp. 2007, 260, 1–8.

167. Kwon, S.M.; Kim, H.S.; Myung, S.J.; Jin, H.J. Poly(methyl methacrylate)/multiwalled carbon nanotube microspheres fabricated via in-situ polymerization. J. Polym. Sci. B Polym. Phys. 2008, 46, 182–189.

168. Song, W.H.; Ni, Q.P.; Zheng, Z.; Tian, L.Y.; Wang, X.L. The preparation of biodegradable polyurethane/carbon nanotube composite based on in situ cross-linking. Polym. Adv. Tech-nol. 2009, 20, 327–331.

169. Kwon, J.; Kim, H. Comparison of the properties of waterborne polyurethane/multiwalled carbon nanotube and acid-treated multiwalled carbon nanotube composites prepared by in situ polymerization. J. Polym. Sci. A Polym. Chem.2005, 43, 3973–3985.

170. Castro, M.; Lu, J.; Bruzaud, S.; Kumar, B.; Feller, J.F. Carbon nanotube/poly(e-caprolcatone) composite vapor sensors.Carbon 2009, 47, 1930–1942.

171. Biedron, T.; Pietrzak, L.; Kubisa, P. Ionic liquid functionalized polylactide by cationic polymerization: Synthesis and stabilization of carbon nanotube suspensions. J. Polym. Sci. A Polym. Chem. 2011, 49, 5239–5244.

172. Liu, P.; Gong, K.; Xiao, P. Preparation and Characterization of Poly(vinyl acetate)-Intercalated Graphite Oxide. Carbon1999, 37, 2073–2075.

173. Xu, Z.; Gao, C. In situ Polymerization Approach to Graphene-Reinforced Nylon-6 Compo-sites. Macromolecules 2010,43, 6716–6723.

174. Ding, R.; Hu, Y.; Gui, Z.; Zong, R.; Chen, Z.; Fan, W. Preparation and characterization of polystyrene/graphite oxide nanocomposite by emulsion polymerization. Polym. Degrad. Stab. 2003, 81, 473–476.

175. Wang, W.F.; Pan, C.Y. Preparation and characterization of poly(methyl methacrylate)-intercalated graphite oxide/poly(methyl methacrylate) nanocomposite. Polym. Eng. Sci. 2004, 44, 2335–2339.

176. Furuichi, N.; Kurokawa, Y.; Fujita, K.; Oya, A.; Yasuda, H.; Kiso, M. Preparation and properties of polypropylene reinforced by smectite. J. Mater. Sci. 1996, 31, 4307–4310.

177. Tseng, C.R.; Wu, J.Y.; Lee, H.Y.; Chang, F.C. Preparation and crystallization behavior of syndiotactic polystyrene-clay nanocomposites. Polymer 2001, 42, 10063–10070.

178. Lee, D.C.; Jang, L.W. Preparation and Characterization of PMMA-Clay Hybrid Composite by Emulsion Polymerization. J. Appl. Polym. Sci. 1996, 61, 1117–1122.

179. Noh, M.H.; Lee, D.C. Comparison of characteristics of SAN–MMT nanocomposites prepared by emulsion and solution polymerization. J. Appl. Polym. Sci. 1999, 74, 2811–2819.

180. Moraes, R.P.; Santos, A.M.; Oliveira, P.C.; Souza, F.C.T.; Amaral, M.; Valera, T.S.; Demar-quette, N.R. Poly(styrene-co-butyl acrylate)-Brazilian Montmorillonite Nanocomposites, Syn-thesis of Hybrid

Latexes via Miniemulsion Polymerization. Macromol. Symp. 2006, 245, 106–115.

181. Sun, Q.; Deng, Y.; Wang, Z.L. Synthesis and Characterization of Polystyrene Encapsulated Laponite Composites via Miniemulsion Polymerization. Macromol. Mater. Eng. 2004, 289, 288–295.

182. Samakande, A.; Sanderson, R.D.; Hartmann, P.C. Encapsulated clay particles in polystyrene by RAFT mediated miniemulsion polymerization. J. Polym. Sci. A Polym. Chem. 2008, 46, 7114–7126.

183. Teo, W.F.; Ramaseshan, R.; Fujihara, K.; Ramakrishna, S. A dynamic liquid support system for continuous electrospun yarn fabrication. Polymer 2007, 48, 3400–3405.

184. Ko, F.; Gogotsi, Y.; Ali, A.; Naguib, N.; Ye, H.; Yang, G.L.; Li, C.; Willis, P. Electrospinning of Continuous Carbon Nanotube-Filled Nanofiber Yarns. Adv. Mater. 2003, 15, 1161–1165.

185. Yu, J.; Qiu, Y.; Zha, X.; Yu, M.; Yu, J.; Rafique, J.; Yin, J. Production of aligned helical polymer nanofibers by electrospinning. Eur. Polym. J. 2008, 44, 2838–2844.

186. Doshi, J.; Reneker, D.H. Electrospinning process and applications of electrospun fibers. J. Electrost. 1995, 35, 151–160.

187. Li, D.; Xia, Y. Electrospinning of Nanofibers: Reinventing the Wheel? Adv. Mater. 2004, 16, 1151–1170.

188. Jing, X.; Mi, H.Y.; Salick, M.R.; Cordie, T.M.; Peng, X.F.; Turng, L.S. Electrospinning thermo-plastic polyurethane/graphene oxide scaffolds for small diameter vascular graft applica-tions. Mater. Sci. Eng. C 2015, 49, 40–50.

189. Ramakrishnan, S.; Dhakshnamoorthy, M.; Jelmy, E.J.; Vasanthakumari, R.; Kothurkar, N.K. Synthesis and characterization of graphene oxide-polyimide nanofiber composites. RSC Adv. 2014, 4, 9743–9749.

190. Panzavolta, S.; Bracci, B.; Gualandi, C.; Focarete, M.L.; Treossi, E.; Kouroupis-Agalou, K.; Rubini, K.; Bosia, F.; Brely, L.; Pugno, N.M.; et al. Structural reinforcement and failure analy-sis in composite nanofibers of graphene oxide and gelatin. Carbon 2014, 78, 566–577.

191. Qi, Y.Y.; Tai, Z.X.; Sun, D.F.; Chen, J.T.; Ma, H.B.; Yan, X.B.; Liu, B.; Xue, Q.J. Fabrication and characterization of poly(vinyl

alcohol)/graphene oxide nanofibrous biocomposite scaf-folds. J. Appl. Polym. Sci. 2013, 127, 1885–1894.

192. Barzegar, F.; Bello, A.; Fabiane, M.; Khamlich, S.; Momodu, D.; Taghizadeh, F.; Dangbegnon, J.; Manyala, N. Preparation and characterization of poly(vinyl alcohol)/graphene nanofibers synthesized by electrospinning. J. Phys.Chem. Solids 2015, 77, 139–145.

193. Pant, H.R.; Park, C.H.; Tijing, L.D.; Amarjargal, A.; Lee, D.H.; Kim, C.S. Bimodal fiber diameter distributed graphene oxide/nylon-6 composite nanofibrous mats via electrospin-ning. Colloids Surf. A Physicochem. Eng. Asp. 2012, 407, 121–125.

194. Wang, Y.; Tang, J.; Xie, S.; Liu, J.; Xin, Z.; Liu, X.; Belfiore, L.A. Leveling graphene sheets through electrospinning and their conductivity. RSC Adv. 2015, 5, 42174–42177.

195. Bao, Q.; Zhang, H.; Yang, J.; Wang, S.; Tang, D.Y.; Jose, R.; Ramakrishna, S.; Lim, C.T.; Loh, K.P. Graphene-Polymer Nanofiber Membrane for Ultrafast Photonics. Adv. Funct. Ma-ter. 2010, 20, 782–791.

196. Wang, M.; Yu, J.H.; Hsieh, A.J.; Rutledge, G.C. Effect of tethering chemistry of cationic surfactants on clay exfoliation, electrospinning and diameter of PMMA/clay nanocomposite fibers. Polymer 2010, 51, 6295–6302.

197. Islam, M.S.; Yeum, J.H.; Das, A.K. Effect of pullulan/poly(vinyl alcohol) blend system on the montmorillonite structure with property characterization of electrospun pullulan/poly(vinyl alcohol)/montmorillonite nanofibers. J. Colloid Interface Sci. 2012, 368, 273–281.

198. Wang, Y.; Li, M.; Rong, J.; Nie, G.; Qiao, J.; Wang, H.; Wu, D.; Su, Z.; Niu, Z.; Huang, Y. Enhanced orientation of PEO polymer chains induced by nanoclays in electrospun PEO/clay composite nanofibers. Colloid Polym. Sci. 2013, 291, 1541–1546.

199. Neppalli, R.; Wanjale, S.; Birajdar, M.; Causin, V. The effect of clay and of electrospinning on the polymorphism, structure and morphology of poly(vinylidene fluoride). Eur. Polym. J. 2013, 49, 90–99.

200. Dong, Y.; Haroosh, H.; Bickford, T. Development and characterisation of novel electrospun polylactic acid/tubular clay nanocomposites. J. Mater. Sci. 2011, 46, 6148–6153.

201. Kim, G.M.; Michler, G.H.; Ania, F.; Calleja, F.J.B. Temperature dependence of polymorphism in electrospun nanofibres of PA6 and PA6/clay nanocomposite. Polymer 2007, 48, 4814–4823.

202. Liu, Y.; Li, C.; Chen, S.; Wachtel, E.; Koga, T.; Sokolov, J.C.; Rafailovich, M.H. Electrospin-ning of poly(ethylene-co-vinyl acetate)/clay nanocomposite fibers. J. Polym. Sci. B Polym. Phys. 2009, 47, 2501–2508.

203. Kim, G.M.; Michler, G.H.; Potschke, P. Deformation processes of ultrahigh porous multi-walled carbon nanotubes/polycarbonate composite fibers prepared by electrospin-ning. Polymer 2005, 46, 7346–7351.

204. Kedem, S.; Schmidt, J.; Paz, Y.; Cohen, Y. Composite Polymer Nanofibers with Carbon Nanotubes and Titanium Dioxide Particles. Langmuir 2005, 21, 5600–5604.

205. Hou, H.; Ge, J.J.; Zeng, J.; Li, Q.; Reneker, D.H.; Greiner, A.; Cheng, S.Z.D. Electrospun Polyacrylonitrile Nanofibers Containing a High Concentration of Well-Aligned Multiwall Carbon Nanotubes. Chem. Mater. 2005, 17, 967–973.

206. Zhou, W.; Wu, Y.; Wei, F.; Luo, G.; Qian, W. Elastic deformation of multiwalled carbon nanotubes in electrospun MWCNTs–PEO and MWCNTs–PVA nanofibers. Polymer 2005, 46, 12689–12695.

207. Minoo, N.; Tong, L.; Mark, P.S.; Liming, D.; Xungai, W. Electrospun single-walled carbon nanotube/polyvinyl alcohol composite nanofibers: Structure-property relation-ships. Nanotechnology 2008, 19.

208. Dror, Y.; Salalha, W.; Pyckhout-Hintzen, W.; Yarin, A.; Zussman, E.; Cohen, Y. From Carbon Nanotube Dispersion to Composite Nanofibers, in Scattering Methods and the Properties of Polymer Materials; Springer: Berlin, Germnay; Heidelberg, Germnay, 2005; pp. 64–69.

209. Sundaray, B.; Subramanian, V.; Natarajan, T.S.; Krishnamurthy, K. Electrical conductivity of a single electrospun fiber of poly(methyl methacrylate) and multiwalled carbon nanotube nanocomposite. Appl. Phys. Lett. 2006, 88.

210. Sen, R.; Zhao, B.; Perea, D.; Itkis, M.E.; Hu, H.; Love, J.; Bekyarova, E.; Haddon, R.C. Preparation of Single-Walled Carbon Nanotube Reinforced Polystyrene and Polyurethane Nanofibers and Membranes by Electrospinning. Nano Lett. 2004, 4, 459–464.

211. Saeed, K.; Park, S.Y.; Lee, H.J.; Baek, J.B.; Huh, W.S. Preparation of electrospun nanofibers of carbon nanotube/polycaprolactone nanocomposite. Polymer 2006, 47, 8019–8025.

212. Pan, C.; Ge, L.Q.; Gu, Z. Fabrication of multi-walled carbon nanotube reinforced polyelectro-lyte hollow nanofibers by electrospinning. Compos. Sci. Technol. 2007, 67, 3271–3277.

213. Jeong, J.S.; Jeon, S.Y.; Lee, T.Y.; Park, J.H.; Shin, J.H.; Alegaonkar, P.S.; Berdinsky, A.S.; Yoo, J.B. Fabrication of MWNTs/nylon conductive composite nanofibers by electrospinning. Diam. Relat. Mater. 2006, 15, 1839–1843.

214. Ayutsede, J.; Gandhi, M.; Sukigara, S.; Ye, H.; Hsu, C.; Gogotsi, Y.; Ko, F. Carbon Nanotube Reinforced Bombyx mori Silk Nanofibers by the Electrospinning Pro-cess. Biomacromolecules 2006, 7, 208–214.

215. Podsiadlo, P.; Shim, B.S.; Kotov, N.A. Polymer/clay and polymer/carbon nanotube hybrid organic-inorganic multilayered composites made by sequential layering of nanometer scale films. Coord. Chem. Rev. 2009, 253, 2835–2851.

216. Caruso, F.; Caruso, R.A.; Mohwald, H. Nanoengineering of inorganic and hybrid hollow spheres by colloidal templating. Science 1998, 282, 1111–1114.

217. Kleinfeld, E.R.; Ferguson, G.S. Stepwise formation of multilayered nanostructural films from macromolecular precursors. Science 1994, 265, 370–373.

218. Kotov, N.A.; Dékány, I.; Fendler, J.H. Ultrathin graphite oxide-polyelectrolyte composites prepared by self-assembly: Transition between conductive and non-conductive states. Adv. Mater. 1996, 8, 637–641.

219. Lee, T.; Min, S.H.; Gu, M.; Jung, Y.K.; Lee, W.; Lee, J.U.; Seong, D.G.; Kim, B.S. Layer-by-Layer Assembly for Graphene-Based Multilayer Nanocomposites: Synthesis and Applica-tions. Chem. Mater. 2015, 27, 3785–3796.

220. Decher, G.; Hong, J.D.; Schmitt, J. Buildup of ultrathin multilayer films by a self-assembly process: III. Consecutively alternating adsorption of anionic and cationic polyelectrolytes on charged surfaces. Thin Solid Films 1992, 210–211, 831–835.

221. Picart, C.; Lavalle, P.; Hubert, P.; Cuisinier, F.J.G.; Decher, G.; Schaaf, P.; Voegel, J.C. Buildup Mechanism for Poly(l-

lysine)/Hyaluronic Acid Films onto a Solid Surface. Langmuir 2001, 17, 7414–7424.

222. Bergbreiter, D.E.; Franchina, J.G.; Sussman, L. Polyvalent Hydrogen-Bonding Functionaliza-tion of Ultrathin Hyperbranched Films on Polyethylene and Gold. Macromolecules 2001, 34, 3018–3023.

223. Such, G.K.; Johnston, A.P.R.; Caruso, F. Engineered hydrogen-bonded polymer multilayers: From assembly to biomedical applications. Chem. Soc. Rev. 2011, 40, 19–29.

224. Wang, F.; Ma, N.; Chen, Q.; Wang, W.; Wang, L. Halogen Bonding as a New Driving Force for Layer-by-Layer Assembly. Langmuir 2007, 23, 9540–9542.

225. Shimazaki, Y.; Mitsuishi, M.; Ito, S.; Yamamoto, M. Preparation of the Layer-by-Layer Deposited Ultrathin Film Based on the Charge-Transfer Interaction. Langmuir 1997, 13, 1385–1387.

226. Kohli, P.; Blanchard, G.J. Applying Polymer Chemistry to Interfaces: Layer-by-Layer and Spontaneous Growth of Covalently Bound Multilayers. Langmuir 2000, 16, 4655–4661.

227. Sun, J.; Wu, T.; Liu, F.; Wang, Z.; Zhang, X.; Shen, J. Covalently Attached Multilayer Assemblies by Sequential Adsorption of Polycationic Diazo-Resins and Polyanionic Poly(acrylic acid). Langmuir 2000, 16, 4620–4624.

228. Bergbreiter, D.E.; Chance, B.S. "Click"-Based Covalent Layer-by-Layer Assembly on Polyethylene Using Water-Soluble Polymeric Reagents. Macromolecules 2007, 40, 5337–5343.

229. Rouse, J.H.; Lillehei, P.T. Electrostatic assembly of polymer/single walled carbon nanotube multilayer films. Nano Lett.2003, 3, 59–62.

230. Artyukhin, A.B.; Bakajin, O.; Stroev, P.; Noy, A. Layer-by-Layer Electrostatic Self-Assembly of Polyelectrolyte Nanoshells on Individual Carbon Nanotube Templates. Langmuir 2004, 20, 1442–1448.

231. Zhang, C.; Liu, T.; Lu, X. Facile fabrication of polystyrene/carbon nanotube composite nanospheres with core-shell structure via self-assembly. Polymer 2010, 51, 3715–3721.

232. He, Q.; Cui, Y.; Ai, S.; Tian, Y.; Li, J. Self-assembly of composite nanotubes and their applications. Curr. Opin. Colloid Interface Sci. 2009, 14, 115–125.

233. Mamedov, A.A.; Kotov, N.A.; Prato, M.; Guldi, D.M.; Wicksted, J.P.; Hirsch, A. Molecular design of strong single-wall carbon nanotube/polyelectrolyte multilayer composites. Nat. Ma-ter. 2002, 1, 190–194.

234. Podsiadlo, P.; Kaushik, A.K.; Arruda, E.M.; Waas, A.M.; Shim, B.S.; Xu, J.D.; Nandivada, H.; Pumplin, B.G.; Lahann, J.; Ramamoorthy, A.; Kotov, N.A. Ultrastrong and Stiff Layered Pol-ymer Nanocomposites. Science 2007, 318, 80–83.

235. Halpin, J.C.; Kardos, J.L. The Halpin-Tsai equations: A review. Polym. Eng. Sci. 1976, 16, 344–352.

236. Terrones, M.; Martín, O.; González, M.; Pozuelo, J.; Serrano, B.; Cabanelas, J.C.; Vega-Díaz, S.M.; Baselga, J. Interphases in Graphene Polymer-based Nanocomposites: Achievements and Challenges. Adv. Mater. 2011, 23, 5302–5310.

237. Cadek, M.; Coleman, J.N.; Ryan, K.P.; Nicolosi, V.; Bister, G.; Fonseca, A.; Nagy, J.B.; Szostak, K.; Béguin, F.; Blau, W.J. Reinforcement of Polymers with Carbon Nanotubes: The Role of Nanotube Surface Area. Nano Lett. 2004, 4, 353–356.

238. Haggenmueller, R.; Gommans, H.H.; Rinzler, A.G.; Fischer, J.E.; Winey, K.I. Aligned single-wall carbon nanotubes in composites by melt processing methods. Chem. Phys. Lett. 2000, 330, 219–225.

239. Haggenmueller, R.; Zhou, W.; Fisher, J.E.; Winey, K.I. Production and characterization of polymer nanocomposites with highly aligned single-walled carbon nanotubes. J. Nanosci. Nanotechnol. 2003, 3, 105–110.

240. Chen, X.Q.; Saito, T.; Yamada, H.; Matsushige, K. Aligning single-wall carbon nanotubes with an alternating-current electric field. Appl. Phys. Lett. 2001, 78, 3714–3716.

241. Kumar, M.S.; Lee, S.H.; Kim, T.Y.; Kim, T.H.; Song, S.M.; Yang, J.W.; Nahm, K.S.; Suh, E.K. DC electric field assisted alignment of carbon nanotubes on metal electrodes. Solid-State Elec-tron. 2003, 47, 2075–2080.

242. Kumar, M.S.; Kim, T.H.; Lee, S.H.; Song, S.M.; Yang, J.W.; Nahm, K.S.; Suh, E.K. Influence of electric field type on the assembly of

single walled carbon nanotubes. Chem. Phys.Lett. 2004, 383, 235–239.

243. Martin, C.A.; Sandler, J.K.W.; Windle, A.H.; Schwarz, M.K.; Bauhofer, W.K.; Shaffer, M.S.P. Electric field-induced aligned multi-wall carbon nanotube networks in epoxy compo-sites. Polymer 2005, 46, 877–886.

244. Strobl, C.J.; Schaflein, C.; Beierlein, U.; Ebbecke, J.; Wixforth, A. Carbon nanotube alignment by surface acoustic waves. Appl. Phys. Lett. 2004, 85, 1427–1429.

245. Kamat, P.V.; Thomas, K.G.; Barazzouk, S.; Girishkumar, G.; Vinodgopal, K.; Meisel, D. Self-Assembled Linear Bundles of Single Wall Carbon Nanotubes and Their Alignment and Depo-sition as a Film in a dc Field. J. Am. Chem. Soc. 2004,126, 10757–10762.

246. Camponeschi, E.; Vance, R.; Al-Haik, M.S.; Garmestani, H.; Tannebaum, R. Properties of carbon nanotube-polymer composites in a magnetic field. Carbon 2007, 45, 2037–2046.

247. Garmestani, H.; Al-Haik, M.S.; Dahmen, K.; Tannenbaum, R.; Li, D.; Sablin, S.S.; Hussaini, M.Y. Polymer-Mediated Alignment of Carbon Nanotubes under High Magnetic Fields. Adv. Mater. 2003, 15, 1918–1921.

248. Steinart, B.W.; Dean, D.R. Magnetic field alignment and electrical properties of solution cast PET-carbon nanotube composite films. Polymer 2009, 50, 898–904.

249. Yang, J.; Wang, C.; Wang, K.; Zhang, Q.; Chen, F.; Du, R.; Fu, Q. Direct Formation of Nanohybrid Shish-Kebab in the Injection Molded Bar of Polyethylene/Multiwalled Carbon Nanotubes Composite. Macromolecules 2009, 42, 7016–7023.

250. Bin, Y.; Kitanaka, M.; Zhu, D.; Matsuo, M. Development of highly oriented polyethylene filled with aligned carbon nanotubes by gelation/crystallization from solu-tions. Macromolecules 2003, 36, 6213–6219.

251. Chen, W.; Tao, X. Production and characterization of polymer nanocomposite with aligned single wall carbon nanotubes. Appl. Surf. Sci. 2006, 252, 3547–3552.

252. Wang, Q.; Dai, J.F.; Li, W.; Wei, Z.Q.; Jiang, J.L. The effects of CNT alignment on electrical conductivity and mechanical properties of

SWNT/epoxy nanocomposites. Compos. Sci. Tech-nol. 2008, 68, 1644–1648.

253. Rafiee, M.A.; Rafiee, J.; Srivastava, I.; Wang, Z.; Song, H.; Yu, Z.Z.; Koratkar, N. Fracture and fatigue in graphene nanocomposites. Small 2010, 6, 179–183.

254. Yasmin, A.; Luo, J.J.; Daniel, I.M. Processing of expanded graphite reinforced polymer nanocomposites. Compos. Sci. Technol. 2006, 66, 1182–1189.

255. Usuki, A.; Hasegawa, N.; Kato, M. Polymer-Clay Nanocomposites. Adv. Polym. Sci. 2005, 179, 135–195.

256. Usuki, A.; Kojima, Y.; Kawasumi, M.; Okada, A.; Fukushima, Y.; Kurauchi, T.; Kamigaito, O. Synthesis of nylon 6-clay hybrid. J. Mater. Res. 1993, 8, 1179–1184.

257. Fornes, T.D.; Yoon, P.J.; Keskkula, H.; Paul, D.R. Nylon 6 nanocomposites: The effect of matrix molecular weight.Polymer 2001, 42, 09929–09940.

258. Hu, K.; Gupta, M.K.; Kulkarni, D.D.; Tsukruk, V.V. Ultra-Robust Graphene Oxide-Silk Fibroin Nanocomposite Membranes. Adv. Mater. 2013, 25, 2301–2307.

259. Kulkarni, D.D.; Choi, I.; Singamaneni, S.S.; Tsukruk, V.V. Graphene Oxide-Polyelectrolyte Nanomembranes. ACS Nano 2010, 4, 4667–4676.

260. Kotov, N.A.; Magonov, S.; Tropsha, E. Layer-by-Layer Self-Assembly of Alumosilicate-Polyelectrolyte Composites: Mechanism of Deposition, Crack Resistance, and Perspectives for Novel Membrane Materials. Chem. Mater. 1998, 10, 886–895.

261. Tang, Z.; Magonov, S.; Ozturk, B. Nanostructured artificial nacre. Nat. Mater. 2003, 2, 413–418.

262. Andrews, R.; Jacques, D.; Minot, M.; Rantell, T. Fabrication of carbon multiwall nano-tube/polymer composites by shear mixing. Macromol. Mater. Eng. 2002, 287, 395–403.

263. Xiao, K.Q.; Zhang, L.C.; Zarudi, I. Mechanical and rheological properties of CNT-reinforced polyethylene composites.Compos. Sci. Technol. 2007, 67, 177–182.

264. Gorrasi, J.; Sarno, M.; Di Bartolomeo, A.; Sannino, D.; Ciambelli, P.; Vittoria, V. Incorporation of carbon nanotubes into polyethylene

by high energy ball milling: Morphology and physical properties. J. Polym. Sci. B Polym. Phys. 2007,45, 597–606.

265. Moore, E.M.; Ortiz, D.L.; Marla, V.T.; Shambaugh, R.L.; Grady, B.P. Enhancing the strength of polypropylene fibers with CNTs. J. Appl. Polym. Sci. 2004, 93, 2926–2933.

266. Manchado, M.A.L.; Valentini, L.; Biagiotti, J.; Kenny, J.M. Thermal and mechanical proper-ties of SWCNT-polypropylene composites prepared by melt processing. Carbon 2005, 43, 1499–1505.

267. Dondero, W.E.; Gorg, R.E.A. Morphological and mechanical properties of CNT-polymer composites via melt compounding. J. Polym. Sci. B Polym. Phys. 2006, 44, 864–878.

268. Jose, M.V.; Dean, D.; Tyner, J.; Price, G.; Nyairo, E. Polypropylene/CNT nanocomposite fibers: Process-morphology-property relationships. J. Appl. Polym. Sci. 2007, 103, 3844–3850.

269. Jia, Z.; Wang, Z.; Xu, C.; Liang, J.; Wei, B.; Wu, D.; Zhu, S. Study on poly(methyl methacry-late)/carbon nanotube composites. Mater. Sci. Eng. A 1999, 271, 395–400.

270. Velasco-Santos, C.; Martinez-Hernandez, A.L.; Fisher, F.T.; Ruoff, R.; Castano, V.M. Improvement of thermal and mechanical properties of carbon nanotube composites through chemical functionalization. Chem. Mater. 2003, 15, 4470–4475.

271. Gorga, R.E.; Cohen, R.E. Toughness enhancements in PMMA by addition of oriented MWCNTs. J. Polym. Sci. B Polym. Phys. 2004, 42, 2690–2702.

272. Blond, D.; Barron, V.; Ruether, M.; Ryan, K.P.; Nicolosi, V.; Blau, W.J.; Coleman, J.N. Enhancement of Modulus, Strength, and Toughness in Poly(methyl methacrylate)-Based Composites by the Incorporation of Poly(methyl methacrylate)-Functionalized Nanotubes. Adv. Funct. Mater. 2006, 16, 1608–1613.

273. Hou, Y.; Tang, J.; Zhang, H.; Qian, C.; Feng, Y.; Liu, J. Functionalised few-walled carbon nanotubes for mechanical reinforcement of polymeric composites. ACS Nano 2009, 3, 1057–1062.

274. Bokobza, L.; Belin, C. Effect of strain in the properties of a styrene-butadiene rubber filled with MWCNTs. J. Appl. Polym. Sci. 2007, 105, 2054–2061.

275. Zhang, W.D.; Shen, L.; Phang, I.Y.; Liu, T. CNT reinforced nylon-6 composite prepared by simple melt compounding.Macromolecules 2004, 37, 256–259.

276. Liu, T.; Phang, I.Y.; Shen, L.; Chow, S.Y.; Zhang, Y.D. Morphology and mechanical proper-ties of MWCNT reinforced nylon-6 composites. Macromolecules 2004, 37, 7214–7222.

277. Ga, J.; Zhao, B.; Itkis, M.E.; Bekyarova, E.; Hu, H.; Kranak, V.; Yu, A.; Haddon, R.C. Chemical Engineering of the Single-Walled Carbon Nanotube-Nylon 6 Interface. J. Am. Chem. Soc. 2006, 128, 7492–7496.

278. Ge, J.J.; Zhang, D.; Li, Q.; Hou, H.; Graham, M.J.; Dai, L.; Harris, F.W.; Cheng, S.Z.D. Multiwalled Carbon Nanotubes with Chemically Grafted Polyetherimides. J. Am. Chem. Soc. 2005, 127, 9984–9985.

279. Liu, T.; Tong, Y.; Zhang, W.D. Preparation and characterization of CNT-polyetherimide nanocomposite films. Compos. Sci. Technol. 2007, 67, 406–412.

280. Siochi, E.J.; Working, D.C.; Park, C.; Lillehei, P.T.; Rouse, J.H.; Topping, C.C.; Bhattacharyya, A.R.; Kumar, S. Melt processing of SWCNT-polyimide nanocomposite fibers. Compos. B 2004, 35, 439–446.

281. Xia, H.; Song, M. Preparation and characterization of polyurethane-carbon nanotube composites. Soft Matter 2005, 1, 386–394.

282. Jung, Y.C.; Sahoo, N.G.; Cho, J.W. Polymeric nanocomposites of polyurethane block copolymers and functionalized multi-walled carbon nanotubes as crosslinkers. Macromol. Rapid Commun. 2006, 27, 126–131.

283. Penumadu, D.; Dutta, A.; Pharr, G.M.; Files, B. Mechanical properties of blended SWCNT composites. J. Mater. Res.2003, 18, 1849–1853.

284. Zheng, W.; Wong, S.C.; Sue, H.J. Transport behavior of PMMA/expanded graphite nano-composites. Polymer 2002, 43, 6767–6773.

285. Ramanathan, T.; Stankovich, S.; Dikin, D.A.; Liu, H.; Shen, H.; Nguyen, S.T.; Brinson, L.C. Graphitic Nanofillers in PMMA Nanocomposites—An Investigation of Particle Size and Dis-persion

and Their Influence on Nanocomposite Properties. J. Polym. Sci. B Polym. Phys. 2007, 45, 2097–2112.

286. Jang, J.Y.; Kim, M.S.; Jeong, H.M.; Shin, C.M. Graphite oxide/poly(methyl methacrylate) nanocomposites prepared by a novel method utilizing macroazoinitiator. Compos. Sci. Tech-nol. 2009, 69, 186–191.

287. Zheng, W.; Lu, X.; Wong, S.C. Electrical and mechanical properties of expanded graphite-reinforced high-density polyethylene. J. Appl. Polym. Sci. 2004, 91, 2781–2788. [Google Schol-ar] [CrossRef]

288. Liang, J.; Zhang, L.; Wang, Y.; Ma, Y.; Guo, T.; Chen, Y. Molecular-Level Dispersion of Graphene into Poly(vinyl alcohol) and Effective Reinforcement of their Nanocomposites. Adv. Funct. Mater. 2009, 19, 2297–2302.

289. Zhao, X.; Zhang, Q.; Chen, D. Enhanced mechanical properties of graphene-based poly(vinyl alcohol) composites.Macromolecules 2010, 43, 2357–2363.

290. Xu, Y.; Hong, W.; Bai, H.; Li, C.; Shi, G. Strong and ductile poly(vinyl alcohol)/graphene oxide composite films with a layered structure. Carbon 2009, 47, 3538–3543.

291. Lee, Y.R.; Raghu, A.V.; Jeong, H.M.; Kim, B.K. Properties of waterborne polyure-thane/functionalized graphene sheet nanocomposites prepared by an in situ method. Macromol. Chem. Phys. 2009, 210, 1247–1254.

292. Feng, R.; Guan, G.; Zhou, W.; Li, C.; Zhang, D.; Xiao, Y. In situ synthesis of poly(ethylene terephthalate)/graphene composites using a catalyst supported on graphite oxide. J. Mater. Chem. 2011, 21, 3931–3939.

293. Garboczi, E.J.; Snyder, K.A.; Douglas, J.F.; Thorpe, M.F. Geometrical percolation threshold of overlapping ellipsoids.Phys. Rev. E 1996, 52, 819–828.

294. Steurer, P.; Wissert, R.; Thomann, R.; Mülhaupt, R. Functionalized Graphenes and Thermo-plastic Nanocomposites Based upon Expanded Graphite Oxide. Macromol. Rapid Com-mun. 2009, 30, 316–327.

295. Bai, J.B.; Allaoui, A. Effect of the length and the aggregate size of MWNTs on the improve-ment efficiency of the mechanical and

electrical properties of nanocomposites—Experimental investigation. Compos. A Appl. Sci. Manuf.2003, 34, 689–694.

296. Celzard, A.; McRae, E.; Deleuze, C.; Dufort, M.; Furdin, G.; Marêché, J.F. Critical concentra-tion in percolating systems containing a high-aspect-ratio filler. Phys. Rev. B 1996, 53, 6209–6214.

297. Qi, X.Y.; Yan, D.; Jiang, Z.; Cao, Y.K.; Yu, Z.Z.; Yavari, F.; Koratkar, N. Enhanced Electrical Conductivity in Polystyrene Nanocomposites at Ultra-Low Graphene Content. ACS Appl. Mater. Interfaces 2011, 3, 3130–3133.

298. Sandler, J.K.W.; Kirk, J.E.; Kinloch, I.A.; Shaffer, M.S.P.; Windle, A.H. Ultra-low electrical percolation threshold in carbon-nanotube-epoxy composites. Polymer 2003, 44, 5893–5899.

299. Li, J.; Sham, M.L.; Kim, J.K.; Marom, G. Morphology and properties of UV/ozone treated graphite nanoplatelet/epoxy nanocomposites. Compos. Sci. Technol. 2007, 67, 296–305.

300. Bauhofer, W.; Kovacs, J.Z. A review and analysis of electrical percolation in carbon nanotube polymer composites.Compos. Sci. Technol. 2009, 69, 1486–1498.

301. Xie, L.; Xu, F.; Qiu, F.; Lu, H.; Yang, Y. Single-walled carbon nanotubes functionalized with high bonding density of polymer layers and enhanced mechanical properties of compo-sites. Macromolecules 2007, 40, 3296–3305.

302. Verdejo, R.; Barroso-Bujans, F.; Rodriguez-Perez, M.A.; Saja, J.A.D.; Lopez-Manchado, M.A. Functionalized graphene sheet filled silicone foam nanocomposites. J. Mater. Chem. 2008, 18, 2221–2226.

303. Veca, L.M.; Meziani, M.J.; Wang, W.; Wang, X.; Lu, F.; Zhang, P.; Lin, Y.; Fee, R.; Connell, J.W.; Sun, Y.P. Carbon nanosheets for polymeric nanocomposites with high thermal conduc-tivity. Adv. Mater. 2009, 21, 2088–2092.

304. Wang, S.; Tambraparni, M.; Qiu, J.; Tipton, J.; Dean, D. Thermal Expansion of Graphene Composites. Macromolecules2009, 42, 5251–5255.

305. Liu, N.; Luo, F.; Wu, H.; Liu, Y.; Zhang, C.; Chen, J. One-Step Ionic-Liquid-Assisted Electro-chemical Synthesis of Ionic-Liquid-

Functionalized Graphene Sheets Directly from Graph-ite. Adv. Funct. Mater. 2008, 18, 1518–1525.

306. Kayano, Y.; Keskkula, H.; Paul, D.R. Effect of polycarbonate molecular weight and pro-cessing conditions on mechanical behaviour of blends with a core-shell impact modifi-er. Polymer 1996, 37, 4505–4518.

307. Koerner, H.; Hampton, E.; Dean, D.; Turgut, Z.; Drummy, L.; Mirau, P.; Vaia, R. Generating Triaxial Reinforced Epoxy/Montmorillonite Nanocomposites with Uniaxial Magnetic Fields. Chem. Mater. 2005, 17, 1990–1996.

308. Koerner, H.; Jacobs, J.D.; Tomlin, D.W.; Busbee, J.D.; Vaia, R.A. Tuning Polymer Nanocom-posite Morphology: AC Electric Field Manipulation of Epoxy-Montmorillonite (Clay) Suspen-sions. Adv. Mater. 2004, 16, 297–302.

309. Sasaki, T.; Shimizu, A.; Mourey, T.H.; Thurau, C.T.; Ediger, M.D. Glass transition of small polystyrene spheres in aqueous suspensions. J. Chem. Phys. 2003, 119, 8730–8735.

310. Ding, J.; Xue, G.; Dai, Q.; Cheng, R. Glass transition temperature of polystyrene microparti-cles. Polymer 1993, 34, 3325–3327.

311. Forrest, J.A.; Dalnoki-Veress, K.; Stevens, J.R.; Dutcher, J.R. Effect of Free Surfaces on the Glass Transition Temperature of Thin Polymer Films. Phys. Rev. Lett. 1996, 77, 2002–2005.

312. Rittigstein, P.; Priestley, R.D.; Broadbelt, L.J.; Torkelson, J.M. Model polymer nanocomposites provide an understanding of confinement effects in real nanocomposites. Nat. Mater. 2007, 6, 278–282.

313. Liao, K.H.; Aoyama, S.; Abdala, A.A.; Macosko, C.W. Does Graphene Change Tg of Nanocomposites? Macromolecules2014, 47, 8311–8319.

314. Gaur, U.; Wunderlich, B. Study of Microphase Separation in Block Copolymers of Styrene and α-Methylstyrene in the Glass Transition Region Using Quantitative Thermal Analy-sis. Macromolecules 1980, 13, 1618–1625.

315. Roth, C.B.; Dutcher, J.R. Glass transition temperature of freely-standing films of atactic poly(methyl methacrylate).Eur. Phys. J. E 2003, 12, 103–107.

316. Tate, R.S.; Fryer, D.S.; Pasqualini, S.; Montague, M.F.; de Pablo, J.J.; Nealey, P.F. Extraordi-nary elevation of the glass transition temperature of thin polymer films grafted to silicon ox-ide substrates. J. Chem. Phys. 2001, 115, 9982–9990.

317. Keddie, J.L.; Jones, R.A.L.; Cory, R.A. Interface and surface effects on the glass-transition temperature in thin polymer films. Faraday Discuss. 1994, 98, 219–230.

318. Rittigstein, P.; Torkelson, J.M. Polymer-nanoparticle interfacial interactions in polymer nanocomposites: Confinement effects on glass transition temperature and suppression of physical aging. J. Polym. Sci. B Polym. Phys. 2006, 44, 2935–2943.

319. Yuen, S.M.; Ma, C.M.; Lin, Y.Y.; Kuan, H.C. Preparation, morphology and properties of acid and amine modified multiwalled carbon nanotube/polyimide composite. Compos. Sci. Tech-nol. 2007, 67, 2564–2573.

320. Grady, B.P. Effects of carbon nanotubes on polymer physics. J. Polym. Sci. B Polym. Phys. 2012, 50, 591–623.

321. Lee, K.M.; Han, C.D. Effect of hydrogen bonding on the rheology of polycar-bonate/organoclay nanocomposites.Polymer 2003, 44, 4573–4588.

322. Dai, X.; Xu, J.; Guo, X.; Lu, Y.; Shen, D.; Zhao, N.; Luo, X.; Zhang, X. Study on Structure and Orientation Action of Polyurethane Nanocomposites. Macromolecules 2004, 37, 5615–5623.

323. Zhang, X.; Loo, L.S. Study of glass transition and reinforcement mechanism in poly-mer/layered silicate nanocomposites. Macromolecules 2009, 42, 5196–5207.

324. Krishnamoorti, R.; Vaia, R.A.; Giannelis, E.P. Structure and Dynamics of Polymer-Layered Silicate Nanocomposites. Macromolecules 1996, 8, 1728–1734.

325. Messersmith, P.B.; Giannelis, E.P. Synthesis and barrier properties of poly(e-caprolactone)-layered silicate nanocomposites. J. Polym. Sci. A Polym. Chem. 1995, 33, 1047–1057.

326. Yang, Y.H.; Bolling, L.; Priolo, M.A.; Grunlan, J.C. Super gas barrier and selectivity of graphene oxide-polymer multilayer thin films. Adv. Mater. 2013, 45, 503–508. [Google Schol-ar] [CrossRef] [PubMed]

327. Liu, H.; Kuila, T.; Kim, N.H.; Kud, B.C.; Lee, J.H. In situ synthesis of the reduced graphene oxide-polyethyleneimine composite and its gas barrier properties. J. Mater. Chem. A 2013, 1, 3739–3746.

328. Tseng, I.H.; Liao, Y.F.; Chiang, J.C.; Tsai, M.H. Transparent polyimide/graphene oxide nanocomposite with improved moisture barrier property. Mater. Chem. Phys. 2012, 136, 247–253.

329. Morgan, A.B. Flame retarded polymer layered silicate nanocomposites:A review of commer-cial and open literature systems. Polym. Adv. Technol. 2006, 96, 206–217.

330. Schutz, M.R.; Kalo, H.; Lunkenbein, T.; Breu, J.; Wilkie, C.A. Intumescent-like behavior of polystyrene synthetic clay nanocomposites. Polymer 2012, 52, 3288–3294.

331. Bartholmai, M.; Schartel, B. Layered silicate polymer nanocomposites: New approach or illusion for fire retardancy? Investigations of the potentials and the tasks using a model sys-tem. Polym. Adv.Technol. 2004, 15, 355–364.

332. Costache, M.C.; Heidecker, M.J.; Manias, E.; Camino, G.; Frache, A.; Beyer, G.; Gupta, R.K.; Wilkie, C.A. The influence of carbon nanotubes, organically modified montmorillonites and layered double hydroxides on the thermal degradation and fire retardancy of polyethylene, ethylene-vinyl acetate copolymer and polystyrene. Polymer 2007, 48, 6352–6345.

333. May, P.; Khan, U.; O'Neill, A.; Coleman, J.N. Approaching the theoretical limit for reinforc-ing polymers with graphene. J. Mater. Chem. 2012, 22, 1278–1282.

334. Grimmer, C.S.; Dharan, C.K.H. High-cycle fatigue of hybrid carbon nanotube/glass fi-ber/polymer composites. J. Mater. Sci. 2008, 43, 4487–4492.

335. Kim, K.T.; Jo, W.H. Non-destructive functionalization of multi-walled carbon nanotubes with naphthalene-containing polymer for high performance Nylon66/multi-walled carbon nano-tube composites. Carbon 2011, 49, 819–826.

336. Yuan, W.; Chan-Park, M.B. Covalent cum Noncovalent Functionalizations of Carbon Nanotubes for Effective Reinforcement of a Solution Cast Composite Film. ACS Appl. Mater. Interfaces 2012, 4, 2065–2073.

337. Aoyama, S.; Park, Y.T.; Ougizawa, T.; Macosko, C.W. Melt crystallization of poly(ethylene terephthalate): Comparing addition of graphene vs. carbon nanotubes. Polymer 2014, 55, 2077–2085.

338. Calcagno, C.I.W.; Mariani, C.M.; Teixeira, S.R.; Mauler, R.S. The effect of organic modifier of the clay on morphology and crystallization properties of PET nanocompo-sites. Polymer 2007, 48, 966–974.

339. Chou, C.C.; McAtee, J.L. Thermal Decomposition of Organo-Ammonium Compounds Exchanged Onto Montmorillonite and Hectorite. Clay and Clay Miner. 1969, 17, 339–346.

340. Xie, W.; Gao, Z.; Pan, W.-P.; Hunter, D.; Singh, A.; Vaia, R. Thermal Degradation Chemistry of Alkyl Quaternary Ammonium Montmorillonite. Chem. Mater. 2001, 13, 2979–2990.

341. Nanostructured and Amorphous Materials Incorporated. Available online: http:www.nanoamor.com (accessed on 6 December 2015).

342. Cheap Tubes. Available online: http://www.cheaptubes.com/ (accessed on 6 December 2015).

4

CHARACTERIZATION OF NANOTUBE-REINFORCED POLYMER COMPOSITES

Wenjie Wang[1] and N. Sanjeeva Murthy[2]

[1] Ames Laboratory, Iowa State University, Ames, IA, U.S.A.

[2] New Jersey Center for Biomaterials, Rutgers – The State University of New Jersey, NJ, U.S.A.

1. INTRODUCTION

Polymers are more readily processible than metals and ceramics, but are not as strong or as stiff. But the mechanical attributes of the polymer can be significantly improved by reinforcing polymers with high Young's modulus fibers. The resulting polymer composites are an important class of light weight materials with excellent specific mechanical properties. Unlike micron size particles and short fibers used in conventional composites, carbon nanotubes (CNTs) allow polymers to be reinforced at the molecular level and thus impart significantly greater improvement in mechanical properties. This is because for the reinforcement to be effective, it is necessary that the fibers be sufficiently long and the interface between the fiber and the matrix be strong, features that are provided by CNTs. The advantages of polymer-nanotube composites (PNCs) were discussed by Calvert (Calvert, 1999). Besides the economic advantage brought about by mixing costly CNTs and inexpensive polymer matrices, it is possible that a synergy exists between the CNTs and polymer matrices so that it is possible to go beyond the simple rule of mixture, and make full use of CNTs' exceptional mechanical properties to enhance the mechanical properties of the composite.

There are several reviews of PNCs (e.g., Breuer & Sundararaj, 2004, Moniruzzaman & Winey, 2006, and Coleman et al., 2006). The present chapter is different from these excellent reviews in that it focuses on the structure and morphology of the reinforced structures, and the consequence of these structures to their performance. The general features of the PNCs will be further illustrated using the results from our work on the study of reinforcements in polyacrylonitrile (PAN) with multi- and single-wall carbon nanotubes (CNTs).

2. POLYMER NANOCOMPOSITES

2.1. Carbon Nanotubes and Nanofibers

Carbon nanotubes (CNTs), discovered by Iijima in 1991, are seamless cylinders made of rolled up hexagonal network of carbon atoms (Iijima, 1991). Single-wall nanotubes (SWNTs) consist of a single cylindrical layer of carbon atoms. The details of the structure of multi-wall carbon nanotubes (MWNTs) are still being resolved, but can be envisioned as a tubular structure consisting of multiple walls with an inter-layer separation of 0.34 nm. The diameter for inner most tube is on the scale of nanometer, and the length of the tube can be up to ~ μm. In one model, the MWNTs are considered as a single sheet of graphene that is rolled up like a scroll to form multiple walls (Dravid et al., 1993). In another model, MWNT is thought to be made of co-axial walls consisting of hexagons aligned in helical manner, the individual tubes forming a nested-shell structure. Ge and Sattler (Ge & Sattler, 1993) found that as the graphene sheet rolls up and matches at joint line seamlessly, different wrapping angles resulting in different helicities is necessary for tubes to be separated by a uniform spacing. It is also likely that there could be the coexistence of scroll-type (open cylinder) and perfect Russian doll (nested shell) structures (Zhou et al., 1994). These two different structure models have an impact on the mechanical as well as the optical and electronic properties of CNTs. The carbon atoms in the plane of graphene sheet are bound with strong covalent sp^2 hybridized bonds. The interaction between walls can be approximated by weak van de Waals force or π bond. Thus, the sliding of walls with respect to each other might significantly reduce axial modulus of MWNT compared to SWNT, more so in the nested shell structure than in the scroll-type structure.

Vapor-grown carbon nano-fibers (VGCNF), which are also used as reinforcements, are 50-200 nm in diameter, 20-100 μm in length, and bear resemblance to MWNT, except that graphite sheets make an angle (~15º) to the long- axis (Uchida et al., 2006). Compared to the CNTs, VGCNF is less costly to produce.

CNTs have exceptional mechanical, optical, electrical and thermal properties (Chae et al., 2009; Chae et al., 2007; Guo et al., 2005; Koganemaru et al., 2004; Sreekumar et al., 2004; Ye et al., 2004). The elastic modulus of an isolated CNT, estimated using intrinsic thermal vibrations in TEM, is 1.8 TPa. (Treacy et al., 1996), almost an order of magnitude larger than that of carbon fibers. This axial modulus is higher than the 1 Tpa in-plane modulus of a single graphene sheet due to the wrap-up of the graphene sheet into a cylinder. Bower et al. measured the fracture strain of MWNT ≥18% in polymer composites, almost one or two orders of magnitude larger than carbon fibers (Bower et al., 1999). Chae and Kumar suggested that, because the tensile strength of CNTs is 100-600 GPa, about two orders of magnitude larger than that of ordinary carbon fibers, the CNTs might be the ultimate reinforcing materials for strong fibers (Chae & Kumar, 2009).

2.2. Cnt/Polymer Composites

The advantages of using CNTs as reinforcements in polymer matrices can be illustrated with the following empirical relationship between the critical length (*lc*), and fracture stress σf of the reinforcing fiber (Wang et al., 2008a):

$$l_c = \frac{\sigma_f d}{2\tau_c} \tag{1}$$

where *d* is the fiber diameter, and τ_c is the interfacial strength. The critical fiber length is the minimum length of the fiber such that the failure occurs within the fiber, and thus ensures that it acts as an effective reinforcement. This relationship shows that nanotubes with their smaller diameter, and greater interfacial strength when molecularly dispersed serve as effective reinforcements even at smaller lengths.

The importance of bonding between CNTs and matrices for effective load transfer without slipping of contact surfaces has been widely recognized (Calvert, 1999). It is desirable that a large fraction of the load applied to the composites is efficiently transferred to CNTs through the interaction or bonding between the outer surface of the CNTs and the surrounding polymer matrix (Ajayan et al., 1997). It is widely accepted that larger the area of interface between the matrix and nano-fillers, the better the mechanical enhancement. Several load-transfer models have been proposed and examined using atomistic molecular dynamic simulations. Most recent molecular dynamics simulations suggest that weak van der Waals force (Srivastava et al., 2003), as well as thermal expansion mismatch can contribute to strong interfacial interaction (Wong et al., 2003; Zhang & Wang, 2005). Panhuis's model suggests that the polymer chains will adapt their conformations to wrap around a single CNT (Panhuis et al., 2003; Wong et al., 2003). Frankland's model suggests that a small fraction of chemical bonds are formed between polymers and carbon atoms of the CNTs thus cross-links are formed between the matrix and CNTs (Frankland et al., 2002).

Chemical bonding between CNT and polymer is apparent in the adhesion of polymers to CNT on the fracture surface through transmission electron microscopy (TEM) (Bower et al., 1999). However, Raman spectroscopy along with the SEM images of fracture surface by Ajayan et al. showed the sliding of SWNT bundles with respect to the matrix suggesting that the actual load transfer was poor between CNT and the polymer matrix, which rules out the chemical bonding model (Ajayan et al., 2000). A number of experiments have been carried out to test these models for the interaction between polymer matrix and CNTs (SWNTs and MWNTs) through tensile tests, microscopy and x-ray scattering. Deformation of the CNTs observed by x-ray scattering measurements (see section 3.3) shows an interaction between the matrix and CNTs that is sufficient to transfer the load from the matrix to the CNTs (Wang et al.,2008b). Actual measurements of the adhesion between the matrix and CNTs were also carried out using scanning probe microscope to pull out the CNTs from the matrix (Cooper et al., 2002). Coleman et al. found that a layer of polymer attached to the CNTs after pullout was very similar to the crystalline polymer coating nucleated by the MWNTs. The presence of interfacial crystalline layers partly contributes to the mechanical properties of composites (Coleman et al., 2006). Pull-out measurement can also be carried out by atomic force microscopy (AFM). Barber et al. attached one end of a single CNT to the tip of AFM and embedded the other end into the polymer matrix. The critical force needed to pull the nanotube from the polymer matrix was then recorded (Figure 1).

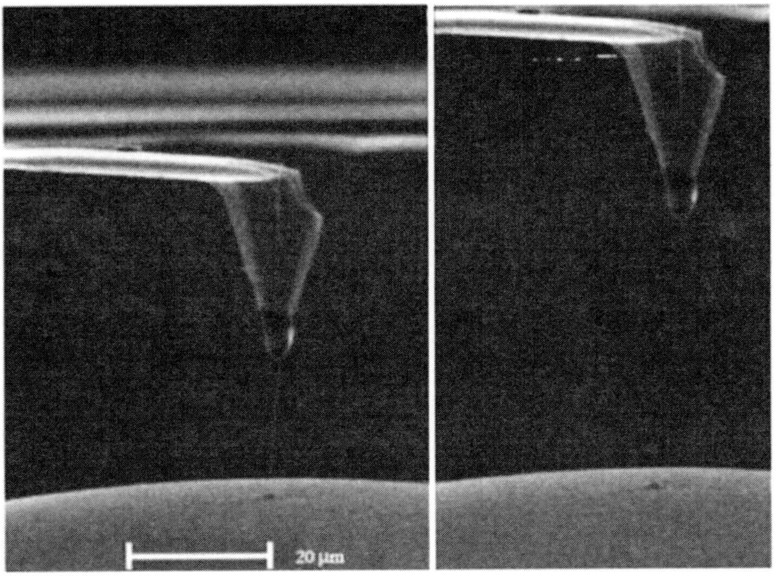

Figure 1. Left panel shows that one end of a single CNT is attached to the AFM tip and the other is embedded in the solid polymer matrix. Right panel shows the process of pull-out of CNT. The critical force needed for pull-out can be recorded by AFM (Reprinted from Philosophical Transactions of the Royal Society A: Mathematical, Physical and Engineering Sciences, vol. 366, W. Wang, P. Ciselli, E. Kuznetsov, T. Peijs and A.H. Barber. "Effective reinforcement in carbon nanotube-polymer composites", **Figure 2**, Copyright 2008, with permission from the Royal Society.).

Because of the interfacial interaction between CNTs and the polymer matrix, the properties of the matrix in the interfacial regions are different from the bulk materials (Schadler, 2007). CNTs dispersed in the polymer matrix, despite the small volume fraction (usually 1-5 vol%), enhance the overall mechanical properties, also affect the structure of the matrix during conventional processing. Many observations show that the presence of CNTs can affect the crystallization behaviors of the polymers (Brosse et al., 2008; Grady et al., 2002; Liu et al., 2004) and alter the morphology of matrix (Brosse et al., 2008; Chatterjee et al., 2007). For instance, instead of the expected spherulites, crystalline lamellae were observed growing perpendicular to the surface of CNTs. (Brosse et al., 2008). Therefore, the microstructure of the polymer matrix should be taken into account when evaluating the performance of nanocomposites (Advani, 2006).

Two other factors that affect the performance of CNT-reinforced nanocomposites are the degree of dispersion and alignment of nanotubes. Poor or nonuniform dispersion causes the fillers to aggregate, acting as defect sites (Advani, 2006). Dispersion process is affected by interactions of CNT-CNT and CNT-matrix. It is controlled by several factors such as filler's size, specific surface area, and interfacial volumes (Moniruzzaman & Winey, 2006). The degree of dispersion of CNTs in polymer matrix can be assessed using small angle x-ray scattering as described in the following section. Rheological study of dispersions of MWNTs in the polymer matrix by Huang et al (Huang et al., 2006) also suggests that a satisfactory dispersion can be achieved if the mixing time exceeds a critical value (characteristic mixing time).

2.3. Processing Of Pncs

CNTs are dispersed into the polymer matrix to form a polymer/CNTs composite, and there are several ways to achieve a good dispersion. Following is a condensed version of the various processing methods as discussed by Coleman et al. (Coleman et al., 2006). Solution processing is the easiest if the polymer can be dissolved in a suitable solvent. Typically, nanotubes well dispersed in a solvent is mixed with a polymer solution with energetic agitation (e.g., sonication), and the solvent is allowed to evaporate in a controlled way leaving a composite film. For insoluble thermoplastic polymers, the polymer is heated above either glass transition (if amorphous) or melt temperature (if semicrystalline) to form a viscous liquid, and the CNTs are mixed into this viscous liquid to be then processed into a fiber or a film using common production processes. For thermosetting polymers such as epoxy, the CNTs are blended when epoxy is in a liquid state, and the mixture is then allowed to cure. Chemical processing of Polymer/CNTs composites aims at formation of chemical bonds between the polymer matrix and the CNTs; this provides for stronger interaction between the polymer and CNTs and better dispersion, and thus results in better mechanical properties. Chemical methods include *in-situ* polymerization and functionalization of CNTs with polymer, also called polymer grafting.

3. CHARACTERIZATION OF CNT/POLYMER COMPOSITES

Although the CNTs' high strength and stiffness as well as high aspect ratio promise significant improvement in mechanical properties of polymer nanocomposites (Treacy et al., 1996), the actual measured enhancement is below these expectations (Schaefer et al., 2003). The major reason might be the intrinsic van der Waals' attraction makes CNTs entangled agglomerates incapable of fully blending into the polymer matrix. Since the four factors important to the reinforcement are CNTs' aspect ratio, alignment, dispersion, and interfacial interactions (Coleman et al., 2006), it is desirable to examine the morphology of CNTs in the polymer matrix. Several methods including x-ray diffraction, Raman spectroscopy, scanning and transmission electron microscopy are employed to characterize the structure, morphology and properties of carbon nanotube-reinforced polymer nanocomposites.

3.1. Electron Microscopy

Transmission electron microscopy (TEM) and scanning electron microscopy (SEM) enable the visualization of the microscopic structure and morphology of CNTs in composites at resolution down to 0.1 nm (Michler, 2008; Sawyer et al., 2010). The limitations of using electron beam to probe the polymer are the low contrast (most polymers are made of light elements) and radiation damage. TEM is used to probe the bulk structure from thin sections of the material, and SEM to examine the surfaces. In addition to imaging, TEM is also used to carry out micromechanical measurements. Examples of these applications will be illustrated below.

3.1.1. Dispersion and Alignment Of The Cnts

One of the primary uses of microscopy is characterizing the dispersion and alignment of the nanotubes in the polymers. Figure 2a is an SEM image from a composite prepared by mixing aqueous solution poly(vinyl alcohol) with CNT dispersions followed by subsequent casting and controlled water evaporation (Shaffer & Windle, 1999). It shows homogeneous dispersion of the nanotubes even at 50 wt%

loading. By examining the fracture surface using SEM, they confirmed the uniform dispersion of CNTs. Figure 2b shows the use of TEM to examine the dispersion of CNTs (Qian et al., 2000). In this experiment Qian et al. used high-energy sonication to disperse MWNTs (predispersed in toluene) into polystyrene (PS) dissolved in toluene, and the solvent was then evaporated to produce 0.4 mm thick films. Figure 2b shows 34 nm diameter MWNTs well dispersed down to the μm length scales. The inset is a plot of the loading of the CNT's at different length scale, and shows that homogeneity increases (decrease in standard deviation) with increase in the length scale. Furthermore, the TEM images also reveal defects that were present as a result of the polymer shrinkage primarily at the end of the nanotubes. Other than the quality of dispersion, both of these nanoscale images show that in PNCs, as pointed out by Fisher et al., the nanotubes are curved instead of being straight. Based on these observations, Fisher et al. illustrated that the reduction of effective modulus of CNT-reinforced materials is due to the curvature of embedded CNTs (Fisher et al., 2002).

Figure 3 illustrates the degree of alignment that can be achieved in PNC composites as seen in a TEM image (Thostenson & Chou, 2002). Unlike the two examples discussed above, in this instance, Thostenson & Chou extruded the polymer (polystyrene, PS) and a master batch of nanotube dispersed in PS using a micro-compounder. The film coming out of a rectangular die was drawn and solidified by passing over a chill roll. TEM was used to characterize the dispersion and alignment of MWNTs in polymer films. Figure 3a shows the large-scale dispersion and overall alignment of the carbon nanotubes and Figure 3b shows the alignment of the individual nanotubes. Examination of the drawn films showed that a draw ratio of five was sufficient to achieve a good nanotube alignment.

3.1.2. Mechanical Properties Of Cnts

TEM is also used to measure the mechanical properties and to study the deformation mechanism in PNCs. In fact, the Young's modulus of CNTs, which is hard to measure by conventional means, was first directly measured using TEM from the thermal vibration amplitudes of CNTs, and thus confirming the predicted axial stiffness on the scale of terapascal (TPa) and the high axial strength due to the in-plane graphitic structure of CNTs (Treacy et al., 1996). Bower et al. (Bower et al., 1999) aligned MWNTs in the thermoplastic polymer (polyhydroxyaminoether, DowChemical Co.) films by stretching above the glass transition temperature and releasing the load at 300 K. The uniaxial alignment of MWNTs was confirmed by x-ray diffraction. The CNTs in the polymer matrix are bent (Figure 4), presumably because of the shrinkage of the polymer matrix as the composite is cooled from 100º C to room temperature. The deformation was reversible at moderate bending, i.e. the CNTs returned to their straight cylindrical shape by heating the polymer matrix. Fracture surfaces in PNC's can be examined to ascertain the strength of the adhesion between the polymer and the nanotube.Figure 5 shows MWNT protruding from polymer matrix attached with a layer of polymer. All these observations point to the effective interactions and intimate contact between CNTs and polymer matrix. In a final example, Figure 6 shows the use of TEM to observe the nucleation and propagation of a crack induced in thin MWNT-PS film by thermal stresses in a TEM (Qian et al., 2000). The nucleation and propagation occur in regions of low CNT density and along the weak CNT-polymer interface. Also, the CNTs tend to align perpendicular to the crack and bridge the gap, and some of them are already pulled out, suggesting that the critical failure force is determined by interaction between polymer matrix and CNTs. These CNTs provide a closure force until a critical value is reached to be pulled out of the matrix.

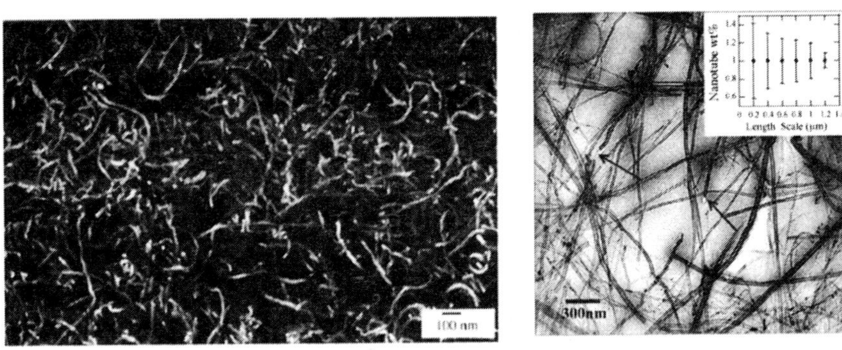

Figure 2. Left Panel (a): A SEM image for 50 wt% CNT/ Poly(vinyl alcohol) composites (S.P. Shaffer and A.H. Windle, "Fabrication and characterization of carbon nanotube/poly(vinyl alcohol) composites". Advanced materials (1999). Vol. 11. 937-941. Copyright Wiley-VCH Verlag GmbH & Co. KGaA. Reproduced with permission.). Right panel (b): An example of TEM image at 1 μm length scale showing homogeneously dispersed MWNT in polystyrene (PS) matrix. Homogeneity of nanotube dispersion was evaluated by assessing the weight fraction of nanotubes in different areas in the TEM images for

different length scale. The inset shows that homogeneity is greater at 1.2 μm length scale (small error bar) than lower length scale. An arrow in the image indicates the defects present at the nanotube ends (Reprinted with permission from D.Qian, E.C. Dickey, R. Andrews and T. Rantell., "Load transfer and deformation mechanisms in carbon nanotube-polystyrene composites", Applied Physics Letters, vol. 76, 2868-2870. Copyright 2000, American Institute of Physics.)

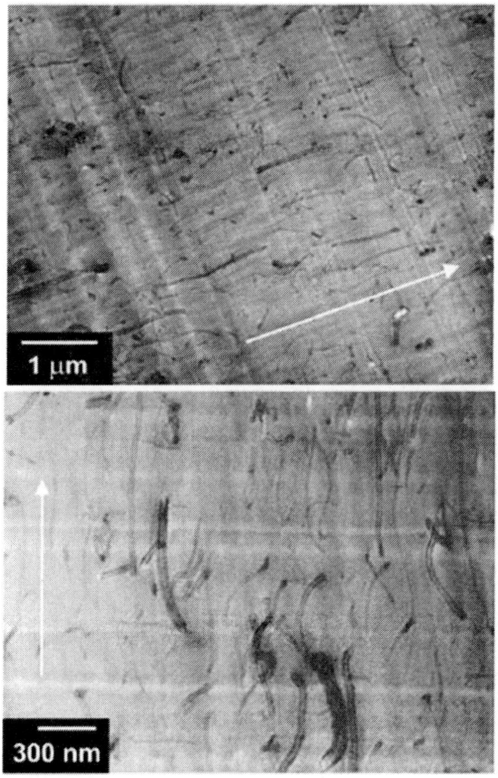

Figure 3. TEM images of dispersion and alignment of CNTs in nanocomposites films. The arrows indicate flow/drawing direction. Left panel (a): TEM at length scale of micrometer. Right panel (b): TEM at length scale of nanometer (Reprinted with permission from E.T. Thostenson and T.W. Chou, "Aligned multi-walled carbon nanotube-reinforced composites: processing and mechanical characterization", J. Phys. D: Appl. Phys., vol. 35, L77-L80. Copyright 2002, Institute of Physics Publishing.).

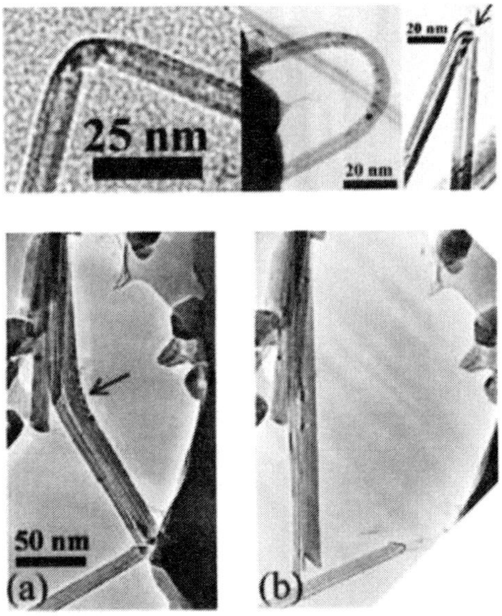

Figure 4. Top panel: TEM images showing the buckling of in bent MWNTs in MWNT/polymer composites. Bottom panel: (a) A bent nanotube before heating; (b) the bent nanotube restored to its straight cylindrical shape after heating (Reprinted with permission from C. Bower, R. Rosen, L. Jin, J. Han and O. Zhou, "Deformation of carbon nanotubes in nanotube-polymer composites", Applied Physics Letters, vol. 74, 3317-3319. Copyright 1999, American Institute of Physics.).

3.2. Raman Spectroscopy

Infrared (IR) and Raman spectroscopy (RS) are the two vibrational spectroscopic techniques widely used to obtain molecular structure information for the identification of polymers, functional groups and stereochemical structures. Both techniques measure the vibrational energies of molecules. However, for a vibrational motion to be IR active, the dipole moment of the molecule must change, and for a transition to be Raman active there must be a change in polarizability of the molecule. In most instances, IR and RS yield complementary information because IR activity suggests RS inactivity and vice versa. Sample requirements for RS are less stringent than for IR. Details of application of RS to CNTs have been discussed thoroughly in the literature (Dresselhaus et al., 2005), and a few examples of the use of RS to PNCs are provided here. For oriented specimen, the intensity of RS spectra line depends on the orientation of molecules relative to the polarization of the incident beam as well as scattering beam. This feature is used to determine the orientation distribution for CNTs in polymer matrix. Polarized RS has also been used to probe the one - dimensional nature of SWNT in the bulk materials (Gommans et al., 2000). These results show that the RS intensity has the same orientation dependence as nanotube's polarizability. A similar work on MWNT has also shown that there is a strong dependence of the graphite-like G band and the disorder-induced D band on the polarization geometry (Rao et al., 2000).

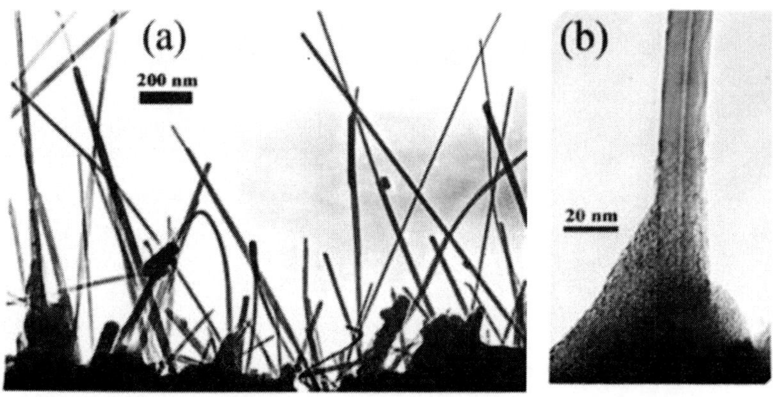

Figure 5. TEM images of the fracture surface showing MWNT protruding from the polymer matrix suggestive of strong adhesion between the MWNT and the polymer matrix (Reprinted with permission from C. Bower, R. Rosen, L. Jin, J. Han and O. Zhou, "Deformation of carbon nanotubes in nanotube-polymer composites", Applied Physics Letters, vol. 74, 3317-3319. Copyright 1999, American Institute of Physics.).

Mathematical description of Raman intensity distribution as a function of SWNT's orientation has been derived using Legendre polynomials by Liu et al. (Liu & Kumar, 2003). They built a relationship between 2nd and 4th orientation parameters of SWNT and Raman scattering intensity and depolarization ratio. This quantitative relation was used by Chae et al. to investigate the orientation of SWNT in polyacrylonitrile fibers (Chae et al., 2006). They measured G-band intensity ratio at about 1592 cm^{-1} for polarization parallel and perpendicular to the fiber axis (Figure 7). By comparing the G-band ratio for conventional spun and gel-spun PAN/SWNT fibers, they were able to conclude that the orientation of SWNT is only slightly improved by the gel-spinning process. They also showed that RS could be used to estimate the relative intensity of disorder band for SWNT, DWNT, MWNT, VGCNF (Chae et al., 2005) (Figure 7). They concluded that SWNT has best perfection while MWNT and VGCNF have more disorder (Chae et al., 2005).

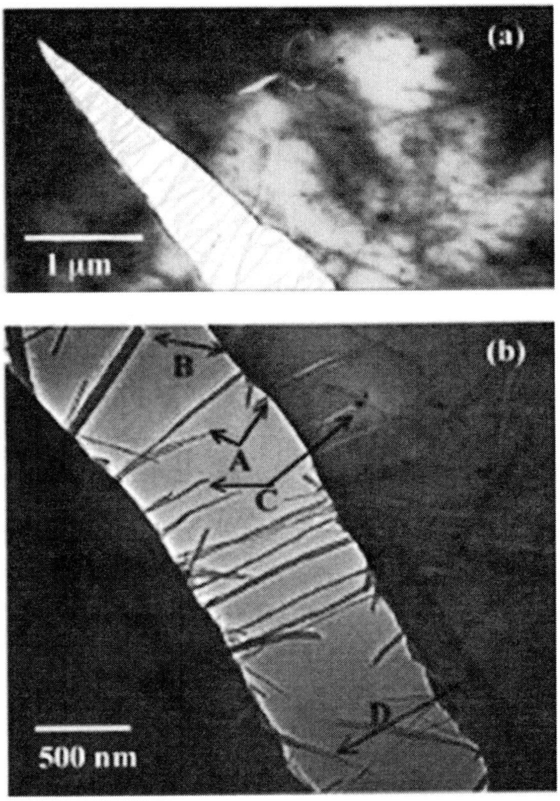

Figure 6. Crack initiated by thermal stress. Upper panel shows the crack initiation and propagation at low CNT density areas. Lower panel shows the CNTs aligned perpendicular to the crack bridging the crack. Arrows labelled A, B, C and D point to CNTs that are broken and pulled out of matrix (Reprinted with permission from D. Qian, E.C. Dickey, R. Andrews and T. Rantell, "Load transfer and deformation mechanisms in carbon nanotube-polystyrene composites", Applied Physics Letters, vol. 76, 2868-2870. Copyright 2000, American Institute of Physics.).

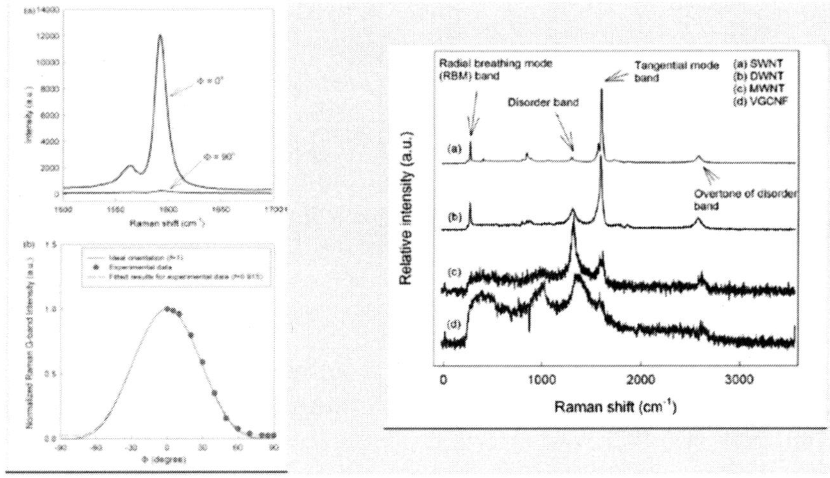

Figure 7. Upper left panel: G-band Raman spectra for PAN/SWNT fiber. The ratio of the peak intensity at 1592 cm^{-1} are compared when the angles between polarizer and the fiber axis are 0 and 90º, respectively. Lower left panel: The normalized G-band Raman intensity as function of angles between polarizer and fiber axis. Experimental data are for PAN/SWNT (1 wt%) at draw ratio 51 (Reprinted from Polymer, Vol.47, H.G. Chae, M.L. Minus and S. Kumar, "Oriented and exfoliated single wall carbon nanotubes in polyacrylonitrile", 3494-4504. Copyright 2006, with permission from Elsevier.) Right panel: Raman spectra for pristine CNT powders (Reprinted from Polymer, Vol.46, H.G. Chae, T.V. Sreekumar, T. Uchida and S. Kumar, "A comparison of reinforcement efficiency of various types of carbon nanotubes in polyacrylonitrile fiber", 10925-10935. Copyright 2005, with permission from Elsevier.).

3.3. X-Ray Scattering

X-ray scattering is an ideal tool to investigate the influence of CNTs on polymer morphology at multiple length scales from nm to sub μm length scales. This is achieved by using variations of the basic diffraction/scattering technique that are usually referred to as x-ray diffraction (XRD), wide-angle x-ray diffraction (WAXD), and small-angle x-ray scattering (SAXS). The scattering or diffraction intensity is expressed as a function of the scattering vector, $q(=4\pi\sin\theta/\lambda)$ or $s(=2\sin\theta/\lambda)$, 2θ being the scattering angle, and λ the x-ray wavelength. The sample preparation is far simpler than that for TEM. Whereas electron microscopy can provide information about the distribution of the CNTs and other structural information within a particular frame of an image, SAXS can provide similar data about the distribution, but averaged over lengths ~ mm in the bulk of the sample. Some of the characteristics of the PNCs that can be measured using x-ray scattering include the structural changes in the polymer due to the incorporation of the CNT, orientation of the polymer chains/crystals and the CNTs, and structure at the interface between the polymer and the CNT, and the changes in these features during deformation. The advent of synchrotron radiation, whose brilliance is ~ 10 orders of magnitude higher than that from a rotating anode tube, makes it possible to carry out *in-situ* magnitude higher than that from a rotating anode tube, makes it possible to carry out *in-situ* dynamical measurements. Some of the data in this section were obtained from CNTs dispersed in polyacrylonitrile (PAN) that was then spun into a fiber. PAN fibers reinforced with CNTs are made by solution- or gel-spinning. PAN/CNTs fibers with significant enhancement in properties suggested good interactions between PAN and CNTs (Chae et al., 2006; Chae et al., 2005; Guo et al., 2005; Sreekumar et al., 2004; Uchida et al., 2006). These studies showed that the improvement in low strain is due to the interaction of PAN and CNTs, while the improvement as high strain is partly due to the CNT length.

3.3.1. Characterization Of The Cnts

Before discussing the results for the PNCs, we will first discuss some of the scattering data relevant to the characterization for the CNTs. Historically, the MWNTs were observed by electron microscopy and were depicted as concentric seamless cylinders (Iijima, 1991). This model was used to interpret XRD data (Pasqualini, 1997), and further direct evidence appeared later to support this concentric cylinder model (Xu et al., 2001). However, there has been evidence that suggest that the local structure is similar to turbostratic graphite or scrolls of graphite sheets (Dravid et al., 1993; Zhou et al., 1994). More recent data suggest that while thin MWNTs could be modelled as concentric 7-15 tubules, thicker MWNTs were found to be mixtures of both scroll-type MWNT and concentric-cylinder (nested Russian doll) MWNTs (Maniwa et al., 2001).

An individual CNT cannot be characterized using SAXS. However, due to the π-π chemical bonding and van de Waals interactions, the CNTs tend to aggregate into bundles while lacking a clear-cut rod-like morphology (Schaefer et al., 2003a), and this typical morphology can be characterized using SAXS. Schaefer et al. combined small-angle light scattering, ultra-SAXS and SAXS to measure morphology and degree of dispersion of SWNTs and MWNTs suspended in water (Schaefer, 2003a;Schaefer, 2003b). They found that the intensities over 5 decades of q are characterized by different power laws. The power-law of intensity from SWNT indicates a branched rope network with fractal characteristics. Porod's law was observed at length scale smaller than inner radius, suggesting the smoothness of interface. The nonuniform dispersion of CNTs in polymers partly explains why the reinforcement of CNTs to polymer composites falls below expectations.

In another experiment (Inada et al., 2005), two segments of different power-law behavior were observed in the intensity distribution (Figure 8). One segment conforms to Porod's law corresponding to smooth interfaces. The other follows a power-law with exponent -1, suggesting scattering is from a rod-like structure. The scattering could be modeled as those from MWNTs represented as hollow tubes with an inner diameter of about 8 nm and a wall thickness of 6 nm, which is in agreement with the reported results from direct measurement through TEM.

3.3.2. Saxs Characterization Of The Voids and The Cnts In The Polymer

Voids are invariably formed in the fibers during the spinning and the drawing process, especially in the presence of reinforcements. Voids, and in general any scattering entities, give rise to SAXS that can be analyzed to obtain information about the length, diameter, and orientation of these entities as well the information about the nature of the interface. In the work with PAN fibers, the degree of orientation of the voids measured by SAXS was found to be same as those of the crystalline domains observed in WAXS. Both increased reversibly upon stretching, indicating that the voids are integral parts of the polymer matrix and closely associated with the crystalline domains in the fibrils (Wang et al., 2009). SAXS data also showed large changes in the size distribution of the voids during elongation in the solution spun PAN fibers, but not the composite fibers. Furthermore, heating the fiber above the glass transition temperature (Tg) decreased the volume fraction of the voids considerably as indicated by a decrease in the scattered intensity. Figure 9 shows examples of the decay in equatorial scattering with scattering angle. The rate of this decay, as indicated by the slope of the log-log plot (called the exponent) provides valuable insight into the nature of the void-polymer interface. In the plot of the

PAN with MWNT, we see the exponent is 4.4 (Figure 9a). An exponent of 4.0 would indicate a sharp, smooth interface. A value higher than 4 suggests that in the PAN/MWNT sample, there is a diffuse interface between the polymer and the void as indicated schematically in the figure. An exponent between 3 and 4 would indicate a rough void-polymer interface, which was observed in gel-spun fibers without any CNT's (Figure 9b). An exponent between 2 and 3 indicates that the structure within the void is similar to a mass-fractal. It is speculated that because of the nature of gel-spinning process, the voids are partially filled with polymer, and perhaps CNT, and as a result the scattering exponent falls below 3 as indicated in Figure 9c. It is possible that a similar analysis can be carried out to probe the nanotube-polymer interface provided there is a contrast between the CNTs and the matrix.

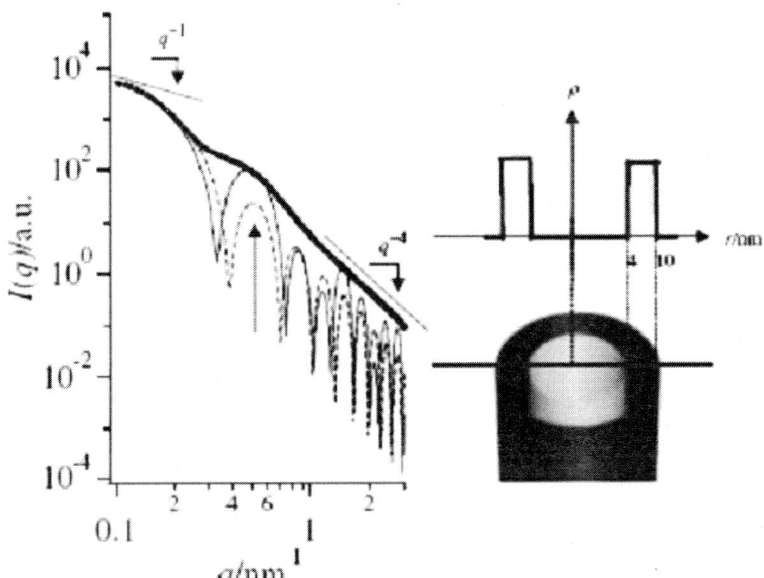

Figure 8. Left: Measured SAXS intensity as function of scattering vector. Two power laws were presented indicating a one-dimension cylinder structure and sharp interface, respectively. Straight line and dotted line are calculated for a core-shell and hard cylinder models, respectively. Right: The best fit parameter suggest the structure of CNT is a core-shell structure (Reprinted from T. Inada et al., "Small-angle x-ray scattering from multi-wall carbon nanotubes (CNTs) dispersed in polymeric matrix", Chemistry Letters, Vol. 34, 525 (2005), with permission from The Chemical Society of Japan.)

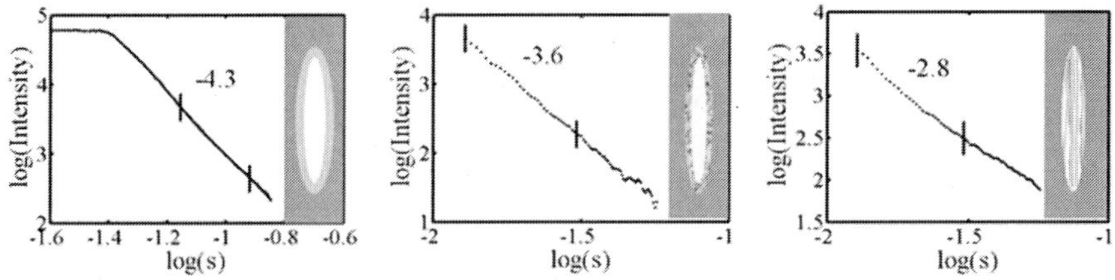

Figure 9. Power law for equatorial intensities. (a) PAN/MWNT for a void with diffuse boundary ; (b) gel-spun PAN for a void with surface fractal; (c) gel-spun PAN/SWNT for a void with mass fractal The linear segments that are fitted to obtain the fractal dimension are shown in the figure (**Wang et al., 2009**).

3.3.3. Waxd Analysis Of The Polymer and The Cnts

Figure 10 shows example of the 2D diffraction patterns and the corresponding 1D scans that can be obtained from polymer composites with CNTs. Because of the large fraction of the polymer (95%) the major features in these scans are due to the polymer, and these features can be used to follow the changes in the polymer structure in terms of the unit cell dimensions, crystallinity, crystallite size and polymer orientation (Wang et al., 2008b). For example, the data show that a monotonic decrease in PAN inter-chain spacing with the fiber strain is accompanied by a reversible change from helix to zigzag conformation, and by changes in the axial repeats of the two conformations. These changes are much larger in gel-spun fibers than in solution-spun fibers, indicating more effective load transfer between the amorphous and crystalline segments in gel -spun fibers. In addition to the intense polymer peaks, it is also possible to see the weak features of the CNT reinforcements in the pattern, especially in the 1D scans. Note the broad feature due to MWNT in Figure 10e, and the sharp features due to VGCNF in Figure 10f. The changes in these features with stress were used to ascertain the modulus of the polymer and the CNT in the composite. Such data are potentially useful in determining the distribution of load between the two components.

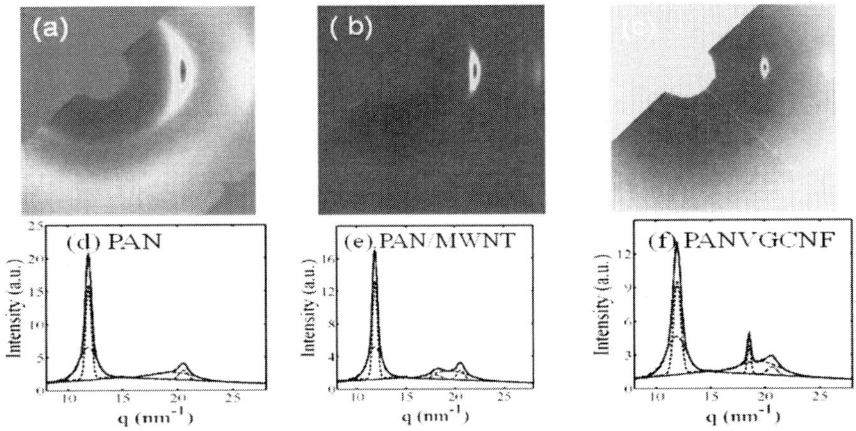

Figure 10. Upper Panel: Two dimensional wide-angle x-ray diffraction pattern for (a) solution-spun PAN ; (b) solution-spun PAN/MWNT; (c) gel-spun PAN/SWNT fibers. Lower Panel: Profile fitted equatorial scans: (d) PAN; (e) PAN/MWNT; (f) PAN/VGCNF. The data for gel fibers are similar to that of PAN. (Reprinted from Polymer, Vol.49, W. Wang, N.S. Murthy, H.G. Chae and S. Kumar, "Structural changes during deformation in carbon nanotube-reinforced polyacrylonitrile fibers", 2133-2145. Copyright 2008, with permission from Elsevier.)

3.3.4. Alignment Of The Cnts and The Polymer

As indicated earlier, to a large extent, the strength, modulus and shrinkage properties of the polymer composite are determined by the distribution in orientation and alignment of both the CNT and the polymer. XRD is an appropriate tool for determining this parameter. (Wang et al., 2008b). For instance, Pichot et al. obtained the orientation of SWNT and PVA chain by azimuthal intensity distribution (Pichot et al., 2006). They related the alignment of SWNT in polyvinyl alcohol (PVA) fibers with different draw ratio and accounted for the improvement of alignment with draw ratio by applying a simplistic continuum mechanical model, i.e. the induced alignment is due to the constant volume. In the work on PAN fibers, by measuring the orientation during loading and unloading of the fibers, it was found that the orientation reversibly increases as the fiber is stressed (Wang et al., 2008b).

Chen et al. (Chen et al., 2006) employed in-situ WAXS/SAXS to investigate the deformation of nanocomposites fiber containing fluorinated MWNT (FMWNT) and fluorinated polyethylene-propylene. Orientation of FMWNT can be characterized by (002) reflection of WAXS and quantitatively described by Hermans' orientation parameter. Similar experiments were carried out byWang et al. (Wang et al., 2008b) on the deformation of PAN and its nanocomposites containing a variety of CNTs (MWNT,VGCNF,SWNT). The effect of CNT reinforcement could be ascertained by monitoring the structural change in PAN crystals. In this work, it was found that CNTs facilitate the orientation of the PAN crystals during deformation (Figure 11) and increase the load-transferred to PAN crystals as evidenced by their increased lateral and axial strain at 75º C.

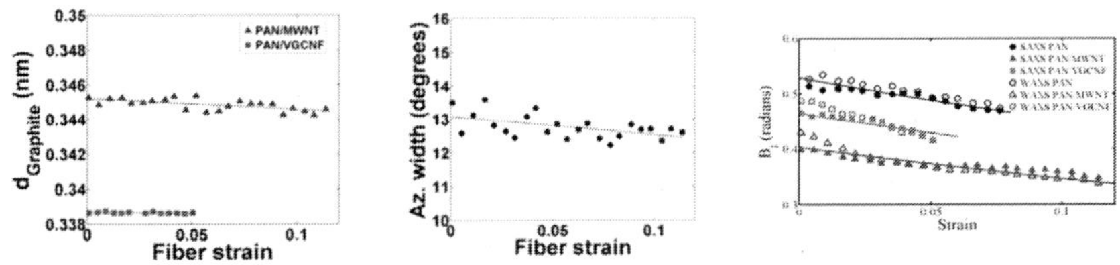

Figure 11. a)-(c) from left to right. Changes in the *d*-spacing of carbon as a function of strain derived from the Bragg's peak of nanotubes at room temperature; (b) Degree of orientation of MWNT in the form of the azimuthal breadth.(Reprinted from Polymer, Vol.49, W. Wang, N.S. Murthy, H.G. Chae and S. Kumar, "Structural changes during deformation in carbon nanotube-reinforced polyacrylonitrile fibers", 2133-2145. Copyright 2008, with permission from Elsevier.) (c) Changes in the extent of misorientation (B_f) for scattering entities at room temperature. Black circle, blue triangle and red square represent PAN, PAN/MWNT, and PAN/VGCNF, respectively. Filled and empty symbols represent SAXS and WAXS results, respectively. Straight line segments are calculated based on affine deformation model. (**Wang et al.,. 2009**).

3.3.5. Deformation Of The Cnts In The Composite

The diffraction patterns shown in Figure 10 can be used to characterize the deformation of CNTs under stress by monitoring the changes in the (002) peak of CNTs. This interlayer distance d_{002} is used to ascertain the changes in the lattice dimension under strain (Figure 11). This figure also shows the changes in the azimuthal width of these reflections with strain as well as the crystallite size corresponding to this reflection. The changes in these parameters show that CNTs deform under load and thus contributes to the increased modulus, and also alter the response of the PAN matrix to stress, thus enhancing the performance of the composite.

4. CONCLUSIONS

Among the many techniques that are currently available, electron microscopy, Raman spectroscopy and x-ray scattering have proved to be most valuable to study the structure in polymer nanocomposites. These techniques complement each other by covering overlapping length scales from sub nm to μm. These techniques are valuable in assessing the dispersion, alignment of the nanotubes in the matrix, to probe the adhesion and the interface between the nanotube and the polymer, and to study the structural changes that occur due to the interaction between the two components as well as during deformation.

REFERENCES

1. S. G. Advani, 2006 Processing and Properties of Nanocomposites: World Scientific.

2. P. M. Ajayan, P. Redlich, M. Ruhle, 1997 Structure of carbon nanotube-based nanocomposites. Journal of Microscopy, 185 275282 .

3. P. M. Ajayan, L. S. Schadler, C. Giannaris, A. Rubio, 2000 Single-Walled Carbon Nanotube-Polymer Composites: Strength and Weakness. Advanced Materials, 12(10), 750-753.

4. C. Bower, R. Rosen, L. Jin, J. Han, O. Zhou, 1999 Deformation of carbon nanotubes in nanotube-polymer composites. Applied Physics Letters, 74(22), 3317-3319.

5. O. Breuer, U. Sundararaj, 2004 Big Return From Small Fibers: A Review of Polymer/Carbon Nanotube Composites. Polymer Composites, 25(6), 630-645.

6. A. Brosse, C. , S. Tence-Girault, P. M. Piccione, L. Leibler, 2008 Effect of multi-walled carbon nanotubes on the lamellae morphology of polyamide-6. Polymer, 49 46804686 .

7. P. Calvert, 1999 Nanotube composites: A recipe for strength. Nature, 399 210211 .

8. H. G. Chae, Y. H. Choi, M. L. Minus, S. Kumar, 2009 Carbon nanotube reinforced small diameter polyacrylonitrile based carbon fiber. Composites Science and Technology, 69 406413 .

9. H. G. Chae, S. Kumar, 2009 Making Strong Fibers. Science, 319(5865), 908-909.

10. H. G. Chae, M. L. Minus, S. Kumar, 2006 Oriented and exfoliated single wall nanotubes in polyacrylonitrile. Polymer, 47(10), 3494-3504.

11. H. G. Chae, M. L. Minus, A. Rasheed, S. Kumar, 2007 Stabilization and Carbonization of gel spun polyacrylonitrile/single Wall nanotube Composite Fibers. Polymer, 48 37813789 .

12. H. G. Chae, T. V. Sreekumar, T. Uchida, S. Kumar, 2005 A comparison of reinforcement efficiency of various types of carbon nanotubes in polyacrylonitrile fiber. Polymer, 46 1092510935 .

13. T. Chatterjee, C. A. Mitchell, V. G. Hadjiev, R. Krishnamoorti, 2007 Hierarchical Polymer-Nanotube Composites. Advanced Materials, 19 38503853 .

14. X. Chen, C. Burger, D. Fang, I. Sics, X. Wang, W. He, et al. 2006 In-Situ X-ray Deformation Study of Fluorinated Multiwalled Carbon Nanotube and Fluorinated Ethylene-Propylene Nanocomposite Fibers. Macromolecules, 39 54275437 .

15. J. N. Coleman, U. Khan, Y. K. Gun'ko, 2006 Mechanical Reinforcement of Polymers Using Carbon Nanotubes. Advanced Materials, 18 689705 .

16. C. A. Cooper, S. R. Cohen, A. H. Barber, H. D. Wagner, 2002 Detachment of nanotubes from a polymer matrix. Applied Physics Letters, 81(20), 3873-3875.

17. V. P. Dravid, X. Lin, Y. Wang, X. K. Wang, A. Yee, J. B. Ketterson, et al. 1993 Buckytubes and Derivatives: Their Growth and Implications for Buckyball Formation. Science, 259(5101), 1601-1604.

18. M. S. Dresselhaus, G. Dresselhaus, R. Saito, A. Jorio, 2005 Raman spectroscopy of carbon nanotubes. Physics Reports, 409 4799 .

19. F. T. Fisher, R. D. Bradshaw, L. C. Brinson, 2002 Effects of nanotube waviness on the modulus of nanotube-rein. Applied Physics Letters, 80, 4647--4649.

20. S. J. V. Frankland, A. Caglar, D. W. Brenner, M. Griebel, 2002 Molecular Simulation of the Influence of Chemical Cross-Links on the Shear Strength of Carbon Nanotube-Polymer Interfaces. J. Phys. Chem. B, 106 30463048 .

21. M. Ge, K. Sattler, 1993 Vapor-Condensation Generation and STM Analysis of Fullerene Tubes. Science, 260(5107), 515-518.

22. H. H. Gommans, J. W. Alldredge, H. Tashiro, J. Park, J. Magnuson, A. G. Rinzler, 2000 Fibers of aligned single-walled carbon nanotubes: Polarized Raman spectroscopy. Journal of Appied Physics, 88 25092514 .

23. B. P. Grady, F. Pompeo, R. L. Shambaugh, D. E. Resasco, 2002 Nucleation of Polypropylene Crystallization by Single-Walled Carbon Nanotubes. Journal of Physical Chemistry B, 106 58525858 .

24. H. Guo, T. V. Sreekumar, T. Liu, M. Minus, S. Kumar, 2005 Structure and Properties of Polyacrylonitrile/single wall carbon nanotube composite films. Polymer, 46 30013005 .

25. Y. Y. Huang, S. V. Ahir, E. M. Terentev, 2006 Dispersion rheology of carbon nanotubes in a polymer matrix. Physical Review B, 73, 125422.

26. S. Iijima, 1991 Helical microtubules of graphitic carbon. Nature, 354 5658 .

27. T. Inada, H. Masunaga, S. Kawasaki, M. Yamada, K. Kobori, K. Sakurai, 2005 Small-angle X-ray Scattering from Multi-walled Carbon Nanotubes (CNTs) Dispersed in Polymeric Matrix. Chemistry Letters, 34(4), 524-525.

28. A. Koganemaru, Y. Bin, Y. Agari, M. Matsuo, 2004 Composites of Polyacrylonitrile and Multiwalled Carbon Nanotubes Prepared by Gelation/Crystallization from solution. Advanced Functional Materials, 14(9), 842-850.

29. T. Liu, S. Kumar, 2003 Quantitative characterization of SWNT orientation by polarized Raman spectroscopy. Chemical Physics Letters, 378 257262 .

30. T. Liu, I. Y. Phang, L. Shen, S. Y. Chow, W. Zhang, D. , 2004 Morphology and Mechanical Properties of Multiwalled Carbon Nanotubes Reinforced Nylon-6 Composites. Macromolecules, 37 72147222 .

31. Y. Maniwa, R. Fujiwara, H. Kira, H. Tou, E. Nishibori, M. Takata, et al. 2001 Multiwall carbon nanotubes grown in hydrogen atmosphere: An X-ray study. Physical Review B, 64 073105073104 .

32. G. H. Michler, 2008 Electron Microscopy of Polymers (1st ed.): Springer.

33. M. Moniruzzaman, K. I. Winey, 2006 Polymer Nanocomposites Containing Carbon Nanotubes. Macromolecules, 39 51945205 .

34. M. i. h. Panhuis, A. Maiti, A. B. Dalton, A. v. d. Noort, J. N. Coleman, B. Mc Carthy, et al. 2003 Selective Interaction in a Polymer-Single-Wall Carbon Nanotube Composite. J. Phys. Chem. B, 107 478482 .

35. E. Pasqualini, 1997 Concentric carbon structure. Physical Review B, 56(13), 7751-7754.

36. V. Pichot, S. Badaire, P. A. Albouy, C. Zakri, P. Poulin, P. Launois, 2006 Structural and mechanical properties of single-wall carbon nanotube fibers. Physical Review B, 74(245416), 245416.

37. D. Qian, E. C. Dickey, R. Andrews, T. Rantell, 2000 Load Transfer and deformation mechanisms in carbon nanotube-polystyrene composites. Applied Physics Letters, 76(20), 2868-2870.

38. A. M. Rao, A. Jorio, M. A. Pimenta, M. S. S. Dantas, R. Saito, G. Dresselhaus, et al. 2000 Polarized Raman Study of Aligned Multiwalled Carbon Nanotubes. Physical Review Letters, 84 18201823 .

39. L. Sawyer, D. T. Grubb, G. F. Meyers, 2010 Polymer Microscopy (3rd ed.): Springer.

40. L. Schadler, 2007 Nanocomposites: Model Interface. Nature Materials, 6 257258 .

41. D. W. Schaefer, J. M. Brown, D. P. Anderson, J. Zhao, K. Chokalingam, D. Tomlin, et al. 2003a Structure and dispersion of carbon nanotubes. J. Appl. Cryst., 36 553557 .

42. D. W. Schaefer, J. Zhao, J. M. Brown, D. P. Anderson, D. W. Tomlin, 2003b Morphology of dispersed carbon single-walled nanotubes. Chemical Physics Letters, 375 369375 .

43. M. S. P. Shaffer, A. H. Windle, 1999 Fabrication and characterization of carbon nanotube/poly(vinyl alcohol) composites. Advanced Materials, 11 937941 .

44. T. V. Sreekumar, T. Liu, B. G. Min, H. Guo, S. Kumar, R. H. Hauge, et al. 2004 Polyacrylonitrile Single-Walled Carbon Nanotube Composite Fibers. Advanced Materials, 16(1), 58-61.

45. D. Srivastava, C. Wei, K. Cho, 2003 Nanomechanics of carbon nanotube and composite. Appl. Mech. Rev 56(2), 215-230.

46. E. T. Thostenson, T. Chou, W. , 2002 Aligned multi-walled carbon nanotube-reinforced composites: processing and mechanical characterization. J. Phys. D: Appl. Phys., 35, L77 -L80.

47. M. M. J. Treacy, T. W. Ebbesen, J. M. Gibson, 1996 Exceptionally high Young's modulus observed for individual carbon nanotubes. Nature, 381 678680 .

48. T. Uchida, D. Anderson, M. Minus, S. Kumar, 2006 Morphology and modulus of vapor grown carbon nano fibers Journal of Materials Science, 41(18), 5851-5856.

49. W. Wang, P. Ciselli, E. Kuznetsov, T. Peijs, A. H. Barber, 2008a Effective reinforcement in carbon nanotube-polymer composites. Philosophical Transactions of the Royal Society A: Mathematical, Physical and Engineering Sciences, 366 16131626 .

50. W. Wang, N. S. Murthy, H. G. Chae, S. Kumar, 2008b Structural changes during deformation in carbon nanotube-reinforced polyacrylonitrile fibers. Polymer, 49(8), 2133-2145.

51. W. Wang, N. S. Murthy, H. G. Chae, S. Kumar, 2009 Small-Angle X-ray Scattering Investigation of Carbon Nanotube-Reinforced Polyacrylonitrile Fibers During Deformation. Journal of Polymer Science: Part B: Polymer Physics.

52. M. Wong, M. Paramsothy, X. J. Xu, Y. Ren, S. Li, K. Liao, 2003 Physical interactions at carbon nanotube-polymer interface. Polymer, 44 77577764 .

53. G. Xu, Z.-c. Feng, Z. Popovic, J.-y. Lin, J. J. Vittal, 2001 Nanotube structure Revealed by High-resolution X-ray Diffraction. Advanced Materials, 13(4), 264-267.

54. H. Ye, H. Lam, T. Nick, Y. Gogotsi, K. Frank, 2004 Reinforcement and rupture behavior of carbon nanotubes-polymer nanofibers. Applied Physics Letters, 85(10), 1775-1777.

55. Y. C. Zhang, X. Wang, 2005 Thermal effects on interfacial stress transfer characteristics of carbon nanotubes/polymer composites. International journal of solids and structures, 42 53995412 .

56. O. Zhou, R. M. Fleming, D. W. Murphy, C. H. Chen, R. C. Haddon, A. P. Ramirez, et al. 1994 Defects in Carbon Nanostructure. Science, 263 17441747 .

5

MICROSTRUCTURE AND PROPERTIES OF POLYPROPYLENE/CARBON NANOTUBE NANOCOMPOSITES

Dimitrios Bikiaris *

Laboratory of Polymer Chemistry and Technology, Department of Chemistry, Aristotle University of Thessaloniki, 541 24, Thessaloniki, Macedonia, Greece;

ABSTRACT

In the last few years, great attention has been paid to the preparation of polypropylene (PP) nanocomposites using carbon nanotubes (CNTs) due to the tremendous enhancement of the mechanical, thermal, electrical, optical and structural properties of the pristine material. This is due to the unique combination of structural, mechanical, electrical, and thermal transport properties of CNTs. However, it is well-known that the properties of polymer-based nanocomposites strongly depend on the dispersion of nanofillers and almost all the discussed properties of PP/CNTs nanocomposites are strongly related to their microstructure. PP/CNTs nanocomposites were, mainly, prepared by melt mixing and *in situ* polymerization. Young's modulus, tensile strength and storage modulus of the PP/CNTs nanocomposites can be increased with increasing CNTs content due to the reinforcement effect of CNTs inside the polymer matrix. However, above a certain CNTs content the mechanical properties are reduced due to the CNTs agglomeration. The microstructure of nanocomposites has been studied mainly by SEM and TEM techniques. Furthermore, it was found that CNTs can act as nucleating agents promoting the crystallization rates of PP and the addition of CNTs enhances all other physical properties of

PP. The aim of this paper is to provide a comprehensive review of the existing literature related to PP/CNTs nanocomposite preparation methods and properties studies.

KEYWORDS

nanocomposites; polypropylene; carbon nanotubes; physical properties

1. INTRODUCTION

Isotactic polypropylene (iPP) is one of the most important commodity thermoplastics, accounting for about 20% of the total world polyolefin production [1]. Its high isotacticity, enhanced mechanical properties, narrowmolecular weight distribution and clarity lead to a balance of physical and mechanical properties, and its environmental friendliness (recyclability) and low cost give it an extra advantage. Due to its low cost, low density, high thermal stability and resistance to corrosion, iPP is widely used in many applications, such as fibers, films for food packaging, production of bottles and tubes, *etc.* However, despite all these advantages, there is also a drawback in its applications. Although its resistance to crack initiation is very high, in crack propagation the resistance of the iPP matrix is very low; therefore, when a crack or mechanical failure exists, the matrix can break very easily. This is especially an issue at low temperatures. Therefore, a great deal of effort has been made to modify its mechanical properties such as blending iPP with inorganic fillers in the form of nanoparticles. Furthermore, these inorganic nanofillers, such as talc, silicon dioxide, carbon black, clay, and carbon nanotubes, act as nucleating agents for PP crystallization as well as they enhance its physical properties.

With the rapid development of nanotechnologies and nanomaterials since 1990s, studies on polymer-based nanocomposites have been extensively carried out in order to find promising alternatives to traditional composites, though these mainly focused on general mechanical and multifunctional properties and filler dispersion. In the case of PP nanocomposites, the most used filler is carbon nanotubes (CNTs). This is due to the unique combination of structural, mechanical, electrical, and thermal transport properties of carbon nanotubes (CNTs) [2], obtaining advanced multifunctional composite materials [3,4].

Carbon nanotubes are categorized as single-walled nanotubes (SWCNTs), multi-walled nanotubes (MWCNTs) and double-walled nanotubes (DWCNTs), which are of minor importance (Figure 1). Most SWCNTs have a diameter of close to 1 nm, with a tube length that can be many thousands of times longer. The structure of SWCNTs can be conceptualized by wrapping a one-atom-thick layer of graphite, called graphene, into a seamless cylinder. SWCNTs are a very important variety of CNTs because they exhibit important electric properties that are not shared by the MWCNTs variants. The MWCNTs consist of multiple layers of graphite rolled in on themselves to form a tube shape. MWCNTs, however, have a lower tensile strength and modulus than SWCNTs. Furthermore, SWCNTs have a shorter diameter (from 0.4 up to 5.6 nm) compared with MWCNTs, which is in the range of 10-100 nm. MWCNTs are also more rigid, because their section is much larger compared to that of SWCNTs.

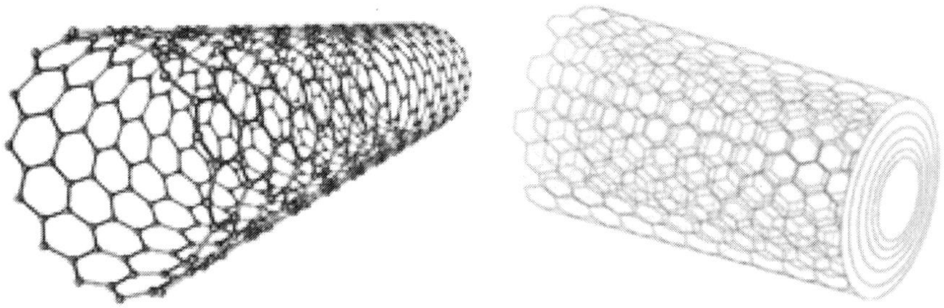

Figure 1. Illustration of SWCNTs (left) and MWCNTs (right).

MWCNTs and SWCNTs have been considered as unique reinforcements for different polymers due to their exceptional electrical, thermal, chemical and mechanical properties. The tensile strength, tensile modulus and Poisson ratio for SWCNTs have been reported to be in the range of 37–100 GPa, 640 GPa to 1–2 TPa [5] and 0.14–0.28 [6], respectively. For comparison, the modulus of graphite fibers and steel are in the range 300–800 and 200–400 GPa, respectively, and the tensile strengths of these materials are about 5 and 10–20 GPa, respectively, whereas the density of SWCNTs is one-sixth that of steel. Furthermore, this is coupled with approximately 500-times more surface area per gram (based on equivalent volume fraction of typical carbon fiber) and an aspect ratio of around 100-1000.

Investigations of PP nanocomposites fabricated with CNTs have been reported by many researchers [7,8,9,10]. However, it is well-known that the properties of polymer-based nanocomposites are strongly dependent on the dispersion of nanofillers.

The microstructure of the PP/CNTs nanocomposites is discussed in the present paper, while emphasis is given to how this microstructure affects the properties of the final material.

2. SYNTHESIS OF CNTS

CNTs are sheets of graphite rolled into tubes and have excellent properties due to their symmetric structure. For the synthesis of CNTs there are three main synthetic methods used: Arc-discharge, laser ablation and catalytic methods such as chemical vapor deposition [11]. Synthesis procedures yielding high purity nanotubes of defined structure are still restricted to carbon arc discharge and pulsed laser ablation, allowing gram per day rather than kilograms per day production quantities. Higher quantities can be manufactured by chemical vapor deposition processes, having reasonable potential for extended industrial production but yielding materials with higher defect density.

2.1. Arc-discharge

Arc-discharge is the easiest and most common method of producing CNTs. This technique involves the growth of CNTs on carbon (graphite) electrodes during the direct current arc-discharge evaporation of carbon in the presence of an inert gas such as helium or argon. The method is similar to that used for the fullerene synthesis. Plasma is created between anode and cathode. The temperature of this plasma typically reaches temperatures of approximately 3700-4000 °C. At this temperature, the carbon on the anode is vaporized and deposits onto the cathode. It should be noted that the diameter of the anode is usually smaller than the diameter of the cathode and both electrodes are water-cooled. The carbon needles, ranging from 4 to 30 nm in diameter and up to 1 mm in length, were grown on the negative end of the carbon electrode used for the direct current arc-discharge evaporation of carbon in an argon-filled vessel.

2.2. Laser ablation

According to this method, in a 1200 °C furnace a pulsed or a continuous laser is used to vaporize a target consisting of a mixture of graphite and metal catalysts, such as cobalt or nickel, in the presence of 500 Torr (67 kPa) helium or argon gas. As the target is vaporized, a cloud of C_3, C_2, C and catalyst vapors is formed rapidly. As the cloud cools, the carbon species with a small molecular weight then combine to form larger molecules. The vaporized catalysts condense slowly and adhere to carbon clusters to prevent their closing into cage structures. The nanotube grows until too many catalyst atoms aggregate on the end of the nanotube. The large particles either detach or become over-coated with sufficient carbon to poison the catalysis. This allows the tube to terminate with a fullerene-like tip or with a catalyst particle. With this method can achieve high yields (>70%) of SWCNTs.

2.3. Chemical Vapor Deposition (CVD)

This method is capable of controlling growth direction on a substrate and synthesizing a large quantity of nanotubes. The CVD technique involves the decomposition of a hydrocarbon in the presence of a catalyst. In this process a mixture of hydrocarbon gas, acetylene, methane or ethylene and nitrogen is introduced into the reaction chamber. During the reaction, nanotubes are formed on the substrate by the decomposition of the hydrocarbon at temperatures 700–900 °C and atmospheric pressure. The most common catalysts used for CVD are iron, nickel or cobalt. Carbon has a low solubility in these metals at high temperatures and thus the carbon will precipitate to form nanotubes.

3. PP/CNTS NANOCOMPOSITES

There are three main processing methods used for the preparation of PP nanocomposites with nanotubes:

a) The solution method; where both the polymer and nanoparticles are dispersed in an organic solvent and nanocomposites are produced in the form of film after solvent evaporation.

b) *In situ* polymerization; where both nanoparticles and monomers are mixed from the beginning of polymerization.

c) Melt mixing; where nanoparticles are added in the molten polymer at temperatures of 30-60 °C above the melting point of the polymer.

Since the solubility of polypropylene is too low in most organic solvents, solution casting is not applied for the preparation of nanocomposites and melt mixing as well as *in situ* polymerization are the most used methods. Melt mixing can be done in a variation of different types of machines and is a very easy and most common method. However, the nanocomposites prepared by melt mixing or *in situ* polymerization have completely different characteristics as well as properties, which will be discussed in following.

3.1. Microstructure of PP/CNTs nanocomposites

Due to the CNTs intrinsic poor solubility, they are often self-assembled into bundles, which limits their applications. Bulk agglomerates of nanotubes up to several tens of microns in diameter act as macro-particle like centres of friction, unable to be compensated by regions exhibiting proper dispersion of nanotubes within the composite where physical properties are strongly enhanced. CNTs have high aspect ratio and surface reactive groups which prompt the interactions between nanotubes and thus very difficult can be separated and intercalated into polymer matrix. Good control over various mixing parameters during melt mixing is essential to obtain homogeneous composite materials. The microstructure and CNTs dispersion into PP matrix is very crucial in order a high performance material to be achieved. Thus, there are many research groups that focus on study of the PP/MWCNTs microstructure using different analytical techniques such as Scanning Electron Microscopy (SEM), Transmission Electron Microscopy (TEM), Atomic Force Microscopy (AFM), Polarizing Optical Microscopy (POM), micro-Raman spectroscopy, *etc.*

The microstructure of iPP/MWCNTs nanocomposite fractured surfaces studied by SEM analysis reveals different filler dispersion efficiency achieved in various composite systems [12,13]. From the micrographs in Figure 2, it can be seen that composites with different MWCNTs contents exhibited different MWCNTs dispersion state. At low MWCNT content, most of the MWCNTs dispersed individually in the iPP matrix. At relatively high MWCNTs contents, MWCNTs aggregates appeared in the polymer matrix, and the size of the aggregates increased with increasing MWCNTs content. [10,14]. Solid-state shear pulverization in conjunction with melt mixing was found that can produce well-dispersed polymer/MWCNTs nanocomposites [15,16].

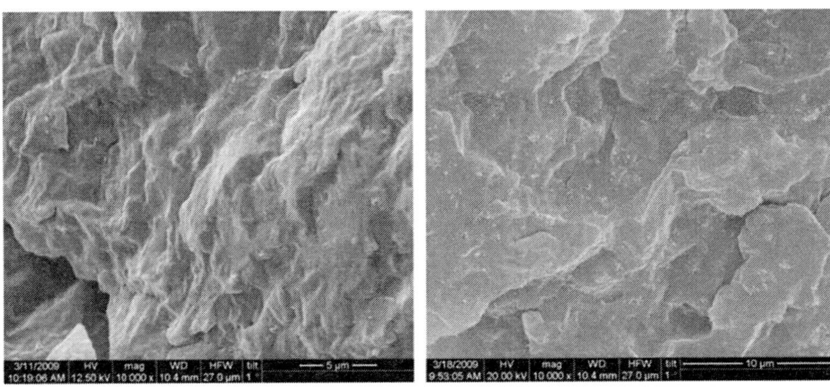

Figure 2. SEM micrographs for the nanocomposites containing 1 wt % (left) and 3 wt % (right) MWCNTs. Reproduced with permission from ref. [13].

The agglomeration of nanotubes in clusters or clumps in the nanocomposite can be seen more clearly in an optical microscopy micrograph (Figure 3) [17,18]. As can be seen CNTs aggregates are observed as dark spots with a wide particle size distribution. Particles with sizes 0.5 up to 20 μm are visible. The particle sizes are mainly depended on the CNTs concentration. At low MWCNTs content they formed agglomerates are in lower magnitude and a better dispersion can be seen. The mean particle sizes are in the range of 0.2 up to 10 μm, while as MWCNTs content increases nanotubes forms larger agglomerates. This is due to the tendency of nanotubes to interact each other due to surface forces. CNTs during their formation have surface carboxyl groups and maybe some hydroxyl groups. These interact through hydrogen bonding or due to weak Van der Wall's forces creating clusters and agglomerates. These are very difficult to break down during nanocomposites fabrication, since the involved forces during melt extrusion are not enough to break these agglomerates.

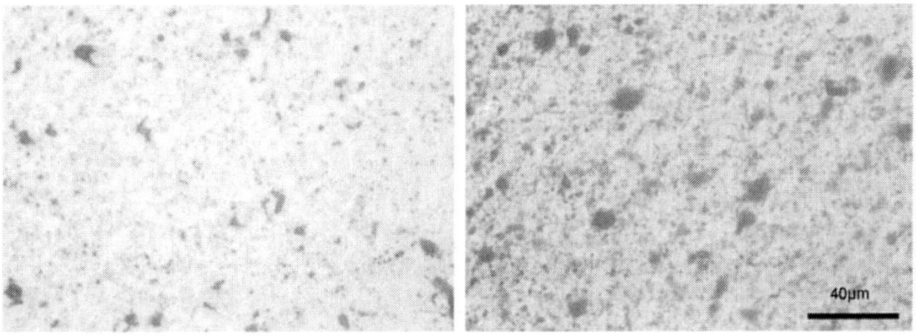

a b

Figure 3. Optical micrographs of PP/MWCNTs nanocomposites containing different MWCNTs content. a) 0.5 wt % and b) 2.5 wt %.

Visualization of the MWCNTs aggregates was also attempted by means of confocal micro-Raman spectroscopy [18]. Their identification is based on the absorption of PP and MWCNTs at completely different wavenumbers. As can be seen in Figure 4a, the MWCNTs spectrum exhibits a distinct pair of peaks at 1573 (major) and 1602 (shoulder) cm^{-1}, which does not conflict with peaks from iPP. In the nanocomposites both the characteristic peaks of MWCNTs and PP are recorded. When plotted with respect to the peak intensity at 1573 cm^{-1}, 2D micro-Raman scans acquired from iPP/MWCNTs samples can provide information on the size of the nanotube aggregates. Red colors indicate very high concentration of MWCNTs while the blue color very low. As can be seen in Figure 4b there are a lot of areas that red color has high intensity indicating that MWCNTs formed aggregates with different sizes. Aggregates with sizes in the area of 0.5 μm are visible as well as particle sizes with diameters till 5 μm. These observations were consistent for all the samples examined and in agreement with POM that was discussed previously and those from TEM images. Thus, the supporting information provided from confocal micro-Raman spectroscopy, which enabled scanning of larger areas and along various planes within the sample, lead us to believe that the findings of the TEM examination can be considered representative of the samples studied.

From the above presented images, it could be wrongly concluded that CNTs are dispersed into PP matrix always in the form of aggregates, while individual CNTs do not exist. However, this is not true, since SEM or POM techniques have low resolution, and thus TEM is more appropriate to study efficiently the microstructure and the CNTs dispersion in a polymer matrix. TEM-images of the PP/MWCNTs nanocomposites containing 0.5 and 1.5 wt % MWCNTs (Figure 5) reveal indeed that these agglomerates observed by SEM and POM techniques are formed from large numbers of nanotubes [19]. However, as can be seen in all agglomerates, the powdery nanotube aggregates (prior to mixing) were infiltrated by the polymer matrix during the melt-mixing procedure and, thus, the diffusion of polymer chains have widened the distance between the individual nanotubes. Furthermore, areas with higher CNTs concentration are seen next to nearly CNTs free regions in the TEM figures at low magnification indicating partially incomplete distribution, though qualitatively there is a relative good state of dispersion at low MWCNTs contents (Figure 5a and c). The dispersion of MWCNTs as individual nanotubes into polymer matrix is very difficult, since the involved interactions between PP macromolecules and nanotubes should be stronger than the interactions between nanotubes. Some individual CNTs can be seen only at low CNTs concentration (0.5 wt %) since the tendency for agglomeration is increased at higher MWCNTs contents (>1.5 wt %).

From the above TEM images it can be seen that during mixing with PP, carbon nanotubes have a small tendency to be separated and dispersed homogenously into PP matrix at low CNTs concentration. This is due to the small interfacial adhesion between the carbon nanotubes and PP matrix as was confirmed by SEM microscopy. From SEM micrograph (Figure 6) corresponding to the fractured surfaces of nanocomposite with 1 wt % of MWNTs, it is observed that the nanotubes appear to have higher diameters at their broken/debonded ends as compared to other parts, indicating that the polymer chains prior to failure are stretched to their maximum followed by recoiling and balling-up around the tip of the broken end of the nanotubes. The outer polymer sheath, in such a case, gets contracted and balled up during fracture process [19].

The dispersion of CNTs into PP matrix may be also affected from the applied draw during the melt procedure [8]. Figure 7shows TEM images of a 3 wt % MWCNTs in PP both un-oriented (no draw) and oriented at 12:1 draw ratio. These images depict the cross section of the fiber parallel to the drawing direction. Comparing the two images reveals how melt drawing the nanocomposite orients the nanotubes. In the un-oriented image (Figure 7a), whole nanotubes lying at various oblique angles are observed; whereas, the image of the drawn sample (Figure 7b) shows nanotubes that run parallel to the fiber axis (as shown by the arrow), therefore indicating flow induced orientation along the fiber axis.

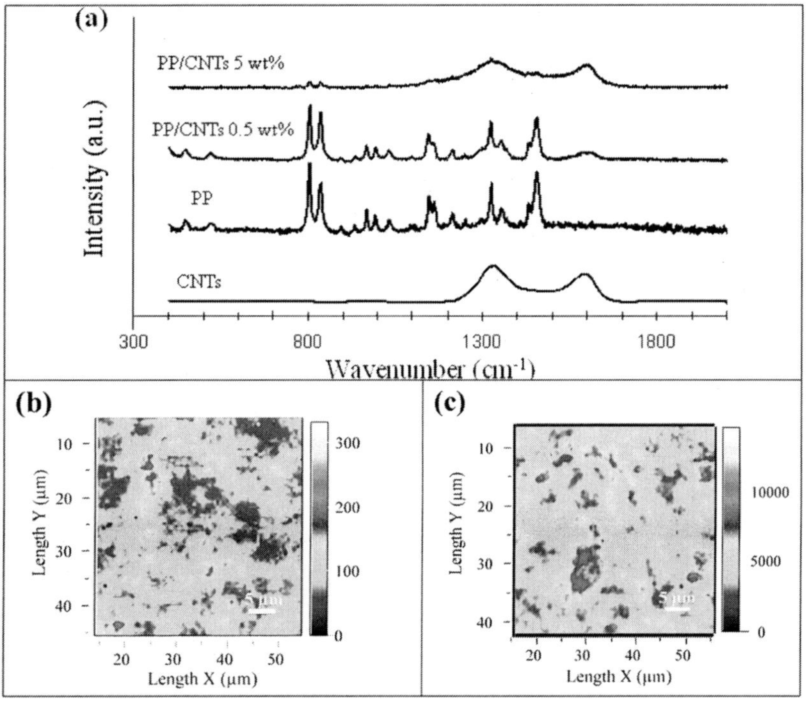

Figure 4. (a) Raman spectra of CNTs, neat iPP, and composites at 0.5 and 5 wt % CNTs content. (b) and (c) Micro-Raman maps from samples containing 0.5 wt % and 5 wt % CNTs, respectively. The color bar indicates the intensity of the peak at 1594 cm^{-1}.

Except for the MWCNTs content, the kind of the PP used also plays an important role for MWCNTs dispersion into PP matrix. In a recent study, both individual nanotubes and nanotube aggregates were present in iPP and sPP matrices [12]. However, a much higher degree of CNT agglomerate exfoliation has been attained in iPP/MWCNTs nanocomposites. sPP-based nanocomposites exhibit strong filler aggregation, while well-separated individual MWCNTs prevail of use iPP/MWCNTs materials. For the sPP/0.4 wt.% MWCNTs sample, predominant formation of relatively compact nanotube clusters integrating 6–8 individual nanotubes takes place, but at filler contents higher than 1.5 wt.% strongly agglomerated CNT structures with micrometer dimensions are formed in sPP matrix. Due to considerable nanotube aggregation, the sPP/MWCNTs materials have substantially lower interfacial area as compared to the respective iPP/MWCNTs nanocomposites.

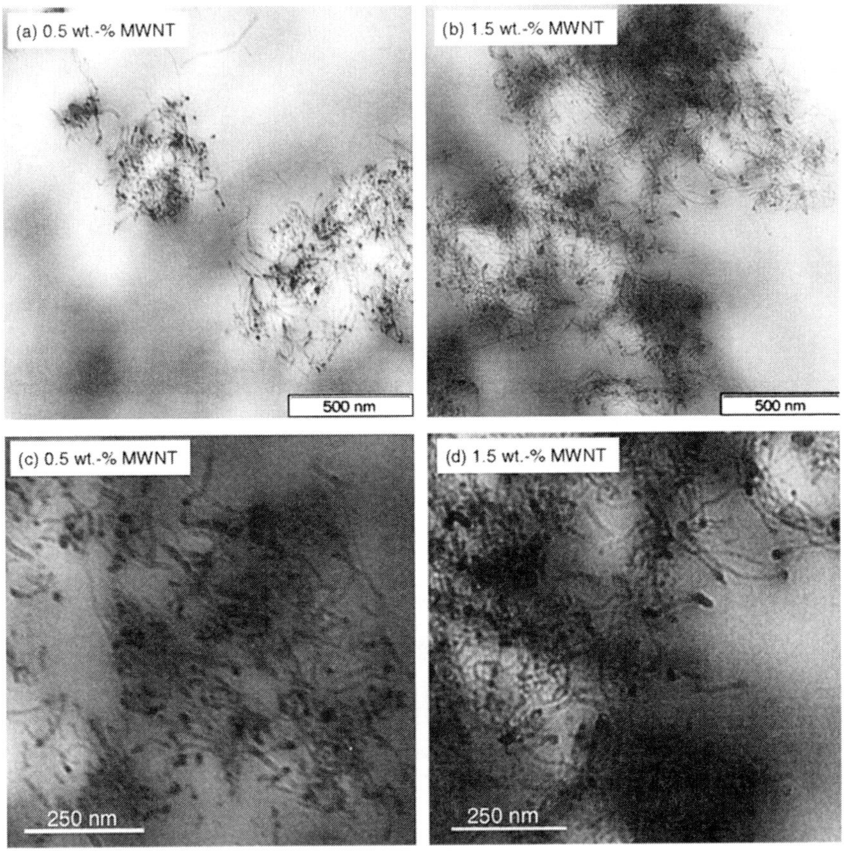

Figure 5. Dispersion of the MWCNTs: TEM image of the nanocomposite with a) 0.5 and b) 1.5 wt.% of MWCNTs at low magnification and c) TEM image of the nanocomposite with 0.5 and (d) 1.5 wt.% of MWCNTs at high magnification. Reproduced with permission from ref. [19].

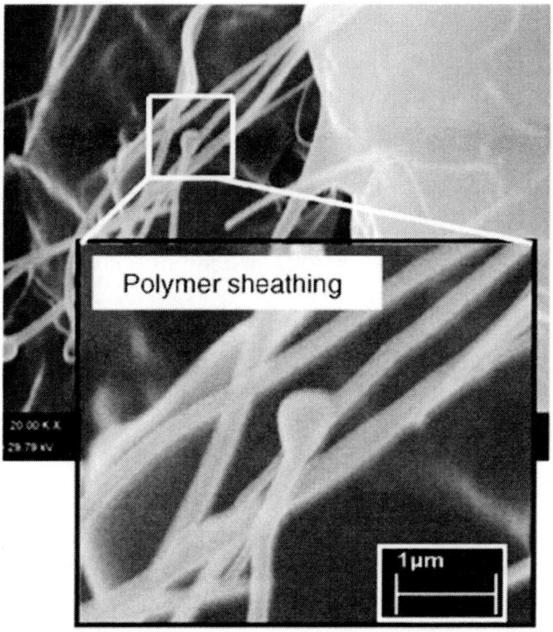

Figure 6. SEM micrograph of "polymer sheathing" phenomena of the fractures surface of PP/MWCNTs nanocomposite with 1 wt % of MWCNTs. Reproduced with permission from ref. [19].

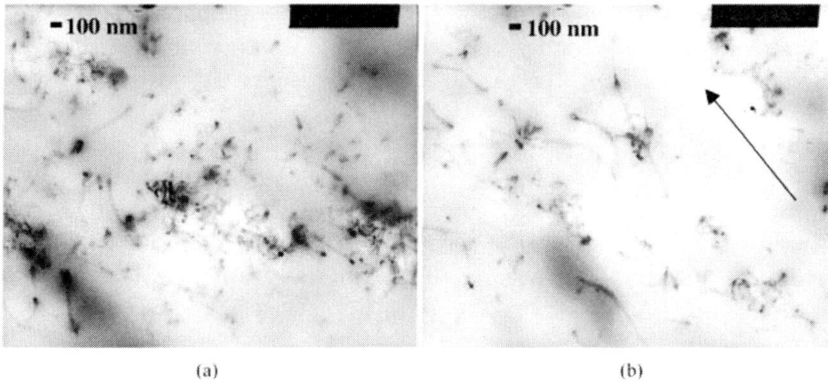

Figure 7. TEM Images of 3 wt % MWNTs in PP a) un-oriented; b) oriented, 12:1 draw ratio. Arrow indicates the direction of orientation. Reproduced with permission from ref. [8].

All the above mentioned studies concerning the microstructure of PP/CNTs nanocomposites revealed that CNTs can form aggregates into PP matrix. These are very difficult to destroy during melt mixing and CNTs dispersed intercalated into PP matrix. It seems that the applied shear forces are not strong enough to destroy the nanotube agglomerates. This microstructure directly affects most of the physical properties of PP/CNTs composites and mainly the mechanical properties. At low CNTs concentration an enhancement is observed, while higher CNTs concentrations cause deterioration. This will be discussed in the following.

3.2. Mechanical properties

The combination of enormous modulus (~1 TPa) and the high aspect ratio (100–1000) of CNTs makes them potentially effective as reinforcement for polymers. Thus, the incorporation into PP matrix will enhance its mechanical properties [20,21]. Due to the nanoscale dimensions of the carbon nanotubes the interfacial regions surrounding the tubes are also in nanometer dimensions and the applied load can be easily transferred from matrix to CNTs. There are three synergizing aspects of interaction in carbon-nanotube-based nanocomposites, *i.e.*, polymer–nanotube interaction, nanotube–nanotube and intra-polymer interactions in polymer-carbon nanotube interactions. It was observed that continuous nanotubes most effectively enhance the buckling resistance of the composites [22]. Despite an increasing thrust in the area of carbon-nanotube-filled polymer nanocomposites, the mechanism and the magnitude of load transfer between polymer matrices and the nanotubes are unclear and research in this area is still in its infancy.

Typical tensile stress–strain curves of nanocomposite samples with respect to their nanotube content are shown in Figure 8. The stress–strain curves of PP/MWCNTs exhibit distinctly regions of elastic, yielding and plastic deformation accompanied by cold drawing, whereas continued deformation after the yield point resulting in necking along the gauge length. For PP, the typical behavior of a ductile material is observed with a very high elongation at break (620%). The strain hardening region appears at about 400% of elongation and after which tensile strength increases almost linearly with strain until fracture eventually occurs. Also, in case of nanocomposites with lower loading (1 wt % MWCNTs) a ductile behavior is observed with necking, but with lower strain at fracture and without strain hardening. Nanocomposites bearing higher nanotube content (2 wt % MWCNTs) showed a brittle behavior with breaking after the yield point [23,24,25].

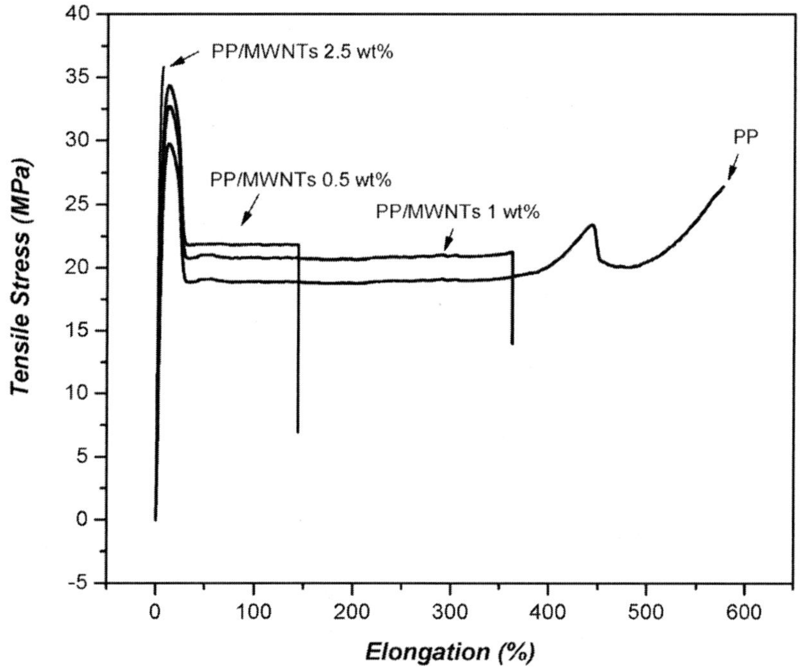

Figure 8. Stress–strain curves of PP/MWCNTs nanocomposites containing different MWNT content.

From stress-strain curves tensile strength and elastic modulus can be calculated. As can be seen in Figure 9, tensile strength and modulus of PP/MWCNTs were increased with increasing MWCNTs content, but after the formation of some aggregates these properties were decreased [26,27]. This is because the size of aggregates increases by increasing the nanotube content. At low CNTs, content nanotubes act as reinforcement agents, but at higher contents the formed aggregates act as mechanical failure concentrators. Thus, PP nanocomposites exhibit their higher mechanical performance at concentration 2-2.5 wt % while with further increase of MWCNTs, the tensile strength of PP/MWCNTs nanocomposites decreases. At low MWCNTs content, partial tensile strain can be transferred to MWCNTs embedded in PP matrix under tensile stress, which leads to the increase of tensile strength. With further addition of MWCNTs, more agglomerates of MWCNTs form in PP matrix and many defects are introduced into the polymer matrix due to the difficulty of homogeneously dispersing MWCNTs by melt mixing. These defects lead to the decrease of tensile strength. However, in all the cases the mechanical properties are higher from that of neat PP indicating the reinforcement effect of MWCNTs on PP matrix. Only elongation at break decreases continuously by increasing CNTs content. This was also verified from stress-strain curves discussed above.

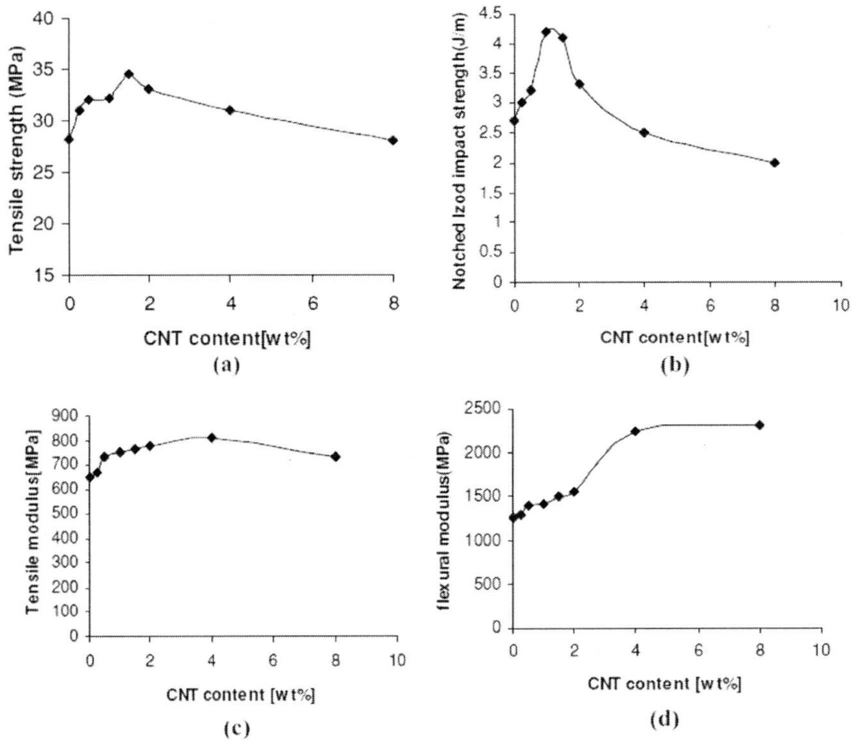

Figure 9. Mechanical properties of PP/CNTs nanocomposites. a) Tensile strength, b) notched Izod impact strength c), tensile modulus and d) flexural modulus. Reproduced with permission from [26].

Evidences of stress transfer in polymer–nanotube composites are largely based on enhancement of modulus. Micromechanical analysis of the effective elastic properties based on the Mori–Tanaka approach was also applied, which was later compared with the results from finite-element modelling [28]. Based on these it was concluded that all the axial and transverse elastic properties are dominated by nanotubes and polymer matrix, respectively, while on the other hand the interphase effects greatly impact the transverse properties at low volume fractions. Composites containing 15 wt % fiber filling showed an improvement of the Young's modulus of about 90%, which means in figures an increase from 410 N/mm^{2}for the pure PP to 780 N/mm^2 for the nanofiber reinforced composite [29].

Impact strength has also a similar behavior with tensile strength as was reported by Seo *et al.*, [30]. At low MWCNTs concentration, there is a steadily increase reaching the maximum value at 1.5-2 wt % MWCNTs, and gradually decreasing after that load. The extended agglomeration at such high MWCNTs content has a negative effect since the applied load into PP matrix cannot be transferred to MWCNTs. Figure 10 shows the impact strength of PP/MWCNTs composites as a function of MWCNTs content in the PP matrix resins.

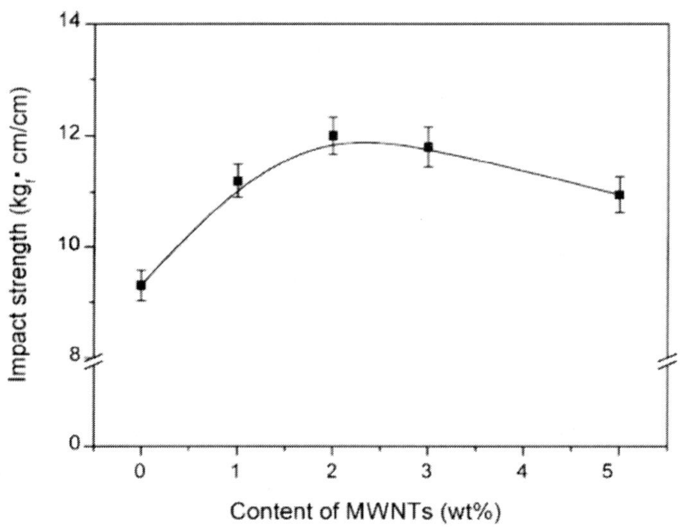

Figure 10. Impact strength of MWCNTs/PP composites as a function of MWNTs content. Reproduced with permission from ref. [30].

The impact strength is improved considerably by the inclusion of just 1 wt % of MWCNTs in the PP/MWCNTs composites and increased with an addition to MWCNTs content, compared with the composites without

MWCNTs. In general, the composites with MWCNTs greatly affect the impact damage condition. During induced impact-damage in the composites, the damage zone acts to magnify stress locally because of matrix cracking. However, this failure behavior by impact loading is hindered by the presence of MWCNTs in the composites. Therefore, an increasing fraction of MWCNTs in the composites improves the impact strength, at least up to 10 wt %, which is also largely related to the fracture toughness of the composites. In other words, the stronger the interfacial adhesions, the higher the impact strength. Moreover, the fracture strain of MWCNTs is estimated to be 10–30%, allowing extensive bending and buckling. This high flexibility most probably provides an additional source of higher impact strength to the composites.

Raman spectroscopy has been used to analyze the load transfer from the matrix to the nanotubes [31]. Frequency of SWCNTs D*-band is sensitive to stress and is often used as a stress sensor [27]. Figure 11 shows the spectra of D*-band of the sample containing 1 wt % SWCNTs with no strain and with 2% strain. The mode shifts to lower frequency under tensile stress. The D*-band peak was fit by a Lorentzian curve to estimate the position of its maximum. The peak position of the D*-band is seen to shift to a lower value on application of a tensile strain. The initial slopes are approximately ~200 cm^{-1} per strain. These data demonstrate clear transfer of the load to SWCNTs. The data also show change in the slope at strain above ~0.8–1.0%.

A ductile-to-semiductile transition in the crack resistance behavior of PP/MWCNTs composites is discussed by Satapathy*et al.* [32] using an essential work of fracture approach based on a post yield fracture mechanics concept and its possible interrelation to the structural attributes studied by TEM, SEM, and WAXD. A maximum in the non-essential work of fracture is observed at 0.5 wt.% MWCNTs content, which demonstrates the enhanced resistance to crack propagation compared to pure PP, followed by a sharp decline with the increase in MWCNTs content to 1.5 wt %, which reveals a ductile-to-semiductile transition (Figure 12).

The incorporation of 0.5 wt % of MWCNTs to PP intrinsically causes α-nucleation, which favors the formation of a large number of smaller spherulites instead of larger spherulitic domains as in pure PP. Thus, the inter-spherulitic boundaries are larger in pure PP than in the nanocomposites, which facilitate the smooth advancement of the fracture once it initiates from a highly stress-localized region by the slippage of the large crystal planes of the spherulites, which causes extensive ductile yielding. On the other hand, in the nanocomposite with 0.5 wt % of MWCNTs, the excessive nucleation narrows the interspherulitic boundary and hence the extensive ductile yielding is arrested to a large extent. This causes an increase in non-essential work of fracture (βw_p) in the case of nanocomposites with 0.5 wt % of MWCNTs. Interestingly, as the MWCNTs content is further increased up to 1.5 wt. %, the nanocomposites tend to show characteristics of forming an interconnecting network. When such a network density becomes higher at higher MWCNTs concentrations, they can potentially lead to intense strain localization and hence cause partial embrittlement of the nanocomposites. The strain localization around the dense network of nanotubes ultimately causes matrix cracking because of a severe modulus mismatch between the polymer and nanotubes, and hence ductile yielding behavior is substantially reduced, *i.e.*, it becomes semiductile behavior.

Creep resistance of PP can be significantly improved by the addition of CNTs. [33]. Three possible mechanisms of load transfer were considered that could contribute to the observed enhancement of creep resistance, which are: (1) fairly good interfacial strength between MWCNTs and polymer matrix, (2) increasing immobility of amorphous regions due to nanotubes acting as restriction sites, and (3) high aspect ratio of MWCNTs. A schematic deformation model under uniaxial stress is demonstrated in Figure 13.

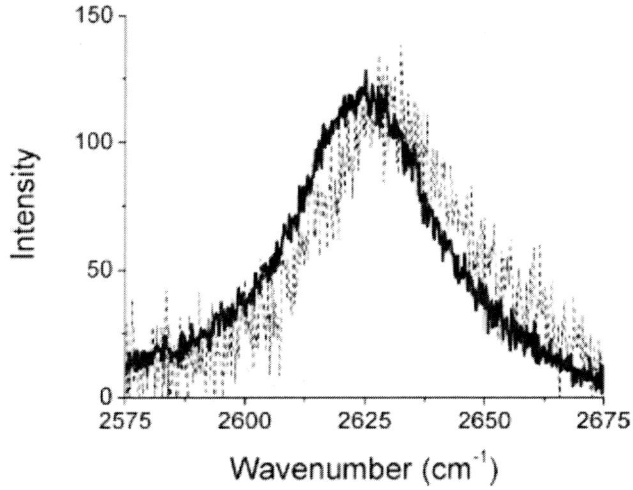

Figure 11. The Raman spectra of D*-mode in PP-1 wt % SWCNTs composite with no strain (dashed line) and with 2% strain (solid line). Reproduced with permission from ref. [27].

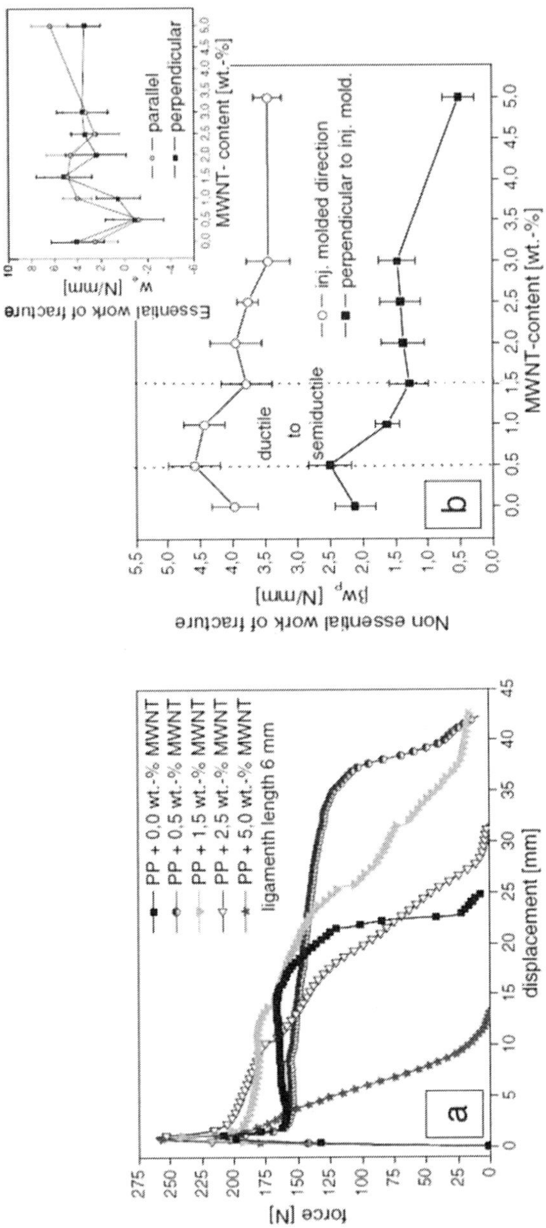

Figure 12. a) Force-displacement diagrams of nanocomposites with different concentrations of MWCNTs for a fixed ligament length (≈6mm). b) Dependence of non-essential work of fracture (βw_p) on MWCNTs content. Essential work of fracture (w_e)*versus* MWNTs content (insert). Reproduced with permission from ref. [32].

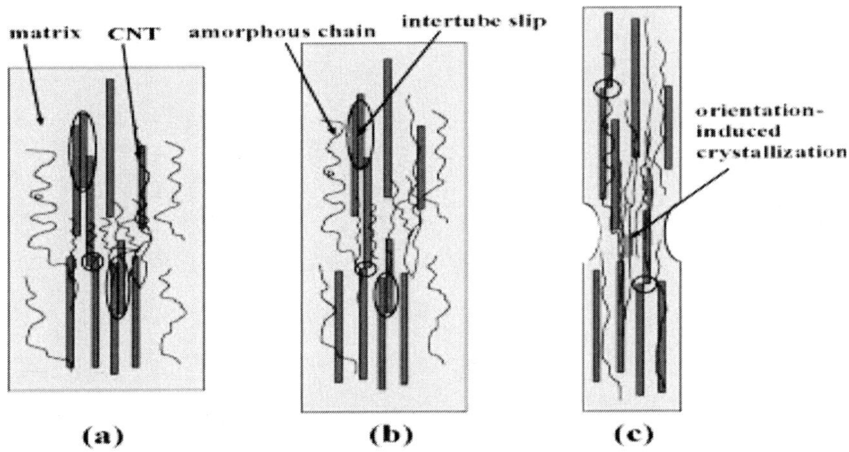

Figure 13. Schematic deformation model and orientation-induced crystallization of PP/CNT composite under uniaxial stress: (a) before loading, (b) under small load levels, and (c) under high load levels. Reproduced with permission from ref. [33].

In fact, MWCNTs are not ideally dispersed in PP, as represented in Figure 13a with some bundles or entanglements. Consequently, the existence of intertube sliding or stick-slip may give rise to an increase in creep rate under low tensile stress, as depicted in Figure 13b. Under high tensile stress level, the large deformation of matrix may introduce high interfacial shear stress to nanotubes across good interfacial bonding. Most of the bundled tubes in the matrix will extensively slip in a short time period (compare the circled parts in Figure 13). The influence of intertube sliding becomes thereafter minor and more nanotubes are effectively transferred to sustain the external tensile load, as shown in Figure 13c with possible necking at matrix-rich parts. The significant reduction of creep rate and deformation is then achieved. The load-bearing capability is enhanced and the secondary creep stage is prolonged under high stress level.

Tensile as well impact strength have a similar behavior in all nanocomposites containing CNTs. However, the most characteristic enhancement in mechanical properties gains the Young's modulus. Figure 14 shows a comparison between the measured Young's modulus of PP/CNTs composites and those predicted by the model [34].

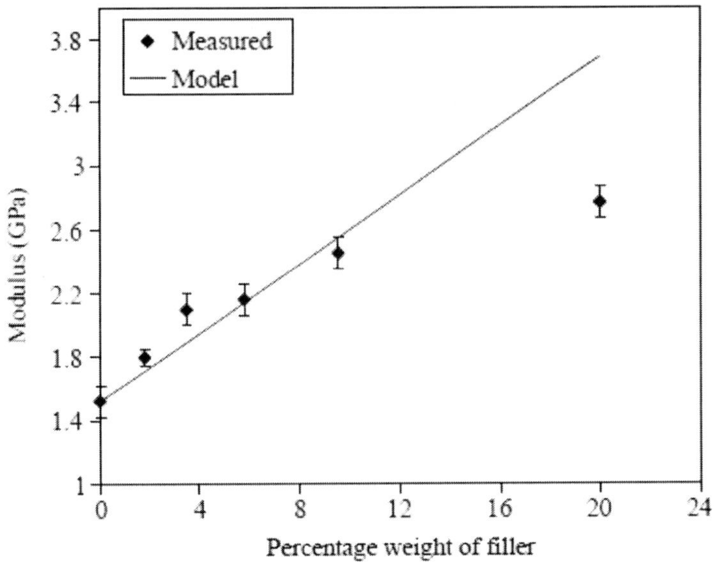

Figure 14. A comparison between measured and model predicted Young's modulus of PP/CNTs composites. Reproduced with permission from ref. [34].

Young's modulus increases continuously with the increase of CNTs content, revealing the reinforcement effect of the nanotubes. It can be seen that for the sample at which the nanotube aspect ratio was actually measured, 6 wt %, the agreement between the measurement and the model is excellent. However, for samples with a lower percentage of nanotubes, the measured values fall above the predicted, while for weight percentages greater than this the predictions are lower; substantially so for the 20 wt % sample. It can be surmised that for the lower percentage samples, which spend less time in the extruder, the fiber attrition could be less and hence the modulus would be higher. For the higher weight percentage, the lower values could be due either to higher fiber attrition, or perhaps more likely to aggregation of fibers into bundles (*i.e.,* poor dispersion). A similar increase was also mentioned for nanocomposites containing higher CNTs content [29]. Composites containing 15 wt % nanotubes filling showed an improvement of the Young's modulus of about 90%, which means in figures an increase from 410 N/mm^2 for the pure PP to 780 N/mm^2 for the nanofiber reinforced composite.

The study of PP/MWCNTs nanocomposites microstructure reveals the reinforcement effect of CNTs and was investigated in fractured surfaces in order to evaluate their effect on matrix deformation and the calculated mechanical properties [19]. As can be seen from the SEM images (Figure 15a) the matrix deformation is taking place with the investigation of fibrils. Furthermore, the network-density of the formed fibrils decreases in the nanocomposite with 0.5 wt % of MWCNTs when compared to pure PP, which indicates a drastic hindrance of the yielding process by MWCNTs despite plastic deformation, while still partly retaining the ductile nature of PP. With further increase of the MWCNTs content, the nanotube-rich areas become bigger and tend to show characteristics of forming an interconnecting network. When such network-density becomes higher at higher MWCNTs concentrations, they can potentially lead to intense strain localization because of hindered plastic deformation/strain of the matrix. The strain localization around the dense network of nanotubes, ultimately causes matrix cracking due to severe modulus mismatch between polymer and nanotubes and hence ductile yielding behavior is substantially reduced, *i.e.*, semi-ductile behavior. Strikingly, when the MWCNTs content was increased further to 1.5 wt %, the network density of fibrils was almost negligible/absent and instead crack-growth on the matrix surface was seen to be spanned by nanotube bridging across the cracks assisted by network of fibrils. However, the formation of a rheologically percolated network above 1 wt % of MWCNTs completely changes the matrix behavior and, thus, the crack toughness of the nanocomposite with 1.5 wt % of MWCNTs becomes a dynamic interplay of percolation and nanotube bridging leading to a semi-ductile behavior. Thus, the drastic hindrance of the ductile yielding in the polymer is controlled by the state of dispersion of the MWCNTs, the interface and MWCNTs-induced structural reorganization (nucleation).

These observations prompt first the understanding that the polymer wrapping phenomenon is being largely confined to the amorphous regions and secondly the fact that the immobilized polymer layer at the polymer–nanotube interface allows the low enthalpy mobility of the free chain ends to take place withgreater ease. The i-PP exists in a 3_1-helix conformational arrangement with a radial dimension of 2.096 nm about its mean chain axis (c-axis). Based on these findings, a qualitative structural model about the state of the relative organization of the PP chains and nanotube has been proposed and is shown schematically in Figure 15b. The model clearly adheres to all the above detailed illustrations, though not exactly up to the scale, and fundamentally complements the understanding of the macro-scale properties correlation to nanoscale organization in the case of two component polymer based systems [19].

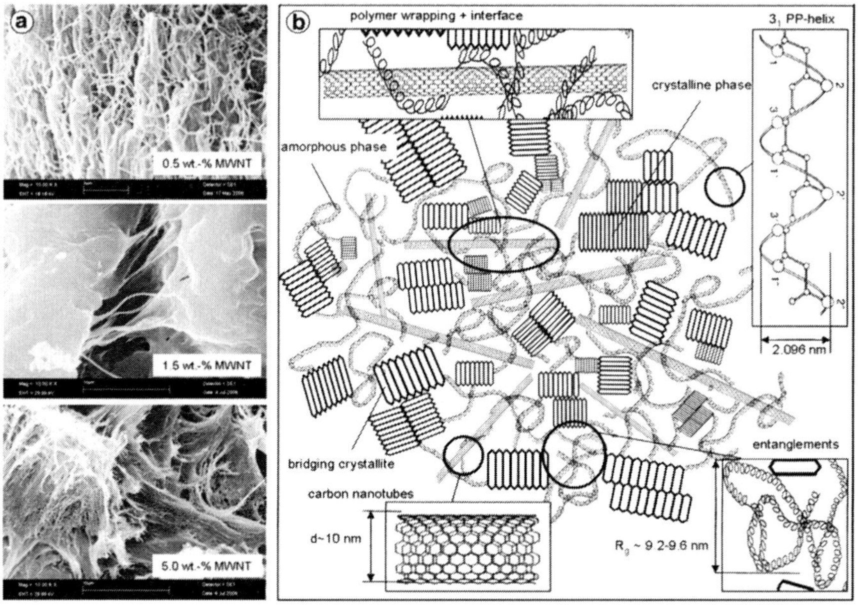

Figure 15. a) SEM investigation: fracture surfaces of nanocomposites with 0.5, 1.5 and 5.0 wt. % MWCNTs, and b) proposed schematic of the state of structural organization and polymer–nanotube interaction in the PP/MWCNTs nanocomposites. Reproduced with permission from ref. [19].

3.3. Dynamic Mechanical Analysis (DMA) of PP/CNTs nanocomposites

The increase in mechanical properties and the reinforcement effect of CNTs can be also measured with Dynamic Mechanical Analysis (DMA). DMA of polypropylene shows three relaxations at about -80 °C (α), 8 °C (β) and 100 °C (γ), respectively. The γ peak is generally attributed to the relaxation of a few chain segments in the amorphous regions. The β-relaxation represents the glass transition, and α-relaxation is attributed to the lamellar slip and rotation in the crystalline phase. The storage modulus (G') and the mechanical loss factor (tanδ) *versus* temperature curves at 5 Hz of PP/SWCNTs are shown in Figure 16 [35]. It can be clearly observed in Figure 16a that the storage modulus of iPP is increased with the addition of SWCNTs due to their stiffening effect. However, it was found that the modulus decreases at the highest SWCNTs concentration (1 wt %). Figure 16b displays the tanδ peaks of the neat iPP and its composites filled with different SWCNTs. A displacement of the peak at higher temperatures is observed by the addition of nanotubes. Furthermore, the peak shift is more evident as higher is the nanotube percentage in the composite. This increase in the T_g is attributed to the reinforcing effect of SWCNTs, hindering the mobility of the chains. However, in PP/MWCNTs the glass transition temperature as determined from DMA is decreasing in the case of the nanocomposites compared to the pure iPP matrix [19,36,37]. Dynamic mechanical properties of PP/CNTs nanocomposites were found that depends also from the type of the used CNTs [38].

3.4. Reological behavior of PP/CNTs nanocomposites

PP/CNTs at low CNTs content exhibited Newtonian behaviour, while by increasing the CNTs content (>2 wt %) non-Newtonian behavior was observed at all shear rate ranges [39]. Similar results were also mentioned from other studies [37,40,41]. Figure 17a and b shows changes of storage modulus, G', and loss modulus, G", of iPP/CNTs at different CNT concentrations with sweeping frequency, respectively. Note that the measuring temperature is 200 °C, which is above the melting point of iPP [42].

At a fixed frequency, both G' and G" increase with increasing CNTs concentration. In addition, at a fixed CNTs concentration, both G' and G" increase with increasing frequency. Moreover, it can be found that G' seems to pleateau at low frequencies for iPP/CNTs with CNTs concentrations of 7.4 wt % and 9.1 wt %. This is indicative of a transition from liquid-like to solid-like viscoelastic behavior. This non-terminal behavior at low frequencies for iPP/CNTs with high CNTs concentrations can be attributed to the formed CNTs network, which restrains long-range motions of polymer chains. It also further infers that the CNTs network may restrain diffusion of polymer chains in the undercooled melt during crystallization, leading to changes of crystallization kinetics of polymer matrix. The network formation usually is accompanied by a gelation behavior. Frequency dependence of loss tangent (tanδ) can reflect more clearly about the appearance of gelation. It has been predicted that in the pre-gel regime, tanδ monotonically decreases with increasing frequency, which is a typical behavior for a viscoelastic liquid, while in the post-gel regime, a moderate increase of tanδ at low frequencies appears with increasing frequency, indicating a dominant elastic response in the composite. Frequency dependence of tanδ for iPP/CNTs is shown in Figure 17c. It can be found that for CNTs concentrations of 7.4 wt % and 9.1 wt % moderate increases of tanδ at low frequencies do occur with increasing frequency, while for the lower CNTs concentrations monotonic decreases of tanδ with increasing frequency are always observed. Therefore, the critical gelation CNTs concentration for iPP/CNTs can be estimated to be around 7.4 wt %.

3.5. Effect of CNTs on PP crystallinity

CNTs cause also a serious effect on the crystallization rate of PP and this can affect most of its properties. For this reason a lot of researchers have studied extensively the effect of CNTs on PP crystallinity [30,43,44,45,46,47,48,49,50,51,52,53,54]. iPP has normally a slow crystallization rate, forming large spherulites. Nucleation can be homogeneous or heterogeneous. Homogeneous nucleation is the process in which nuclei form spontaneously in the PP melt as it cools down. This crystallization will not occur before the melt is super cooled well below the equilibrium melting temperature. On the other hand, during heterogeneous nucleation the crystallite growth is initiated by foreign bodies within the molten phase. In this case, crystallization tends to occur at higher temperatures, closer to the melting point of PP. Therefore, heterogeneous nucleation is amply used to enhance the mechanical properties and/or to provide consistent optical properties.

Isothermal crystallization behaviors of iPP/CNTs with CNTs concentrations less than 2.0 wt % at several selected temperatures between 136 °C and 144 °C were studied by using optical microscopy [42]. Typical optical micrographs taken at 136 °C for certain times are presented in Figure 18.

It is obviously seen that the sizes of spherulites in neat iPP are much larger than that in iPP/CNTs with CNTs concentration of 0.2 wt %. Nucleation density in the former is also obviously lower than that in the latter. Apparently, heterogeneous nucleation event is considered to occur mainly in the latter, due to nucleating effect of CNTs on primary crystallization of iPP [55,56]. It is also found that the sizes of spherulites and nucleation density of iPP/CNTs shows no obvious changes with increasing CNTs concentration from 0.2 wt % to 1.0 wt %, which infers that the nucleating effect of CNTs may saturate at least at CNTs concentration of 0.2 wt %. This implies that a further elevation of CNTs concentration does not further improve the nucleation effect [52,57].

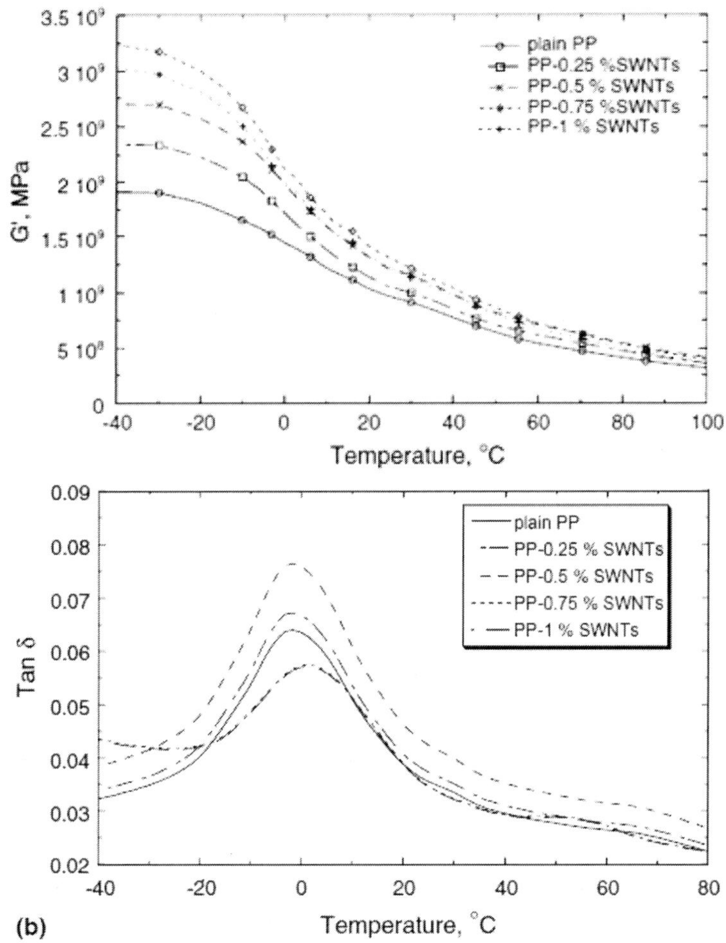

Figure 16. a) Storage modulus for pristine PP and its composites filled with different SWCNTs content and b) Tanδ for pristine PP and its composites as a function of SWCNTs content. Reproduced with permission from ref. [35].

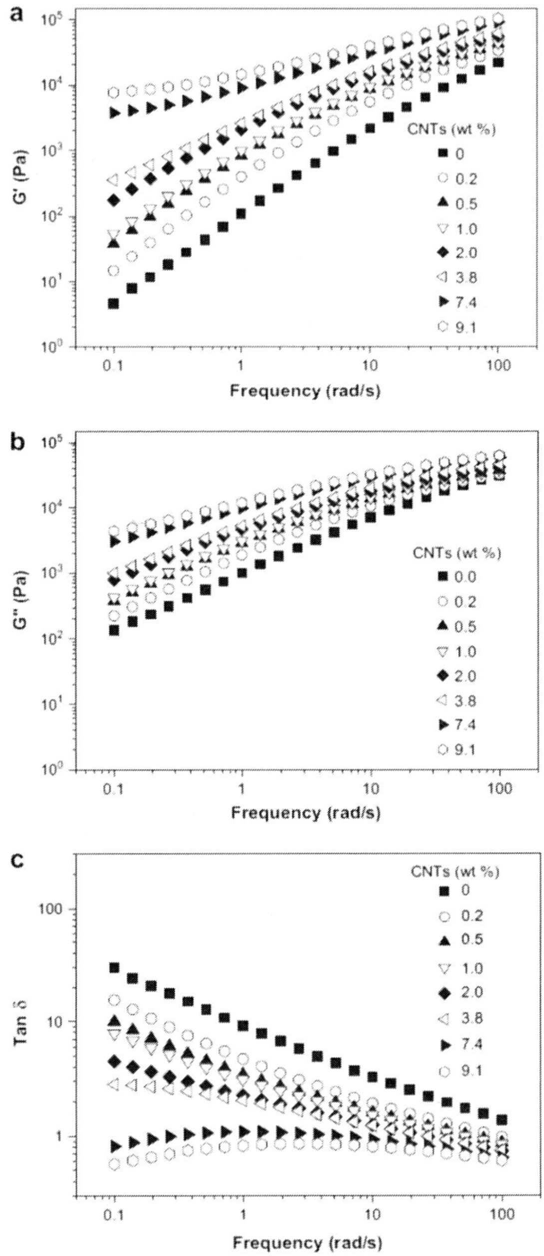

Figure 17. Storage modulus (G') (a), loss modulus (G") (b), and loss tangent, tanδ (c), as functions of frequency for iPP/CNTs with different CNT concentrations. Rheological measurements were performed with a strain of 2% at 200 °C. Reproduced with permission from ref. [42].

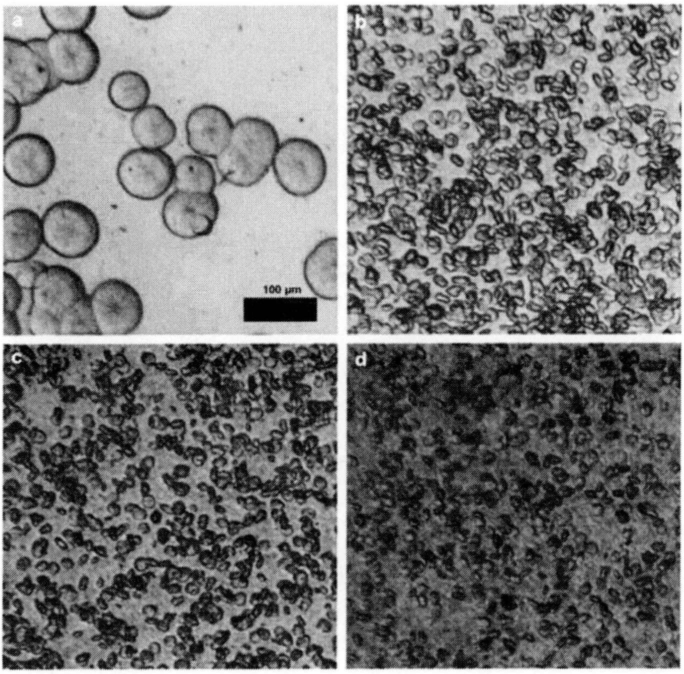

Figure 18. Optical micrographs of iPP/CNTs with CNTs concentrations of a) 0 wt %, b) 0.2 wt %, c) 0.5 wt % and d) 1.0 wt %, during isothermal crystallization at 136 °C for 21 min, 6 min, 6 min, and 6 min, respectively. Optical micrographs were taken in bright field. Reproduced with permission from ref. [42].

Even though a lot of published papers discuss the nucleation ability of CNTs on crystallization of semicrystalline polymers, the reasons why CNTs can be efficient nucleating agents have not been explained clearly. The authors consider that CNTs surfaces might help decrease the energy barrier for nucleation of polymer crystallization due to the evolved interactions between CNTs and PP macromolecules. For a better understanding of these interactions between CNTs and the surrounding iPP matrix, some researchers have performed crystallization experiments in ultrathin films of iPP and CNTs [58]. For this purpose, a droplet of the pure aqueous polypropylene latex dispersion or the CNTs/iPP latex dispersion was deposited on an amorphous carbon-coated TEM grid, heated to 180 °C for 5 min, and subsequently cooled to room temperature so that the iPP can crystallize. Figure 19 shows TEM bright-field images in slight defocus conditions; in contrast to stained samples, in case of phase contrast imaging as applied here, the darker lines correspond to the lamellar crystals. Again, for the neat iPP the common cross-hatched morphology is formed (Figure 19a).

As can be seen from Figure 19b, when the iPP is crystallized in contact with MWCNTs lamellar crystals are formed that grow perpendicular to the long axis of the MWCNTs. It is obvious that the iPP crystallization is nucleated at the surface of the MWCNTs, because the perpendicular orientation of the iPP lamellae follows even narrow curvatures of the MWCNTs [58].

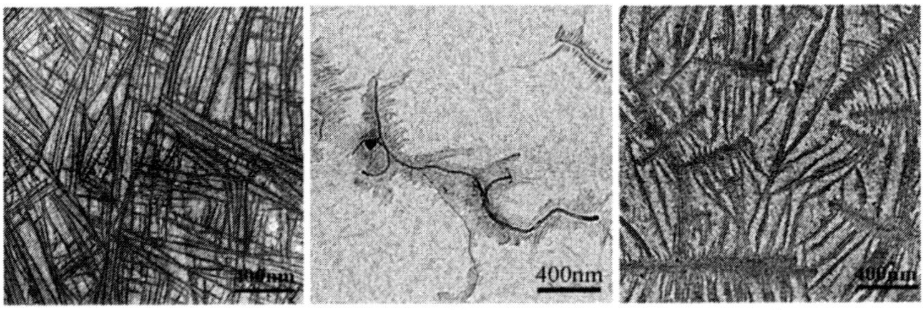

a) b) c)

Figure 19. TEM bright-field images of ultrathin film samples of (a) pure iPP showing cross-hatched morphology; and transcrystalline organization of iPP around (b) MWCNTs and (c) SWCNTs (the SWCNTs are not visible). Reproduced with permission from ref. [58].

Figure 20 shows typical polypropylene morphologies around the carbon fibers in the quiescent melt at 125 °C. The images show oriented crystalline lamellae surrounding the carbon fibers. This supramolecular structure is identified as the transcrystalline interphase. Away from the carbon fibers, the polypropylene spherulites are observed. The impingement lines of the transcrystalline interphase and bulk spherulites are also observed. When two carbon fibers are close to each other, the area between two carbon fibers consists almost exclusively of transcrystalline interphases [59].

PP crystallization is a dynamic process depending on the temperature (as was described before) and on the time. Figure 21 shows a series of optical images taken at 122 °C with different crystallization times. Polypropylene nucleation first occurred at the surface of the carbon fibers within 30s (Figure 21a). This uniform transcrystalline growth front indicates the high nucleating ability of the carbon fibers toward the matrix. When the crystallization time approached 1 min, the spherulite nuclei in the polypropylene matrix were also observed (Figure 21b), which grew into bulk spherulites (Figure 21c, e and f). The growth direction of transcrystals is normal to the carbon fibers axis. This growth is ultimately hindered by the impingement with bulk spherulites or by another transcrystal (Figure 21c, e and f).

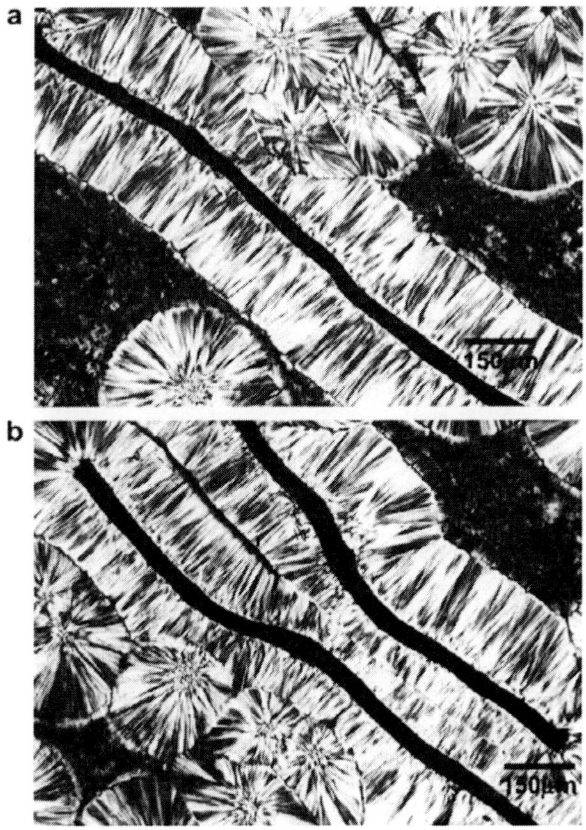

Figure 20. Optical micrographs of transcrystalline interphases for polypropylene surrounding the carbon fibers isothermally crystallized at 125 °C: (a) a single carbon fiber and (b) two carbon fibers. Reproduced with permission from ref. [59].

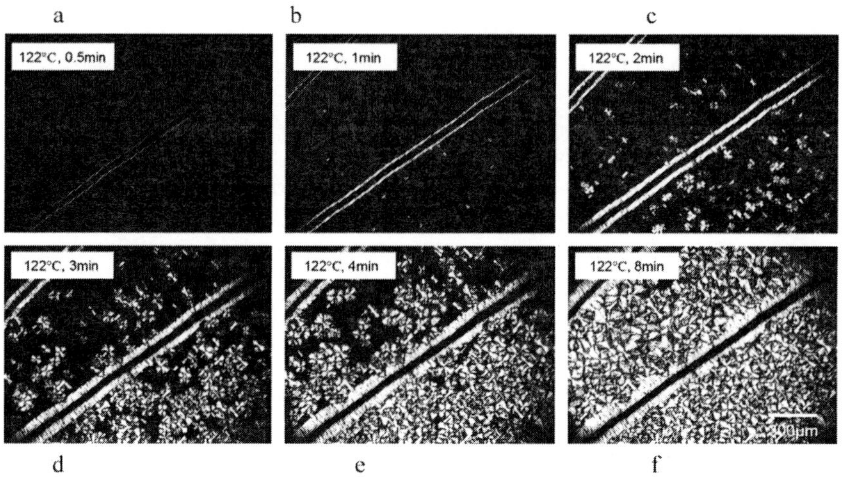

Figure 21. Optical micrographs under cross-polars of transcrystal developments for polypropylene surrounding the carbon fibers at 122 °C for different times: (a) 0.5 min, (b) 1 min, (c) 2 min, (d) 3 min, (e) 4 min, and (f) 8 min. Reproduced with permission from ref. [59].

The mechanism of heterogeneous nucleation is still not precisely known, it has been suggested that, apart from the dispersion of the agent, the mechanism is controlled by the polymer–nucleator interactions, and these can be chemical or physical, or even related to Van der Waals attractions. The carbon nanotubes are able to enhance the crystallization rate by increasing the nuclei sites. As a result, a raise in the crystallization temperature of the polypropylene is observed using different techniques [43,44,45,46,47,48,49,50,51]. From a recent study it was proposed that the orientation of the very first molecules, probably adsorbed at the surface of the CNTs, that induces the formation of a stable nucleus should be perpendicular rather than parallel to the long axis of the CNTs. This situation is sketched in Figure 22: iPP macromolecules are adsorbed at the surface of the CNTs in the melt and are partially wrapped around the CNTs; probably the protruding methyl groups of the iPP interact with the graphite layer of the CNT. This confinement of the macromolecules is kept during cooling of the sample from quiescent melt and causes enhanced and oriented nucleation of iPP lamellar crystals, which have their crystallographic *c*-axis tangential to the graphite shell(s) of the CNTs [58].

CNTs cannot cause only heterogeneous nucleation on PP, but also can increase the crystallization rates. The effects of CNTs on the crystallization of iPP were analyzed with nonisothermal DSC experiments on cooling. Figure 23 shows the dynamic thermograms obtained for neat PP and its composites at a cooling rate 20 °C/min. The observed dynamic crystallization behavior shows the positive effects of the nanotubes on the crystallization kinetics of PP. In particular, Figure 23 values confirm that the addition of CNTs to the polymer matrix produces an increase in T_c. The relative shift of T_c is quite evident at the lowest reinforcement content with a slow but continuous further increase with the CNTs concentration. These results confirm that the addition of a low concentration of nanotubes enhances the nucleation process for iPP crystallization [60].

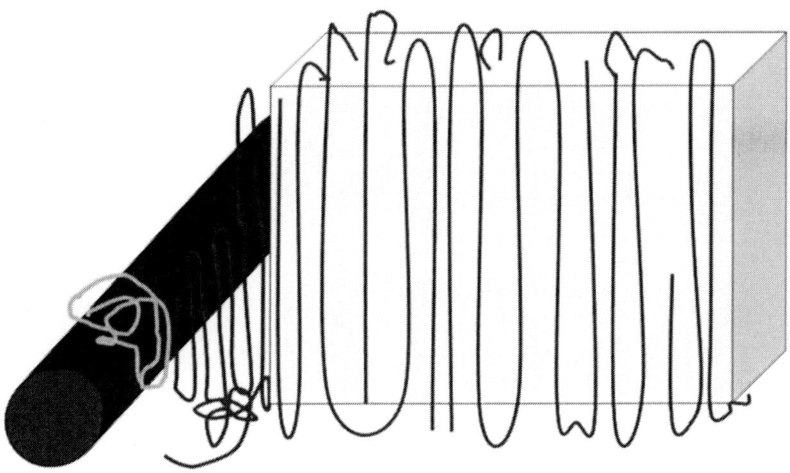

Figure 22. Sketch explaining the possible nucleation mechanism of iPP on the surface of a CNTs; iPP macromolecules initially are partly wrapped around the CNTs (brighter macromolecule in the sketch, dark rod represents a CNTs) and form a nuclei with crystallographic *c*-axis perpendicular to the long axis of the CNTs. Reproduced with permission from ref. [58].

To quantitatively compare isothermal crystallization kinetics of iPP/CNTs with different CNTs concentrations, the Avrami analyses on DSC data were conducted and the half crystallization time, $t_{1/2}$, values were obtained [42]. Changes of $t_{1/2}$ of iPP/CNTs with increasing CNTs concentration at different crystallization temperatures are shown in

Figure 24. An obviously observable phenomenon is an initially rapid decrease of $t_{1/2}$ and subsequent modest decrease of $t_{1/2}$, followed by about constant $t_{1/2}$, with increasing CNTs concentration at each isothermal crystallization temperature. For example, at 127 °C, addition of 0.2 wt % CNTs reduces $t_{1/2}$ to 21% of $t_{1/2}$ of neat iPP, additions of 0.2-3.8 wt % CNTs only reduce $t_{1/2}$ from 21% to 10% of $t_{1/2}$ of neat iPP, and additions of above 3.8 wt % CNTs maintain about constant 10% of $t_{1/2}$ of neat iPP.

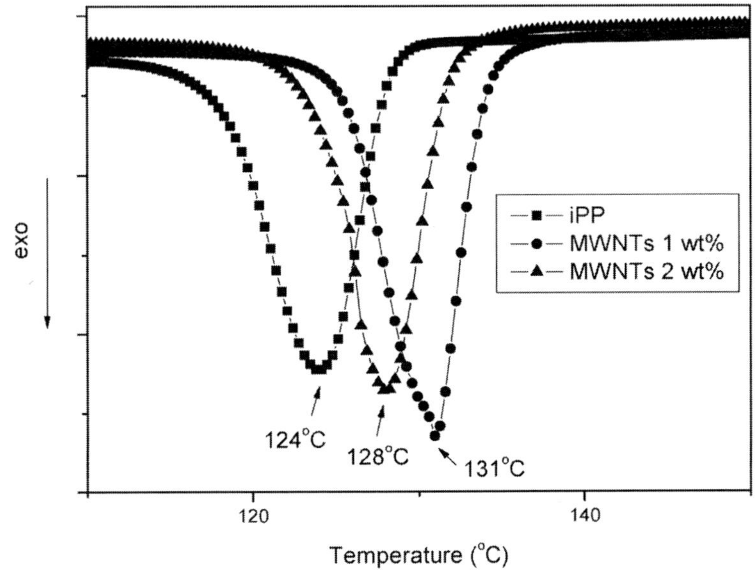

Figure 23. Nonisothermal crystallization curves of PP and PP/MWCNTs composites.

The main question from the above mentioned data is if this heterogeneous nucleation as well as enhancement of crystallization rate can also increase the crystallinity of PP. Crystallinity has a major effect on the mechanical properties of polymers and hence the effect of nanoparticles on nucleation and crystallinity development is of interest. DSC thermographs of PP and PP/MWCNTs nanocomposites with different MWCNTs weight loadings are shown in Figure 25. The melting range of PP as well as its nanocomposites is 160–165 °C. A small reduction was observed in the melting point of nanocomposites compared with neat PP. Furthermore, concerning the effect of MWCNTs on iPP crystallinity, different findings were reported. As the MWCNTs loadings increased, it was reported that the percent of iPP crystallinity could remain stable [47], decrease slightly [54] or increase slightly [49].

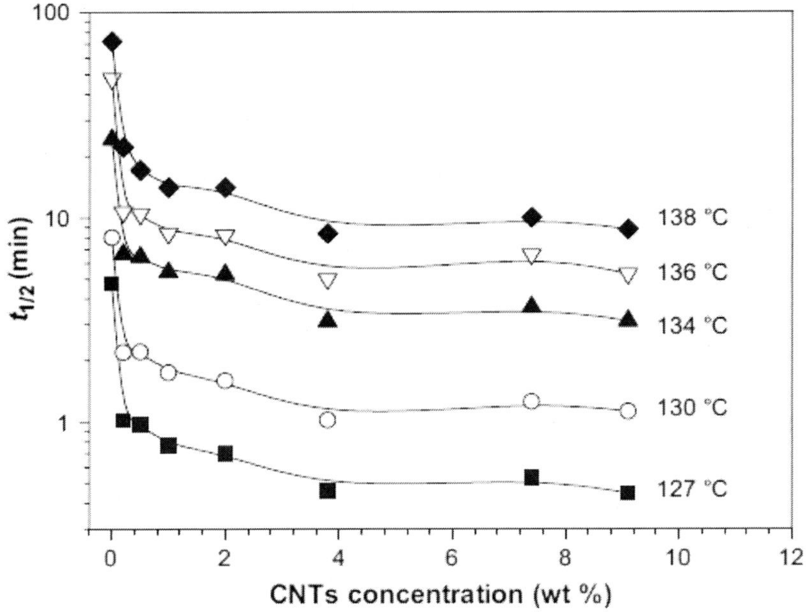

Figure 24. Changes of half crystallization time $t_{1/2}$, of iPP/CNTs with different CNTs concentration at crystallization temperatures of 127 °C, 130 °C, 134 °C, 136 °C and 138 °C. Reproduced with permission from ref. [42].

Some studies have been done in order to determine the equilibrium melting temperature (T^o_m) of PP/MWCNTs nanocomposites [56]. It was found that the T^o_m for neat iPP falls within the 212–215 °C range while the nanocomposites had lower T^o_m than their corresponding base polymer. The presence of CNTs, and the nucleating effect associated with them (vide infra), increases the quantity of amorphous material, especially in between lamellar stacks. It may be noted that the samples prepared with functionalized carbon nanotubes had a higher T^o_m than those filled with pristine MWCNTs. This is indicative of a less disrupting role of PP-g-MWCNTs, with respect to pristine MWCNTs, on the ordered and regular crystallization of the matrix.

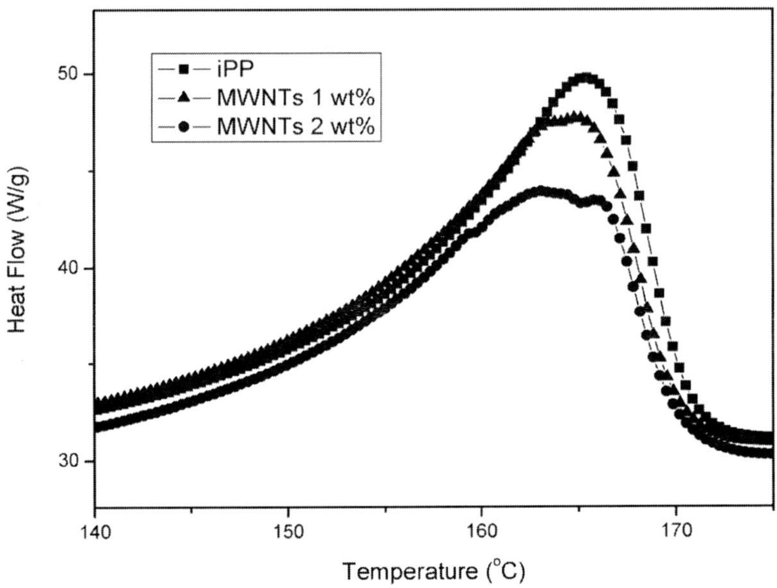

Figure 25. DSC thermographs for different weight percentages of nanotube loadings in PP.

The melting behavior after nonisothermal crystallization of iPP was investigated by modulated temperature DSC (MTDSC) at a heating rate of 2.5 °C/min [61]. Figure 26 shows the melting traces for the iPP nanocomposites containing SWCNTs or MWCNTs. Clearly, the presence of CNTs has a dramatic impact on the melting behavior of the matrix material. Whereas the unfilled matrix unambiguously shows double melting behavior, the shape of the melting transition progressively evolves toward single melting with increasing SWCNTs content. Indeed, the lower melting peak $T_{m,1}$progressively shifts to higher temperature upon increasing the filler loading, until it finally almost coincides with the higher melting peak $T_{m,2}$ at 2 wt % SWCNTs loading. Even though an upward shift in $T_{m,1}$ can also be observed for MWCNTs composites, the effect is less pronounced than in the case of SWCNTs, with double melting behavior prevailing even at higher loading.

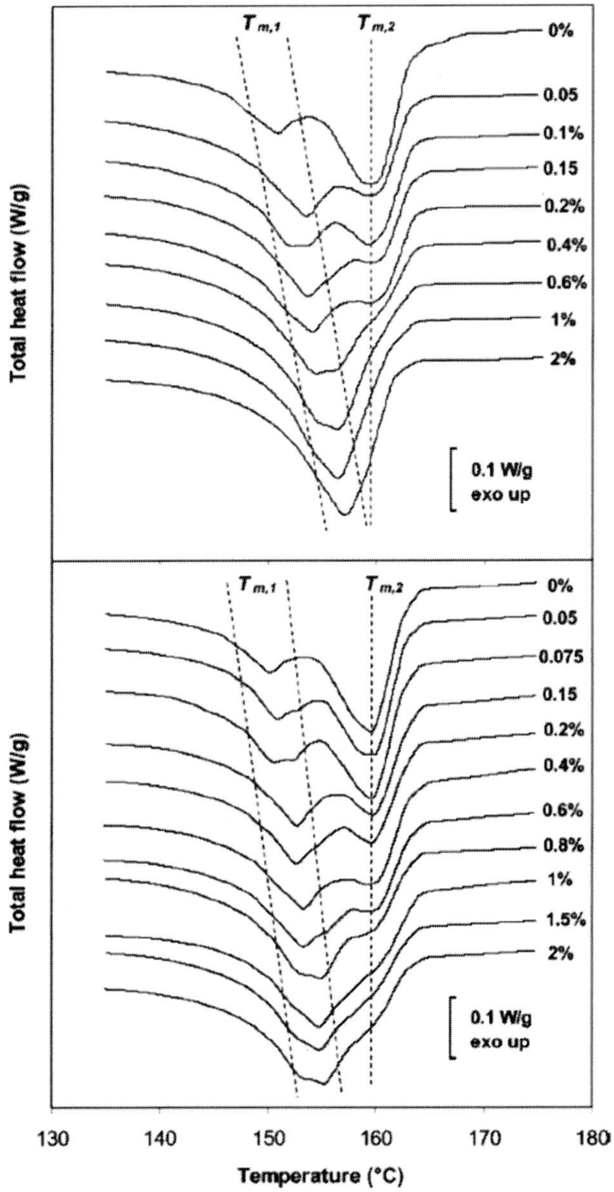

Figure 26. MTDSC total heat flow traces during heating at 2.5 °C/min of iPP nanocomposites containing SWCNTs (top) and MWCNTs (bottom). The dashed lines are guidelines for the eye showing the position of the melting peaks for the different systems ($T_{m,1}$ marked as a broad melting region and $T_{m,2}$ as a single peak). Reproduced with permission from ref. [61].

Multiple melting behavior is generally assumed to result from polymorphism, from the successive melting of crystal populations with distinct degrees of perfection, or from the rapid succession of melting-crystallization-melting or reorganization phenomena. The impact of fillers on the polymorphic behavior of iPP has been extensively reported over the years. Some authors reported that CNTs as well can induce crystallization of iPP in the hexagonal β-polymorph instead of the more common monoclinic α-form. Since both polymorphs display distinct melting temperatures, the observed changes in the shape of the melting transition in the presence of CNTs might potentially be the result of an altered balance between crystal forms simultaneously present in the sample.

From the above studies with DSC, it was revealed that CNTs can have an effect on crystallization rate, melting point as well as the degree of crystallinity of PP matrix. However, it seems the CNTs can not only promote the crystallization rates and the formation of α-iPP crystals, which are the main crystal forms of PP, but can also to induce the formation of β-crystals of iPP. In DSC melting curves of pure iPP and iPP/MWCNTs composites, two obvious peaks at approximately 147 and 154 °C appeared with a MWCNTs concentration of 0.1 wt % [14]. Polypropylene in the β-crystal form melts at lower temperature than the α-crystals, so the appearance of the peaks at 147 and 154 °C can be attributed to the melting of β-crystals or smaller or imperfect α-crystals. WAXD of the iPP/MWCNTs composites confirmed the presence of β-crystals, in agreement with DSC data. Considering that the pure iPP was not able to form the β-structure crystal under identical crystallization condition, it can be concluded that the presence of MWCNTs induced the growth of the β-form crystals in iPP. This result was surprising, since most studies [13,41,45,61] suggested that MWCNTs could only promote the formation of α-iPP crystals, and no literature reported that MWCNTs could induce the formation of β-phase iPP. The characteristic crystalline reflection of α-form iPP at 2θ= 13.8° (110), 1.6° (040), 18.3° (130), and 21.5° (111,131) were observed in the curves of both pure iPP and PP/MWCNTs composites with various MWCNTs contents (Figure 27).

3.6. Electrical conductivity of PP/CNTs

One of the major benefits of CNTs is electrical conductivity and thus their addition into PP matrix can increase the conductivity of PP. Electrically conductive polymer composites (CPCs) based on polypropylene and conductive fillers were investigated in recent studies

[13,46,47,48,49,50,51,52,53,54,55,56,57,58,59,60,61,62,63,64,65,66]. The dispersion of low concentrations (<0.5 vol %) of MWCNTs in PP, resulted in substantial decreases in the electrical surface resistivity of the derived composite material. In PP the inclusion of MWCNTs at the percolation threshold of 0.05 vol %, produced a reduction in resistivity from $>10^{12}$ Ω/square to a value of $\approx 10^5$ Ω/square and higher concentrations produced further reductions to a 100 Ω/square [67]. MWCNTs form a more entangled and stable network compared to carbon black (CB) due to their large aspect ratio [68]. The aggregation of the MWCNTs, as well as CB, into sphere-like agglomerates is considered to be a main process in the formation of the conductive network in a polymer melt. The network reconstruction process during isothermal annealing is explained by re-aggregation of the conductive filler agglomerates.

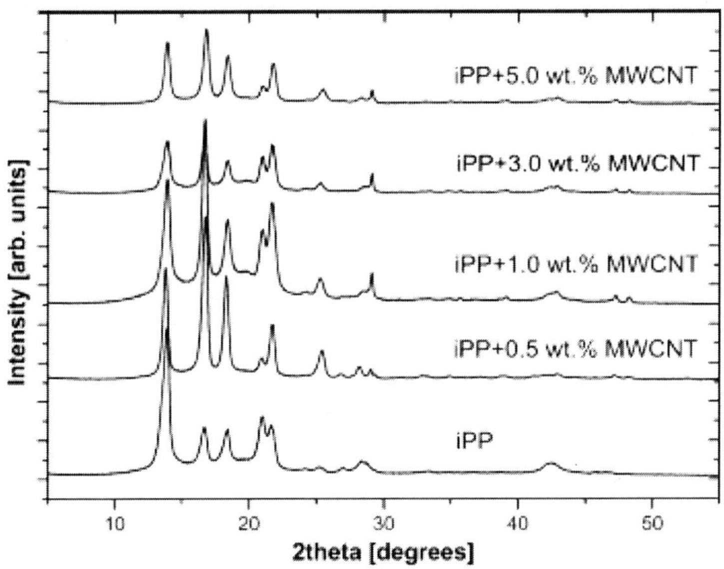

Figure 27. WAXD patterns of pure iPP and iPP/MWCNTs composites with differing MWCNTs contents. Reproduced with permission from ref. [13].

The influence of CNTs contents on electrical properties of CNTs-reinforced PP nanocomposites was studied by many researchers. It was found that the volume resistivity of the composites was decreased with increasing the CNTs content and the electrical percolation threshold was formed between 1 and 2 wt % CNTs, which were caused by the formation of conductive chains in the composites [13,65,69,70]. It is obvious from the conductivity spectra and shape of the curves that the sample containing 2 wt % MWCNTs is close to the percolation threshold, whereas the samples with 3.5 and 5 wt % are above (Figure 28) [71]. The composite at low MWCNTs concentration is an insulator since the conductivity of the polymer matrix is taken to be negligibly small, but a "percolating" network of particles forms at higher concentrations as the randomly positioned and oriented particles begin to "overlap." For a sample containing 2 wt % MWCNTS, the DC conductivity in the sheared melt at 220 °C is less than 10^{-7} S/m, whereas the value in the annealed melt (20 min after stopping the extruder) increased to about 1.4×10^{-3} S/m. The difference in conductivity values can be explained by a strong dependence of the "effective number of conducting bonds" on the arrangement of the nanotubes and clusters in the matrix as well as on the arrangement of the polymer chains in the gaps between the tubes. All these effects depend strongly on the thermo-mechanical prehistory of the sample. Thus, the DC conductivity and the static permittivity were found to increase with annealing (rest time after shear was stopped) of the melt for the sample with 2 wt % MWCNTs.

Electrical conductivity of PP/CNTs is also affected by the kind of CNTs and preparation method [17]. Microwave studies indicate conspicuous difference in the electrical properties of the iPP/MWCNTs nanocomposites as compared to the sPP/MWCNTs. Thus, the iPP/MWCNTs nanocomposites exhibit considerably higher permittivity values in comparison with the sPP/MWCNTs materials [12]. This behavior corresponds to different frequencies in microwave range. Slower growth of the sPP/MWCNTs permittivity with increasing filler content is attributed to the nanotube agglomeration.

Electrical conductivity can be further enhanced when conductive polymers or other fillers are mixing together with MWCNTs. In a recent study, hybrid materials of conducting polyaniline (PANi) and multiwalled carbon nanotubes (CNTs) were synthesized using an *in situ* polymerization method [72]. The resulting CNTs with a thin layer of PANi coating were further compounded with polypropylene to produce electrically conductive composite material. The PANi coating was found to enhance the CNTs dispersion in PP. In a recent study, the fabrication of lightweight and high performance nanocomposite bipolar plates for the application in polymer electrode membrane fuel cells (PEMFCs) was reported [73]. The thin nanocomposite bipolar plates (thickness <1.2

mm) consisting of MWCNTs, graphite powder and PP were fabricated by compression molding. Three types of PP with different crystallinities including high crystallinity PP (HC-PP), medium crystallinity PP (MC-PP), low crystallinity PP (LC-PP) were prepared to investigate the influence of crystallinity on the dispersion of MWCNTs in PP matrix. The optimum composition of original composite bipolar plates was determined at 80 wt % graphite content and 20 wt % PP content based on the measurements of electrical and mechanical properties with various graphite contents. Results also indicate that MWCNTs was dispersed better in LC-PP than other PP owing to enough dispersed regions in nanocomposite bipolar plates. This good MWCNTs dispersion of LC-PP would cause better bulk electrical conductivity, mechanical properties and thermal stability of MWCNTs/PP nanocomposite bipolar plates. Figure 29depicts a proposed model of conductive paths in the MWCNTs/PP nanocomposite bipolar plate with different crystallinities of PP matrices. This model was based on the presented experimental results. Electrical insulated polymers are filled with electrically conductive graphites and carbon nanotubes. The insulting zones between graphites are bridged by the MWCNTs. When voltage is applied, more continuous paths are available for the flow of electrical current due to the MWCNTs homogeneity in LC-PP matrix. Comparing these two conductive-path models, it is assumed that more electrical conductive paths were formed effectively by the better dispersed MWCNTs and graphite particles in LC-PP than that of HC-PP.

Electrical and thermal conductivity of polymers can be also increased by adding conductive metals. Silver exhibits the largest electrical and thermal conductivities among all the metals. Owing to the lack of conductive paths, PP/Ag nanocomposites exhibit poor electrical properties. In order to improve the electrical properties of PP/Ag nanocomposites, MWCNTs having large aspect ratio are compounded with PP and nano-Ag to form hybrid nanocomposites [74]. The dielectric constant and conductivity of PP/Ag/MWCNTs composites increases gradually with MWCNTs content at low filler loading. As the MWCNTs content reaches 2 wt %, the dielectric constant increases sharply by two orders of magnitude, and the conductivity by more than four orders of magnitude, exhibiting the percolation phenomenon (Figure 30).

This reveals that MWCNTs with large aspect ratio favor the formation of conductive pathways in the PP/Ag/MWCNTs hybrid nanocomposites. The nanotubes act as bridges for individual Ag particles, yielding high electrical conductivity for the ultimate composites. This clearly indicates that the Ag particles facilitate the formation of conductive pathways in the presence of MWCNTs. The opposite results were mentioned when an electrically inert additive line calcium carbonate ($CaCO_3$) is used together with MWCNTs. A significant reduction in electrical resistivity of PP/MWCNTs composites was found with the addition of $CaCO_3$ [75].

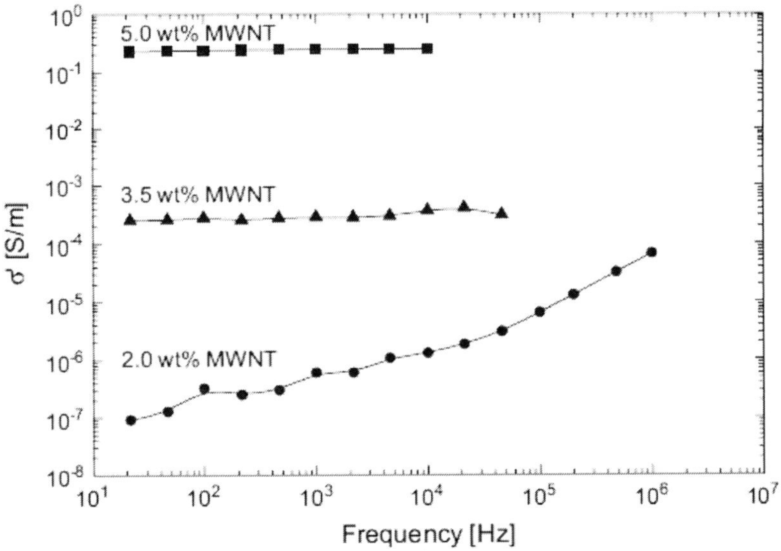

Figure 28. Conductivity spectra for polypropylene carbon nanotube composites with different MWCNTs contents recorded at about 120 s after the extruder was stopped. The temperature of the slit die was about 225 °C. Reproduced with permission from ref. [71].

3.7. Enhancement of PP thermal stability by addition of CNTs

There are many studies concerning the thermal stability enhancement of PP reinforced with CNTs [76,77,78]. CNTs are themselves very stable materials. In air, the main mass loss of MWCNTs starts at 400 °C while the maximum decomposition rate T_{max} is found at about 600 °C. At the end of the analysis (680 °C), the degradation process is almost complete with a residue 1.5 wt %. DWCNTs start to lose mass at about 450 °C while the majority of the material degrades with a single large peak centered at about 525 °C (Figure 31). The degradation is complete at about 600 °C, leaving a 4% residue probably due to catalyst and support from synthesis. This difference can be attributed to the higher defectiveness of DWCNTs due to the synthetic process that makes them more subject to the degradative action of air with respect to MWCNTs [79]. In nitrogen, both nanotubes have similar behavior.

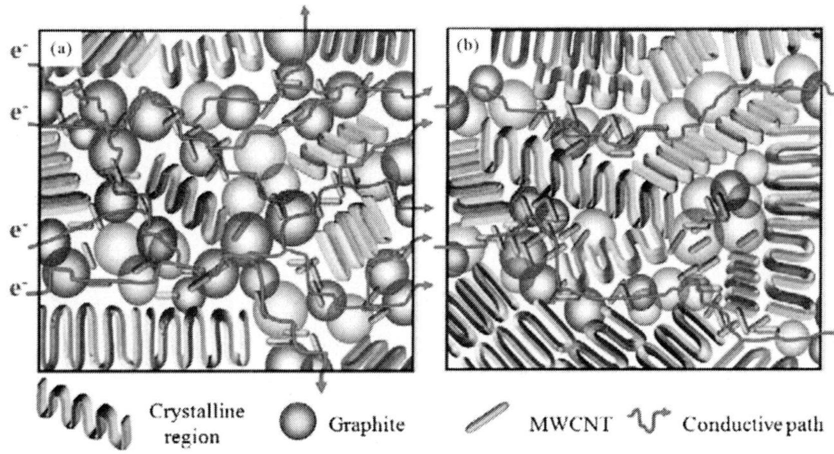

Figure 29. The model of conductive paths in the nanocomposite bipolar plates with (a) better dispersion of MWCNTs in LC-PP matrix (b) MWCNTs aggregation in HC-PP matrix. Reproduced with permission from ref. [73].

The thermogravimetric (TG) and derivative thermogravimetric (DTG) curves for the neat PP and PP/MWCNTs composites at a constant heating rate of 10 °C/min are shown in Figure 32(a) and (b), respectively. [54,78] It is clear that PP degrades in a single step. The step initiates at 220 °C and ends at 410 °C. The TGA curves show no significant weight loss till 250 °C. However beyond this temperature the onset of degradation was notably influenced by the nanotube loading within the polymer. The onset of thermal decomposition values increased with nanotube loading and the nanocomposite attained a maximum onset of 354 °C. Compared with the pure PP, the nanocomposites showed an increase of 16 °C with the addition of 0.5 wt % MWCNTs and 34 °C with the addition of 1.0 wt % in the onset of thermal degradation. The improvement in thermal stability can be attributed to good matrix–nanotube interaction and also due to the thermal conductivity of the nanotubes. As the nanotubes are good thermal conductors the tubes easily take up the heat that is applied to the nanocomposite fibers. The good dispersion of the nanotubes in the polymer matrix allows the spreading of heat uniformly along the fiber. Another factor that can potentially contribute to the thermal stability is the formation of a relatively uniform network-structured layer which covers the entire sample surface without any cracks or gaps forming during heating. This layer re-emits much of the incident radiation back from its hot surface, thereby reducing the heat transmitted to the PP layers below.

The values of the activation energies of degradation at 50% conversion can be seen in Figure 33 [78]. There is a clear trend between the activation energy of degradation and the MWCNTs content of these samples. The activation energy values for 1 wt % PP/MWCNTs nanocomposites are 10 kJ/mol for neat PP, and increasing with increasing MWCNTs content. These results should be attributed to the increasing stability of the radicals formed during the pyrolysis process and the improving dispersion of the particles in the composites.

Thermogravimetric analyses (TGA) of PP-CNTs composites were also investigated in nitrogen and in air atmosphere in order to get a complete picture of their thermal behavior [79]. It was found that addition of DWCNTs in nitrogen increased both $T5\%$ and T_{max} of the pristine polymer with a maximum increase of 32 °C for $T5\%$ and of 18 °C for T_{max}. In air, the difference between the range of temperature for composite mass loss, as compared to PP, was much higher than in nitrogen, and increased with CNTs concentration (maximum ΔT_{max}=18 °C in nitrogen and 101 °C in air). A larger effect is observed for MWCNTs as compared to DWCNTs. The delay in T_{max} induced by a 3% loading of DWCNTs is in fact very similar to that induced by a 1% loading of MWCNTs. The protection action of nanotubes can be attributed to a barrier effect implying a hindered transport of polymer degradation products from the condensed to the gas phase. The shape of the mass loss curve of PP is strongly modified by the CNTs in the presence of air becoming strongly asymmetrical. At 3% MWCNTs loaded sample, the beginning of the degradation can be fixed at 320 °C, as in the case of the other samples, and the mass loss proceeds slowly up to 454 °C. When this temperature is reached an abrupt acceleration of the mass loss is observed. Such behavior can be attributed to the formation of a superficial protecting nanotube network which is destroyed when increasing the temperature up to a limit value. The protection from oxygen action is therefore extremely efficient at low temperatures but the final result is in any case a complete degradation of the polymer. The stabilizing effect of CNTs can be also ascribed to the interfacial interaction between the nanofibres and PP increasing the activation energy of degradation and to retarding the degradation at the surface of the samples owing to the better heat distribution within the heat conductive CNTs-containing samples.

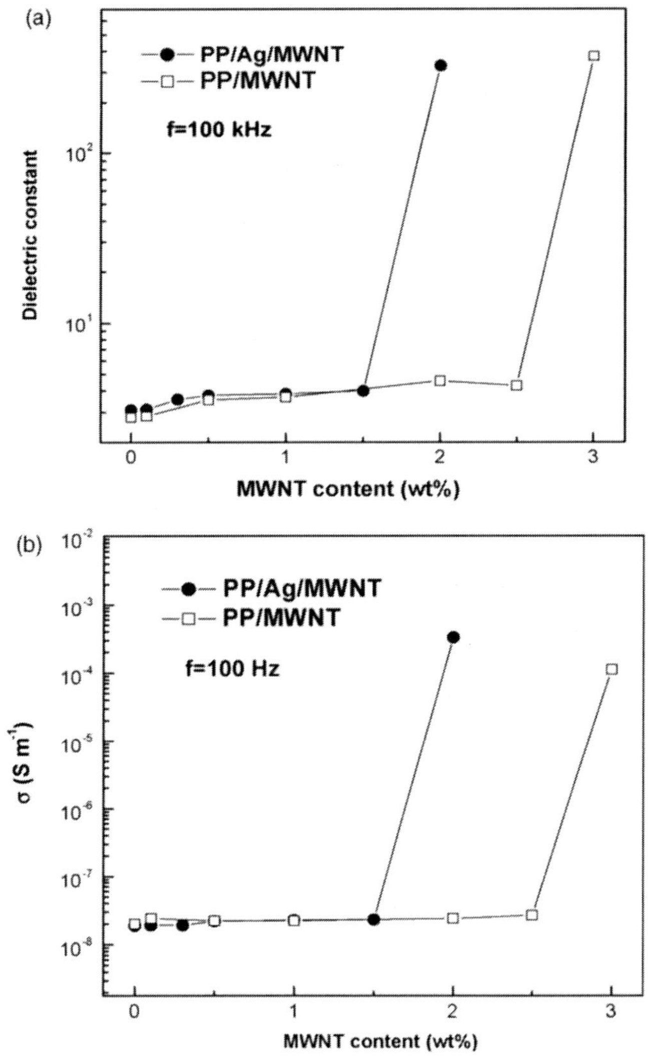

Figure 30. Variations of (a) dielectric constant and (b) conductivity with MWNT content for PP/MWNT and PP/Ag/MWNT nanocomposites. The nano-Ag content is fixed at 10 wt % for PP/Ag/MWNT hybrid nanocomposites. Reproduced with permission from ref. [74].

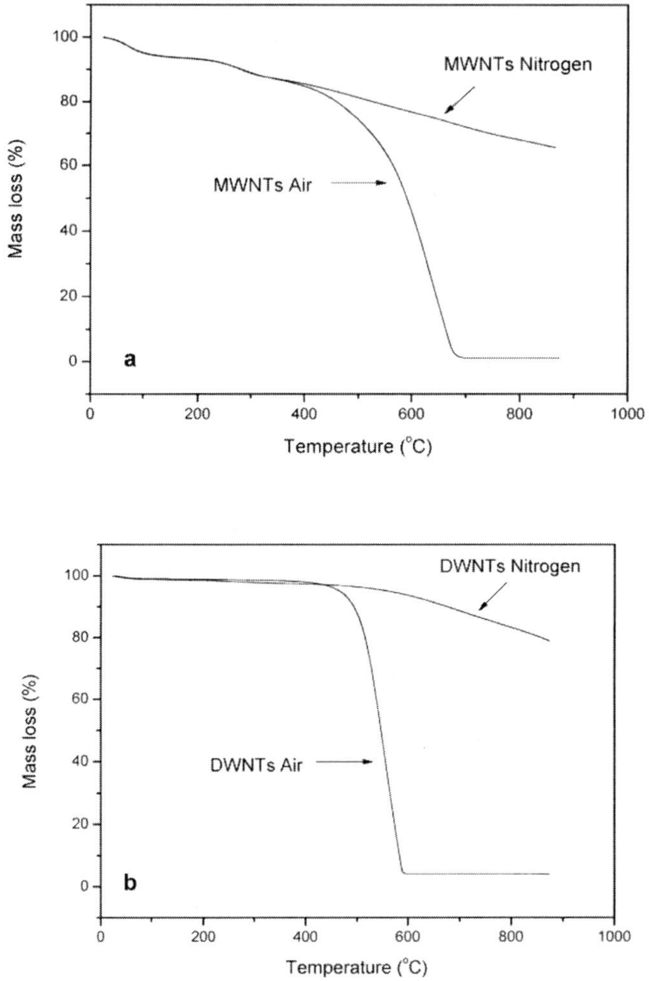

Figure 31. TG and DTG curves in air of MWCNTs and DWCNTs.

3.8. Effect of CNTs on PP permeability

Nanoparticles are well known to reduce also the permeability due to the shielding effect in PP matrices. In the case of PP/MWCNTs, the studies on gas permeability are limited [14]. Neat iPP has an O_2 permeability of 7.35 (mL cm)/(day m^{2}atm), while this is reduced to 5.87 (mL cm)/(day m^2 atm) after the addition of 2.5 wt % MWCNTs (Figure 34). A similar behavior is also observed for N_2 permeability. It is well known that such nanoparticles are considered impenetrable by gas molecules. Thus, it is believed that their addition in the resin matrix would enhance its barrier properties by forcing the gas molecules to follow a more tortuous path as they diffuse through the material, retarding the progress of the phenomenon. This has been observed in many polymer/layered silicate nanocomposites as well as in iPP using fumed silica nanoparticles [80,81]. The presence of MWCNTs in the studied nanocomposites is believed to create a labyrinth, complicating the path that the gas molecules must go through in order to pass through the whole width of the film.

3.9. Thermal expansion of PP/CNTs nanocomposites

Thermal expansion is a very important parameter for many PP products, especially since PP has been used for preparation of pipes for hot and cold water circulation. Measurements of thermal expansion behavior of the isotropic PP/CNTs composites were made on the changes in length between 15 and 25 °C, to provide an average value for a room temperature of 20 °C [34]. In Figure 35, the measurements are compared with model predictions following a similar approach to that described above for the mechanical behavior. It is interesting to note that the thermal expansion behavior mirrors exactly that for the Young's modulus, with the predictions for the 6 wt % composite being in excellent agreement with the measurements and similar discrepancies for the other materials. For thermal expansion, the measured values for the lower weight percentages are lower than predicted and the higher weight percentages are higher, supporting the proposals above regarding the structure of these samples.

3.10. Flammability of PP/CNTs nanocomposites

Some brief papers involving MWCNTs were published showing significant flame retardant effectiveness of polypropylene PP/MWCNTs nanocomposites [82,83,84] even thought the residual amount of iron particles, which are used as catalyst to make the MWCNTs, can affect the thermal and thermo-oxidative stability of nanocomposites [85]. Flammability properties were measured with a cone calorimeter in air and a gasification device in a nitrogen atmosphere [83]. A significant reduction in the peak heat release rate was observed; the greatest reduction was obtained with a MWCNTs content of 1% by mass. The radiative ignition delay time of a nanocomposite having less than 2% by mass of MWCNTs was shorter than that of PP due to an increase in the radiation in-depth absorption coefficient by the addition of carbon nanotubes. The effects of residual iron particles and of defects in the MWCNTs on the heat release rate of the nanocomposite were not significant. The flame retardant performance was achieved through the formation of a relatively uniform network-structured floccule layer covering the entire sample surface without any cracks or gaps.

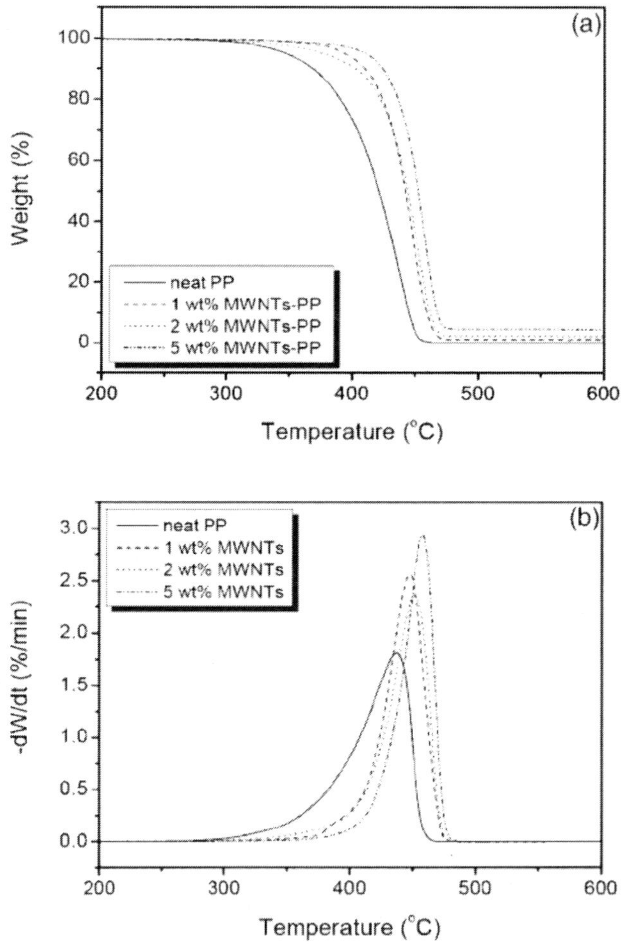

Figure 32. (a) TGA and (b) DTG curves of different contents of PP/MWCNTs nanocomposites at a heating rate of 10 °C/min. Reproduced with permission from ref. [78].

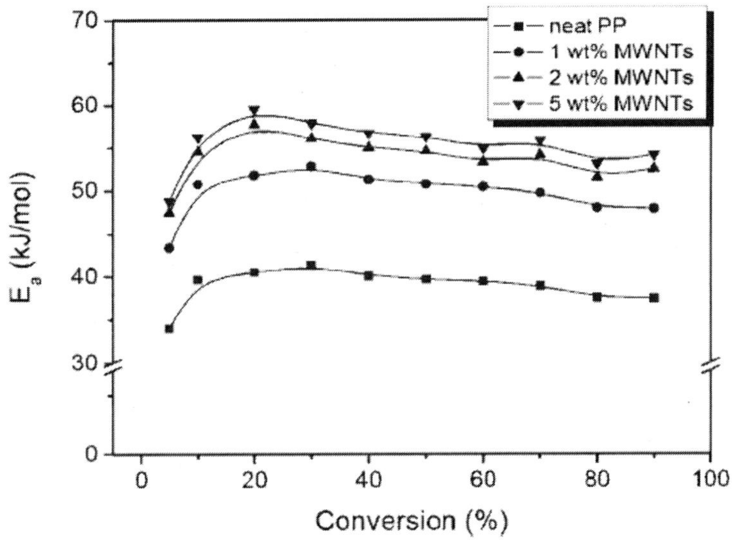

Figure 33. Activation energies for degradation (Ea) of different contents of MWCNTs-filled PP nanocomposites as a function of conversion using the Flynn-Wall-Ozawa method. Reproduced with permission from ref. [78].

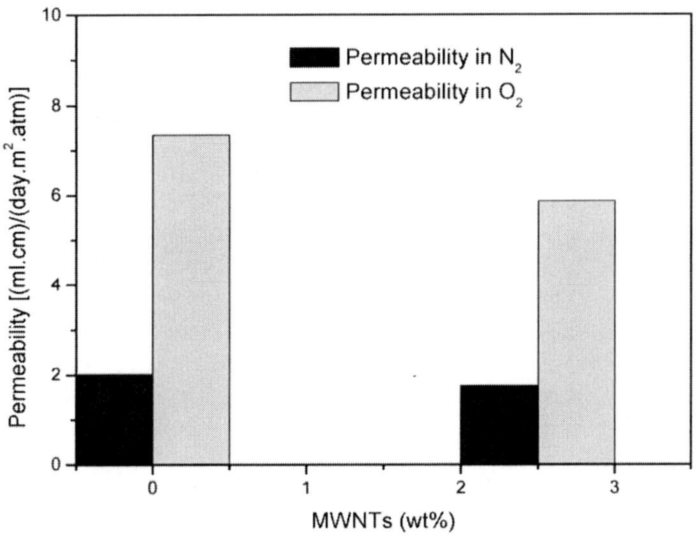

Figure 34. Permeability of O_2 and N_2 from iPP/MWCNTs nanocomposites containing 2.5 wt % MWCNTs.

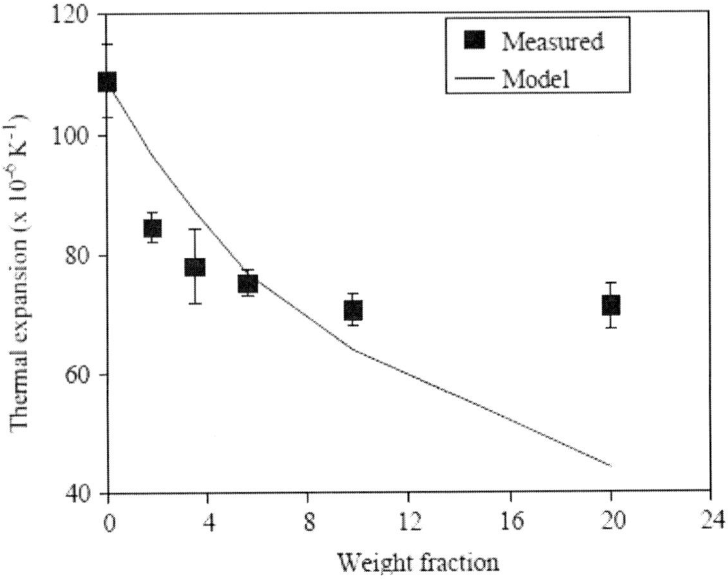

Figure 35. A comparison between measurements and model predictions for thermal expansion. Reproduced with permission from ref. [34].

4. EFFECT OF CNTS FUNCTIONALIZATION ON PP/CNTS NANOCOMPOSITES PROPERTIES

A good dispersion of CNTs in a polymer matrix and a strong interfacial interaction between CNTs and polymers are prerequisites to maximize the advantage of CNTs as effective reinforcing filler in polymer composites [16,10,35,77,86,87,88]. As found from almost all the reports in PP/CNTs by studying their microstructure, CNTs tend to aggregate to form bundles because of strong Van der Waal's force, which makes the dispersion of CNTs in the polymer matrix difficult. CNTs are hydrophilic while PP is a non polar hydrophobic material. Thus, in order to achieve a finer dispersion into PP matrix CNTs should become hydrophobic due functionalization or surface coating. Consequently, techniques such as end group chemical functionalization on the CNTs and modification by grafting reactive groups on the polymer matrix have been used to improve the dispersion of CNTs in the polymer matrix [33,89,90, 91,92,93,94,95].

Silane and alkyl groups are the most important to modify CNTs and gain a hydrophobic surface [90,96]. According to the sol-gel method, the MWCNTs functionalization can be achieved using tetraethoxysilane (TEOS). The MWCNTs were treated with TEOS:H_2O:ethanol = 2:1:4 in a 5:1 volume ratio and the obtained dispersion was sonicated for 2 h making MWCNTs uniformly disperse and avoiding phase separation. The dispersion was maintained by stirring at room temperature overnight to allow the hydrolysis and condensation of TEOS to form silica coating on the surface of MWCNTs. After 24 h of hydrolysis, the dispersion was centrifuged three times to wash out uncoated silica from MWCNTs using ethanol to obtain silica-coated MWCNTs. CNTs can be also coated with methacryloxypropyltrimethoxysilane (3-MPTS). In this case, 1 g of silica-coated MWCNTs was dispersed in 1 L of ethanol, then the dispersion was sonicated for 30 min. 0.5 g of 3-MPTS was slowly added to 500 mL of this MWCNTs dispersion, then the mixture was refluxed for 5 h at 75 °C under constant stirring. After completion of reaction, the product was washed with deionized water and ethanol respectively five times to obtain 3-MPTS functionalized MWCNTs, which were filtrated and dried in a vacuum at 60 °C for two days. According to this procedure, the surfaces of CNTs are coated with a small layer of silica and MPTS (Figure 36).

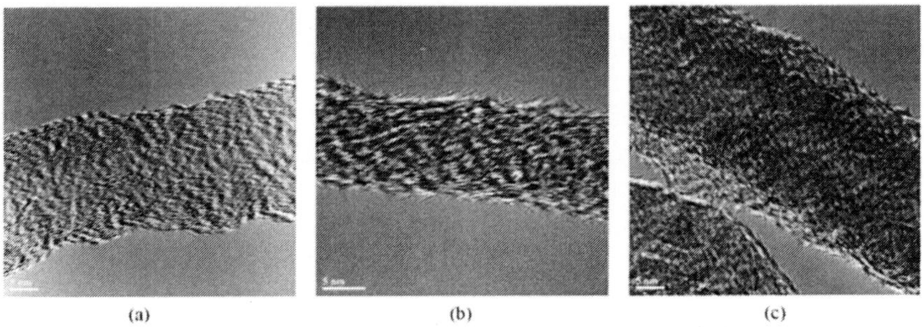

Figure 36. TEM images of (a) the raw MWCNTs, (b) silica-coated MWCNTs, and (c) 3-MPTS functionalized MWCNTs. Reproduced with permission from ref. [96].

Alkyl modified CNTs such as CNT($COOC_{18}H_{37}$)n can also be prepared using a simple method. In the first stage, CNTs can be converted into acid form [CNT(COOH)n] via conisation in 1/3 relative volume ratio of nitric acid/sulfuric acid mixture at 40 °C and with NaOH can be converted into sodium salt form [CNT(COONa)n] [42,93]. The addition of cetyltrimethylammonium bromide (CTAB) and $C_{18}H_{37}Br$ converts CNT(COONa)n into CNT($COOC_{18}H_{37}$)n alkyl-modified CNTs. Undecyl-functionalized MWCNTs (C11-MWCNTs) were also prepared by Koval'chuk *et al.*, as described in Figure 37[97].

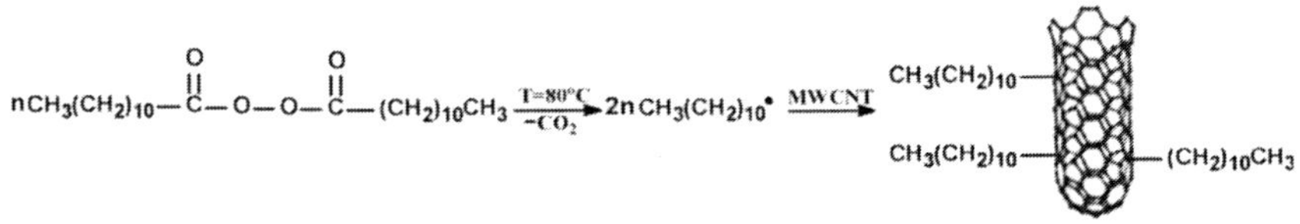

Figure 37. Steps for the preparation of undecyl-functionalized MWCNTs. Reproduced with permission from ref. [97].

The CNTs functionalization results in a finer dispersion of CNTs into PP matrix due to increased interfacial adhesion between the two materials. It is evident from the optical microscope images that CNTs tend to form large, non-uniform network agglomerates when dispersed in PP (Figure 38a) [98]. This is very usual in PP/CNTs nanocomposites. However, octadecyl amide-functionalized CNTs form agglomerates, which are more uniformly shaped and sized, yet still preserve their network-like-structure (Figure 38b). Except surface treatment, surfactants can be also used for finer dispersion of CNTs into PP matrix. To obtain effective steric barriers to aggregation, non-ionic surfactants with a relatively low hydrophile–liphophile balance can be applied in the preparation of MWCNTs/PP composites. While Tergitol NP-9 [poly(ethylene oxide) (9) nonylphenyl ether] surfactant, with an unbranched hydrocarbon tail, only slightly improved the untreated CNTs dispersion (Figure 38c), Triton-100, with the same head group but a branched hydrocarbon tail, generated an aggregate-free system (Figure 38d). The application of Disperbyk-163 (a commercial anionic solution of copolymers with acidic groups) was motivated by the fact that this additive is manufactured as a carbon-black dispersant that has an electrosteric stabilization effect, and was previously used as a polymer viscosity modifier. Figure 38e shows Disperbyk-163 containing functionalized MWCNTs/PP composites, and demonstrates the highly uniform distribution of the modified nanotubes within the polymer.

CNTs functionallization also alters the crystallization rate of PP. Figure 39 shows POM micrographs of PP and PP/MWCNTs composites that have been isothermal crystallized at 150 °C for 1 h [90]. It is clear that spherulites of pure PP are about 100 μm in diameter, larger than those of PP in PP/MWCNTs composites. The spherulites grew perfectly with the maltese cross. The addition of MWCNTs increased the amount of heterogeneous nuclei in the PP matrix, resulting in an obvious decrease in the size of the PP. When the MWCNTs content was 0.5 wt %, the spherulites grown with the maltese cross could be still observed. However, when the content of MWCNTs reached 1 wt %, only a large quantity of small crystal aggregates was visible. After the 3-MPTS functionalization of MWCNTs, for PP/0.5 wt % MWCNTs composite, the size of PP spherulites did not change, and a large quantity of spherulite crystal aggregates could be observed, indicating the 3-MPTS functionalization of MWCNTs increased the number of PP spherulites without changing the size of spherulites. When the content of MWCNTs reached 1 wt %, no difference could be found from the POM micrographs of PP/1 wt % raw MWCNTs and PP/1 wt % 3-MPTS functionalized MWCNTs composites.

The finer dispersion of CNTs into polymer matrix further enhances the mechanical properties of the PP. Such reinforcing effect of functionalized MWCNTs on PP matrix was recently reported [96]. Comparing PP/CNTs nanocomposites with the corresponding containing functionalized MWCNTs with 3-MPTS (PP/3-MPTS) it can be seen that nanomposites with functionalized MWCNTs have higher performance. The effect of MWCNTs and 3-MPTS functionalized MWCNTs content on the tensile properties of PP composites are described in Figure 40. It can be seen that the tensile strength of PP composites increases with the increasing MWCNTs content reaching the highest values at 1 wt % MWCNTs content. The effect of 3-MPTS functionalized MWCNTs on tensile strength of PP composites is the same as raw MWCNTs. However, at the same content, the PP/3-MPTS functionalized MWCNTs composites have higher tensile strength than PP/raw MWCNTs composites.

In order to further illustrate the effect of 3-MPTS functionalization on the MWCNTs reinforced PP, FESEM was used to investigate the fracture surface of PP/MWCNTs nanocomposites [96]. The FESEM micrographs of cryogenically fractured surface for PP/1 wt % MWCNTs composite are shown in Figure 41. As seen in Figure 41a, most raw MWCNTs are well dispersed in PP matrix but some MWCNTs still exist as aggregates. 3-MPTS functionalized MWCNTs have much better dispersion than the raw MWCNTs in the composites (Figure 41c). As seen in Figure 41d, 3-MPTS functionalized MWCNTs were not simply or completely pulled out from PP matrix compared with raw MWCNTs (Figure 41b), which may be accounted from good interfacial interactions between MWCNTs and PP. For PP/3-MPTS functionalized MWCNTs composites, the interfacial interaction between MWCNTs and PP matrix could be improved through the boundary bonding between organic group chains of silane and PP matrix, and the boundary bonding could not be very strong.

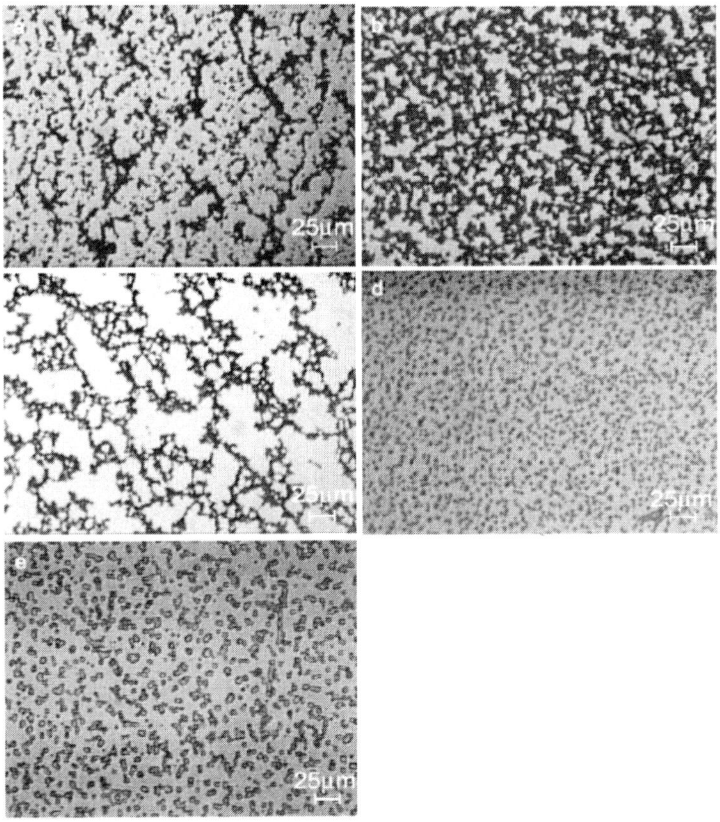

Figure 38. Optical microscope images of 1 wt % MWSNTs/PP composite films. The character of the CNTs dispersion was governed by surface modification via octadecylamide functionalization or surfactant adsorption: a) MWCNTs/PP; b) functionalized MWCNTAS/PP; c) MWCNTS/PP with NP-9; d) MWCNTS/PP with Triton-100, e) functionalized MWCNTS/PP with Disperbyk-163. Reproduced with permission from ref. [98].

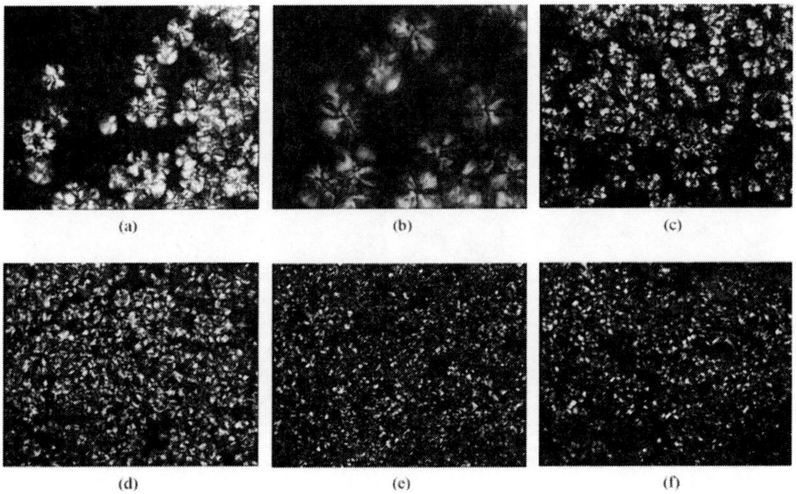

Figure 39. Crystalline morphology of (a) PP, (c) PP/0.5 wt % raw MWCNTs, (d) PP/0.5 wt % 3-MPTS functionalized MWCNTs, (e) PP/1 wt % raw MWCNTs, and (f) PP/1 wt % 3-MPTS functionalized MWCNTs by POM (3200 magnification); (b) PP by POM (3400 magnification). Reproduced with permission from ref. [90].

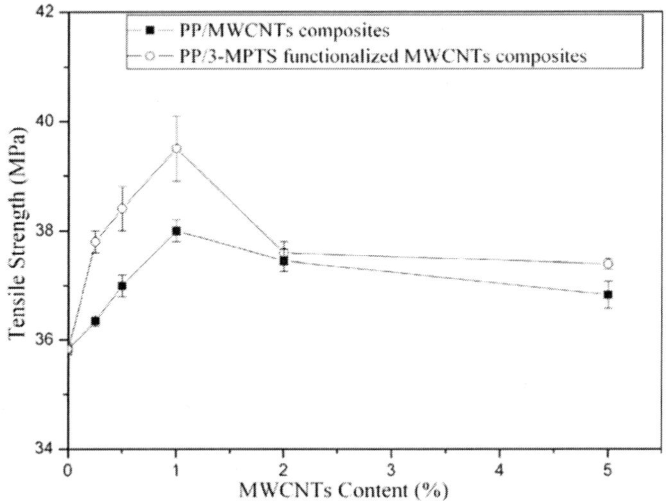

Figure 40. Effect of raw MWCNTs and 3-MPTS functionalized MWCNTs content on tensile strength of the PP composite. Reproduced with permission from ref. [96].

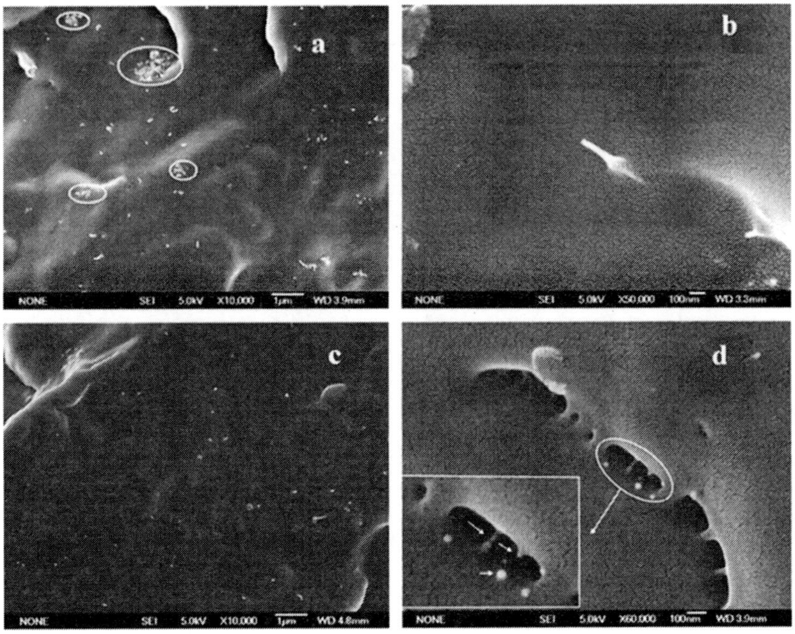

Figure 41. FESEM images of (a and b) PP/1 wt % raw MWCNTs; (c and d) PP/1 wt % 3-MPTS functionalized MWCNTs composites. Reproduced with permission from ref. [96].

In most of the above described techniques functionalization of CNTs was prepared in a previous step before melt mixing with PP. However, chemical functionalization can introduce defects in nanotubes, thereby degrading their mechanical properties [99]. Thus, some researchers have used maleic anhydride grafted PP (PP-g-MA) or maleic anhydride grafted styrene–ethylene/butylene–styrene (MA-g-SEBS) as the compatibilizer to improve dispersion of the purified CNTs without further chemical modification in PP matrix [39,48,100,101]. In this case, functionalization can be achieved *in situ* during melt mixing of PP with MWCNTs. The formation of strong hydrogen bonding between hydroxyl and/or carboxyl groups of the purified CNTs and maleic anhydride groups of PP-g-MA can stabilize the morphology and enhance the interfacial interaction between PP/CNTs and finally promotes homogeneous dispersion of CNTs [100]. This confirms that PP-g-MA acts as a role of compatibilizer to improve dispersion of CNTs. The strong intermolecular van der Waals interactions among the CNTs, in combination with their high surface area and high aspect ratio as well as poor affinity to the nonpolar PP matrix, commonly causes significant agglomeration. The presence of PP-g-MA can lead to strong hydrogen bonding between carboxyl/hydroxyl groups of the CNTs and maleic anhydride groups of PP-g-MA. That noncovalent attachment, which is also called wrapping, can alter the nature of the CNT's surface and make it more compatible with the PP because of the chemical similarity between matrix PP and grafted PP. As a result, the presence of PP-g-MA enhances the interfacial adhesions between CNTs and PP matrix and finally promotes homogeneous dispersion of CNTs, as schematically described in Figure 42 [102].

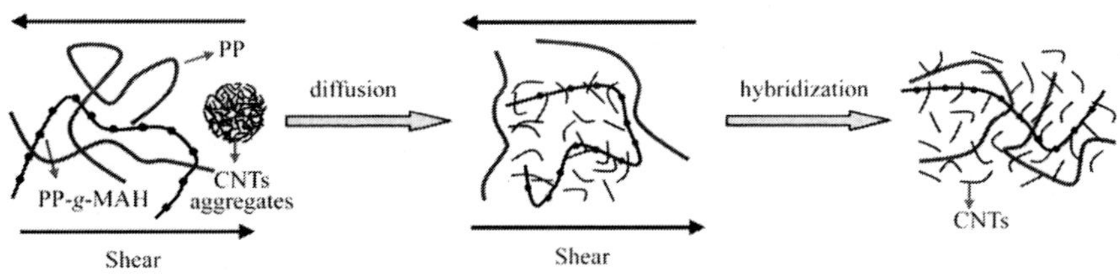

Figure 42. Schematic diagrams of melt compounding process for PP/CNTs nanocomposites using PP-g-MA as a compatibilizer. Reproduced with permission from ref. [102].

CNTs can also be covalently grafted into PP (Figure 43) with an easy procedure including oxidation of MWCNTs with a 3:1 mixture of concentrated sulfuric and nitric acid, reaction with ethylenediamine to afford MWNT–NH_2 and finally melt blending with PP-g-MA at 220 °C [14,103].

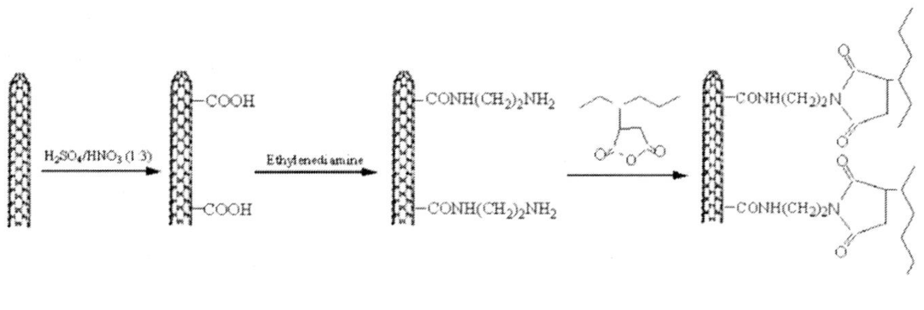

Figure 43. Grafting of PP onto MWCNTs. Reproduced with permission from ref. [103].

The SEM images of composites containing 3.0 wt % MWCNTs with and without PP-g-MA are shown in Figure 44 [104]. As Figure 44a shows, a large amount of self-organized MWCNTs bundles was observed on the cryofractured surface. This indicates that a significant part of the nanotubes was dispersed as nanotubes aggregates due to the imperfect mixing of the masterbatch. But in the PP/MWCNTs/PP-g-MA composites (Figure 44b), reasonably uniform dispersion and good distribution of the MWCNTs was observed with a small amount of aggregates.

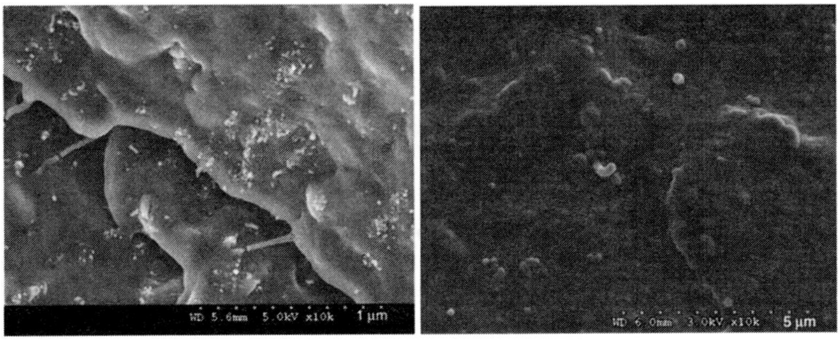

a b

Figure 44. SEM images of a) CNTs-MA-PP composite and b) CNTs/PP composite. Reproduced with permission from ref. [104].

From these observations, one can conclude that the added PP-g-MA promotes the dispersion of individual carbon nanotubes and limits the presence of nanotubes aggregates. Moreover, in the case of the PP/MWCNTs/PP-g-MA composite , the diameter of the MWCNTs has increased to about 150–200 nm, while in case of PP/MWCNTs diameter is about 30-40 nm. The diameter of the nanotubes was determined from different SEM observations at different places for each composite so as to be representative of the average diameter. As the nanotube used for the different composites is the same, this increase in the diameter of the nanotubes is assumed to be due to wrapping of nanotubes by PP-g-MA thanks to hydrogen bonding.

This better dispersion and adhesion of CNTs with PP matrix results in a substantial enhancement of mechanical properties. However, the improvement in the mechanical properties arises from several factors.

Firstly, PP-g-MWCNTs promote the crystallization of PP and the crystallites strengthen the composites. Secondly, PP-g-MWCNTs are well dispersed in the matrix, allowing a more uniform load distribution. Thirdly, the PP chains grafted onto MWCNTs enable a more efficient load transfer from the matrix to the nanotubes. The variation of tensile strength, modulus and elongation at break with MWCNTs content for the PP/MWCNTs and PP/MWCNTs/2 wt % PP-g-MA composites are shown in Figure 45[104]. It is obvious that Young's modulus (Figure 45a) and yield stress (Figure 45b) of PP/MWCNTs/PP-g-MA composites are higher than those of PP/MCWNTs composites, and tends to increase substantially with the MWCNTs content in composites. This was attributed to the enhanced adhesion between MWCNTs and PP matrix and the improvement of MWCNTs dispersion, which also increases the effective average aspect ratio of the nanofillers in the polymer matrix. However, in spite of the better dispersion of MWCNTs to PP matrix thanks to PP-g-MA, elongation at break decreases with increase in MWCNTs content (Figure 45c). This is probably due to the presence of few masterbatch aggregates.

Figure 46 shows the impact properties of notched samples of PP/MWCNTs and PP/MWCNTs/2 wt % PPg-MA composites. Charpy impact strength of the notched specimens slightly increased as the MWCNTs content increased up to 2 wt % for PP/MWNT nanocomposites and up to 3 wt % in case of PP/MWCNTs/2 wt % PP-g-MA nanocomposites. This increase is significantly larger for the samples with PP-g-MA. This is due to the fact that notched impact behavior is controlled to a greater extent by factors affecting the propagation of fracture initiated due to stress concentration at the notch tip. Lowering of impact energy at higher nanotube content is due to the presence fewer master-batch aggregates in the composites.

5. PP/CNTS NANOCOMPOSITES PREPARED BY *IN SITU* POLYMERIZATION

Bulk composites reinforced with CNTs have been investigated by melt mixing, as described before, with respect to mechanical, thermal, electrical, and other properties. However, the main problem in order to gain sufficient enhancement of most of these properties is the insufficient filler dispersion into PP matrix. Especially at high filler contents, which lead to aggregation and intercalation in turn decreasing the mechanical properties. A promising route to resolving this issue is the employment of *in situ* polymerization method [105,106,107,108,109,110,111,112,113] which involves ultrasonication for effective CNT dispersion and polymerization of monomer on nanotube surface in the present of catalysts that promotes intimate contact between polymer chains and nanotubes. By *in situ* polymerization the cocatalyst methylaluminoxane (MAO) can be absorbed or anchored on the surface of the nanofillers, changing the surface to a hydrophobic one [109]. According to this method, CNTs dispersion and subsequent propylene polymerization are accomplished in a bulk of liquid monomer [12].

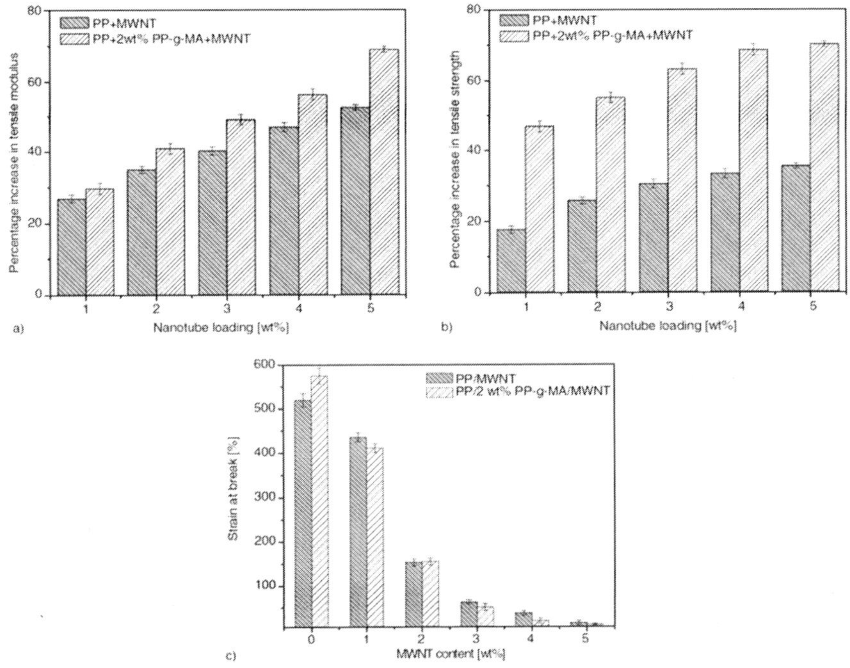

Figure 45. Percentage increase in tensile modulus with respect to PP and PP-g-MA as a function of MWCNTs content for the PP/MWCNTs and PP/MWCNTs/2 wt % PP-g-MA composites (a). Percentage increase in tensile strength with respect to PP and PP-g-MA as a function of MWCNTs content for the PP/MWCNTs and PP/MWCNTs/2 wt % PP-g-MA composites (b). Variation of percentage strain at break with MWCNTs content for the PP/MWCNTs and PP/MWCNTs/2 wt % PP-g-MA composites (c). Reproduced with permission from ref. [104].

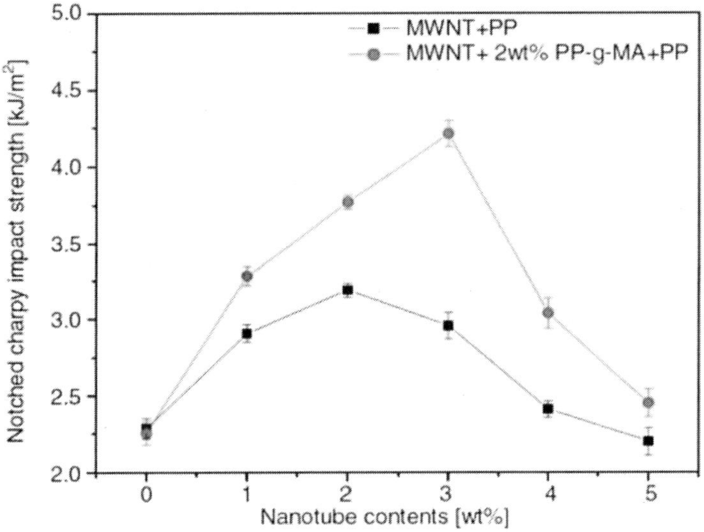

Figure 46. Impact properties of PP/MWCNTs (a) and PP/MWCNTs/2 wt % PP-g-MA (b) nanocomposites. Reproduced with permission from ref. [104].

Koval'chuk, *et al.*, [114] used *in situ* polymerization for the preparation of PP/CNTs composites with distinct features as follows.

(1)Bulk of liquid propylene is used as a medium for CNTs dispersion and subsequent polymerization reaction. Utilization of liquid monomer as a reaction medium allows obtaining of nanocomposites with high yields and high molecular weights (considering matrix polypropylene) and makes the synthesis easily scalable.

(2)CNTs dispersion via ultrasonication proceeds in the presence of the cocatalyst, methylaluminoxane (MAO). This leads to immobilization of MAO molecules on CNT surface by loose ionic interactions and, to a lesser extent, by virtue of covalent bonding (see scheme in Figure 47) to -COOH or -OH groups, which are inherent to partially oxidized CNTs.

(3)The formation of catalytic active sites is accomplished by means of heterogenization of the metallocene catalyst precursor on CNTs surface [77] owing to chemical interaction of metallocene with MAO directly during the initial stage of polymerization at lowered temperature. Covalent bonding between MAO and CNTs has considerable impact on the morphology and properties of nanocomposites; MAO molecules chemically grafted to CNTs surface form catalytic active species, yielding polypropylene chains attached directly to nanotubes.

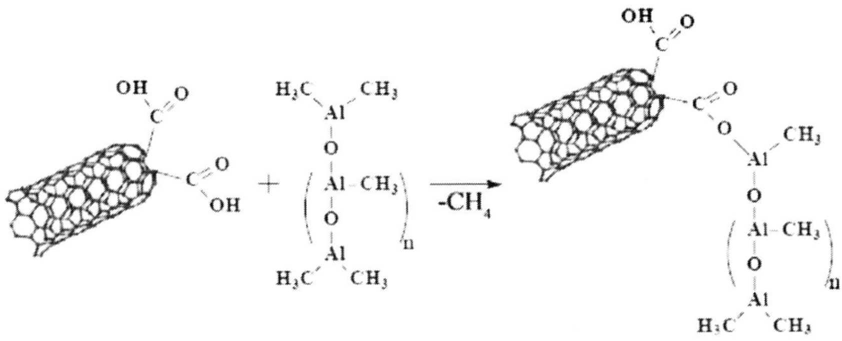

Figure 47. Scheme of chemical interaction between MAO and CNTs. Reproduced with permission from ref. [114].

Due to this reason, the polymerization of the monomer proceeds mainly in the nanotube surface. In Figure 48, CNTs encapsulated by a high molecular weight isotactic polypropylene generated by the combination of such catalyst can be seen. The detailed micrograph on the right shows the different layers (about 20) of the tube with a distance of 0.3 nm. In higher magnification it can be seen that the nanotube was covered with a high molecular weight PP (HMWiPP) film of about 8 nm thickness. Even the open cap of the MWCNTs was encapsulated too. This resulted in an excellent adhesion between tube and polymer.

For nanocomposites prepared by *in situ* polymerization noticeable mechanical reinforcement effect is observed in the iPP and sPP nanocomposites even upon incorporating the lowest nanotube amount [12,109]. At 0.1 wt.% MWCNTs loading, Young's modulus of iPP increases by ~22% from ~1,200 MPa to ~1,465 MPa, and modulus of sPP increases by ~34% from ~380 MPa to ~510 MPa (Figure 49). Further modulus enhancement continues up at higher filler concentrations. Ultimate Young's modulus improvement of iPP is ~37% (from ~1,200 MPa to ~1,650 MPa) at 2.1 wt.% MWCNTs content. Subsequent moderate modulus drop at 3.5 wt.% of MWCNTs is apparently connected with the nanotube agglomeration that is inevitable at such high CNTs concentrations. For the sPP/MWCNTs composites maximal mechanical reinforcement is achieved at 0.4 wt.%

nanotube loading (Young's modulus grows by ~66% from ~380 MPa to ~635 MPa), and posterior modulus reduction at higher nanotube contents signifies considerable impact of MWCNTs aggregation that takes place in this system.

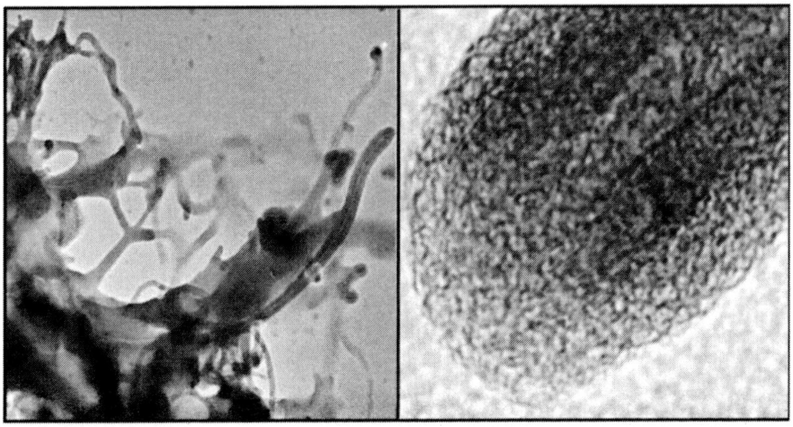

Figure 48. TEM image of polymer encapsulated ox. MWCNTs. Almost every nanotube is covered by a thin and homogenous *in situ* grown HMWiPP film (left). On the right: TEM images (magnification x 400,000) of an ox. MWCNTs, coated by a thin polymer film (~8 nm). Reproduced with permission from ref. [113].

6. SYNDIOTACTIC PP/CNTS NANOCOMPOSITES

Syndiotactic polypropylene has been found to exhibit much superior electrical, thermal and mechanical properties compared with those of conventional isotactic and atactic polypropylene. These excellent characteristics of syndiotactic polypropylene are found to originate from lower crystallinity, much smaller spherulites, different crystal lattice and less residual catalysts than in isotactic polypropylene. However, in order to apply syndiotactic polypropylene to some applications low temperature brittleness and thermal durability should be improved. Furthermore, a drawback for its application is the relatively slow crystallization rate [115].

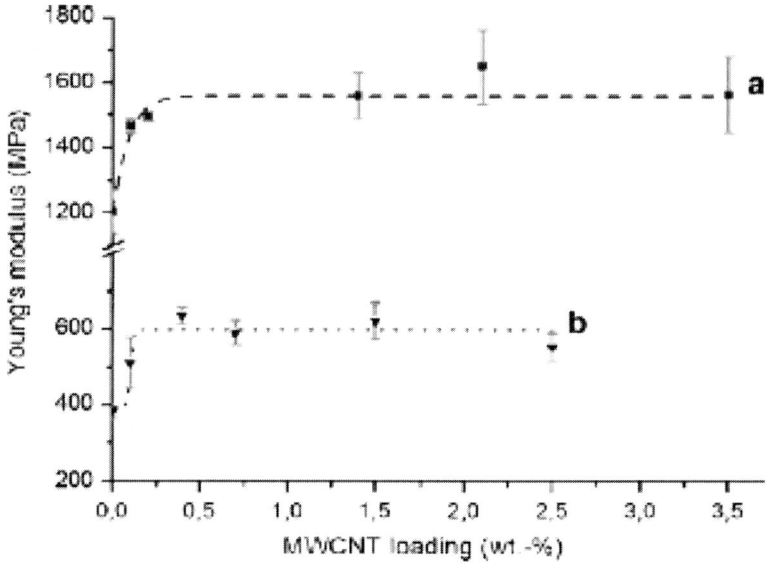

Figure 49. Young's modulus of iPP/MWCNTs nanocomposites (a) and sPP/MWCNTs nanocomposites (b) as a function of MWCNTs loading. Reproduced with permission from ref. [109].

The addition of CNTs was reported that can further enhance the mechanical properties of sPP [116,117]. Figure 50 shows the elastic modulus of all the samples of sPP, as a function of the CNT content.

As can be seen, the modulus increases with increasing CNTs content. The enhancement of the mechanical properties of the composites relies on many factors, the most important being: the morphology of the polymeric matrix, in terms of crystallinity and crystal dimensions, the good dispersion of the nanotubes into the polymeric matrix and the strong interfacial adhesion between the phases, leading to a high degree of load transfer between the matrix and the nanotubes. Furthermore, the modulus of the sample crystallized at 25 °C (25sPPCN0) is lower than that of sample crystallized at 100 °C (100sPPCN0), due to the higher crystallinity degree of the latter sample, which seems to be the predominant factor. However, an inversion is observed for the nanocomposites. The increase of the modulus with CNTs content is more relevant for the nanocomposites quenched at 25 °C, while a less dramatic effect is observed for the samples quenched at 100 °C. This result must be correlated with the morphology of the amorphous phase in both samples. The carbon nanotubes are dispersed in this phase, and their behavior depends both on their properties (dispersion and interaction) and on the amorphous phase interconnection with the crystalline phase. In the samples crystallized at lower temperature (25 °C) the crystals are initially smaller and the amorphous phase is better dispersed between the small crystals, forming a more interconnected matrix.

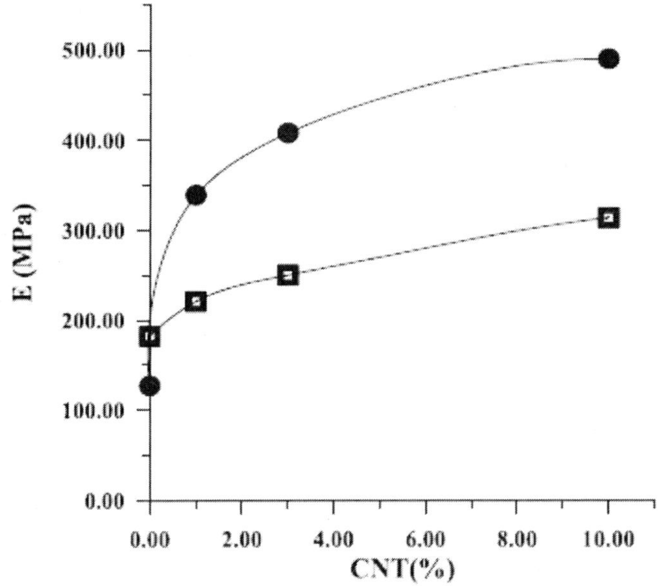

Figure 50. The elastic moduli for samples quenched at 25 °C (●) and at 100 °C (◘). Reproduced with permission from ref. [116].

Thermal properties of sPP are also affected from the addition of CNTs (Figure 51). The main result regarding the glass transition temperature is that, besides small variations on increasing the CNTs content, the glass temperature of 25sPP samples is always consistently higher than that the T_g of the 100sPP samples. This indicates that the higher temperature of crystallization of 100 °C influences the amorphous phase, making the chains in this phase more relaxed with respect to the samples crystallized at 25 °C. Also the melting enthalpy of the mesophase is consistently higher in samples crystallized at 25 °C, and it increases on increasing the CNTs content [116].

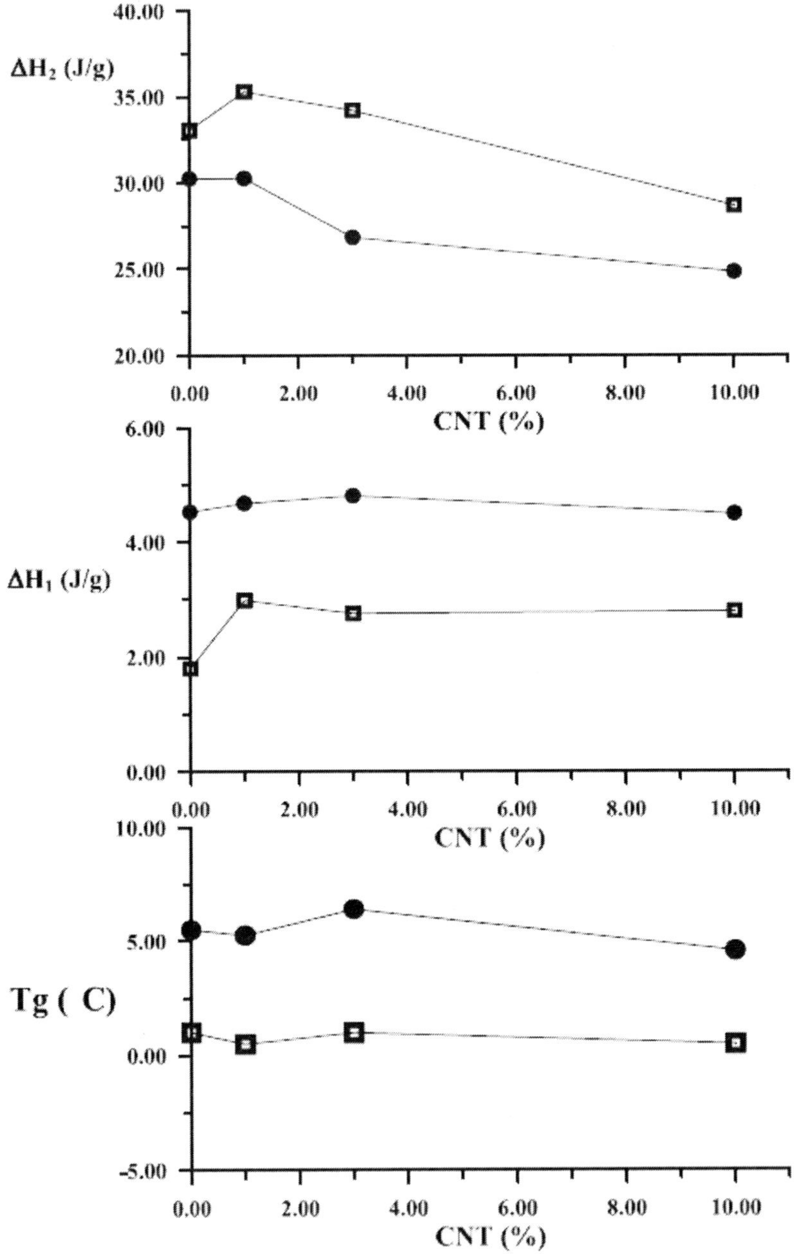

Figure 51. The glass transition temperature (Tg (°C)), the melting enthalpy of the mesophase (ΔH_1 (J g^{-1})), the melting enthalpy of the crystalline phase (ΔH_2 (J g^{-1})) for samples quenched at 25 °C (•) and at 100 °C (□). Reproduced with permission from ref. [116].

Carbon nanotubes also improve the thermal stability of sPP in nitrogen as well as in air, although the complex behavior of the degradation in air was interestingly found to depend on the structural organization and morphology of the composite samples [117]. The thermal degradation of sPP is caused by the scission of C–C bonds followed by hydrogen transfer at the site of scission. Samples containing CNTs (CN-A25) also degrades in a one-loss step, while both the onset of the degradation temperature and the peak temperature of the derivative weight loss (DTG) are shifted to higher values by about 20°C (Figure 52). This result, which is similar to that reported for isotactic polypropylene, could be explained by a barrier effect of carbon nanotubes, which hinders the diffusion of the degraded products from the bulk of the polymer to the gas phase.

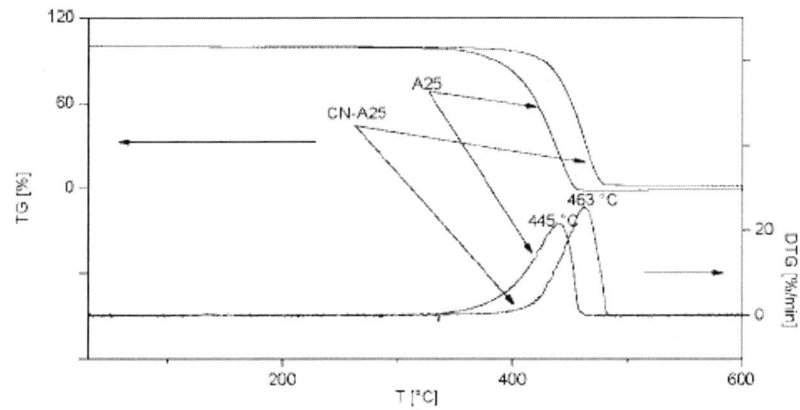

Figure 52. TG-DTG analysis in nitrogen of sPP/CNTs nanocomposites A25; CNA25. Reproduced with permission from ref. [117].

Syntiotactic PP nanocomposites with MWCNTs were also prepared by *in situ* polymerization. However, different pre-treatments before the polymerization were necessary to achieve a homogeneous distribution of MWCNTs [107]. Preliminary experiments showed that MWCNTs stay aggregated if not treated with ultrasound prior to the polymerization. The effect of different ultrasonic amplitudes (which are a measure of the energy input) and various sonication times was investigated. Higher amplitude led to a better dispersion. Figure 53 shows microscopic photographs of sPP/MWCNTs nanocomposites prepared without pre reaction after treatment of the MWCNTs with ultrasonic amplitude of 10 or 30%. It can be seen that a higher amplitude leads to a slightly better dispersion, which is still not satisfactory. To achieve a more homogeneous distribution, the MWCNTs were stirred with MAO solution overnight after sonication (with pre-reaction). The results showed that the pre-reaction with MAO has led to a much more homogeneous dispersion of the fillers in the matrix. During the pre-reaction with MAO, hydroxyl groups on the nanotube surface can react with the MAO by the formation of covalent oxygen-aluminum bonds. The metallocene is then heterogenized indirectly on the MWCNTs. This should lead to polymer growth directly on the filler surface to resulting in a better dispersion compared to nanocomposites prepared without pre-reaction.

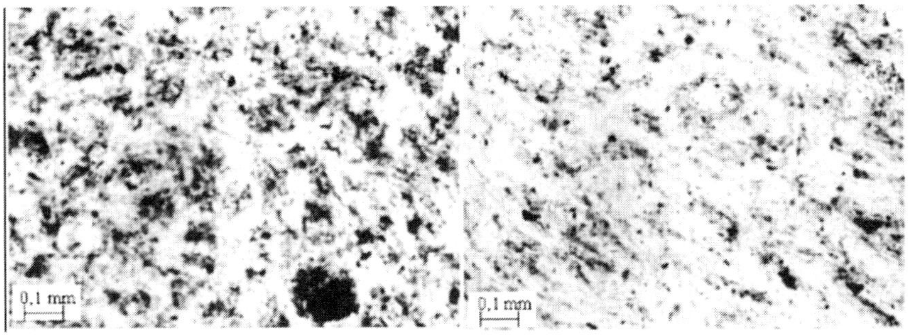

Figure 53. Micrographs of sPP/MWCNTs nanocomposites prepared with prereaction after sonication of the nanotubes with different amplitudes (left: 10%, right: 30%). Reproduced with permission from ref. [107].

7. POLYPROPYLENE POLYMER BLENDS WITH CNTS

Besides the application of CNTs in single polymer matrices, other researchers have shown that CNTs also exhibit apparent roles in influencing the morphologies, mechanical and electrical properties of PP polymer blends [118,119,120]. Most of these blends are immiscible and show poor mechanical properties because of the weak interface between the two phases. The toughening of PP by rubbers, particularly ethylene–propylene copolymers and terpolymers (EPM and EPDM, respectively), was found to be highly effective and can be further enhanced by the addition of carbon nanotubes [121]. For example, although poly(ethylene-co-vinyl acetate) (EVA) has been proved to improve the fracture toughness of PP, the toughening effect is still inconspicuous. The addition of a few amounts of MWCNTs induces the change of fracture toughness of composites in different degrees. The improvement of impact strength is greatly dependent of the contents of EVA and MWCNTs. For PP/EVA 80/20 w/w, the impact strength increases slightly with the increasing content of MWCNTs, indicating that MWCNTs do not influence the fracture toughness of such blend significantly (Figure 54). However, for PP/EVA 60/40 w/w, the impact strength increases greatly with increasing content of MWCNTs. This may provide a simple but efficient way to improve the mechanical properties of such immiscible polyolefin blends and this improvement depends from the morphological characteristics of the blends [122].

For PP/EVA 80/20 w/w with low MWCNTs content, MWCNTs tend to form clusters, which are limited in the EVA phase (Figure 55a); at high f-MWCNTs content, these clusters migrate to PP phase, possibly inducing the local network structure or aggregation of MWCNTs around EVA particles (Figure 55b). For PP/EVA 60/40 w/w, MWCNTs exhibit good dispersion in EVA phase at low load due to the largely increased EVA phase and the good interaction between EVA and MWCNTs (Figure 55c). At high load, MWCNTs form the network structure in EVA phase. Furthermore, some MWCNTs tend to migrate to PP phase, possibly inducing some MWCNTs span the two phases, leading to the bridge effect for PP and EVA (Figure 55d). Khare *et al.* [123] revealed also a co-continuous structure and a refinement in the morphology of polypropylene/acrylonitrile–butadiene–styrene (PP/ABS) blends in the presence of CNTs. Furthermore, lower electrical percolation threshold was observed in continuous PP/ABS blends with CNTs.

8. SUMMARY AND CONCLUSIONS

Carbon nanotubes (CNTs) show promise to revolutionize several fields in material science and are suggested to pave the way into nanotechnology. The discovery of carbon nanotubes and carbon nanotube based materials has inspired scientists for a range of potential applications due to their extraordinary mechanical, electrical, and optical properties. Isotactic polypropylene (iPP) is a thermoplastic polymer widely used in many fields and its applications are increasing last years. It has been reported that carbon nanotubes in iPP produce the following improvements in properties:

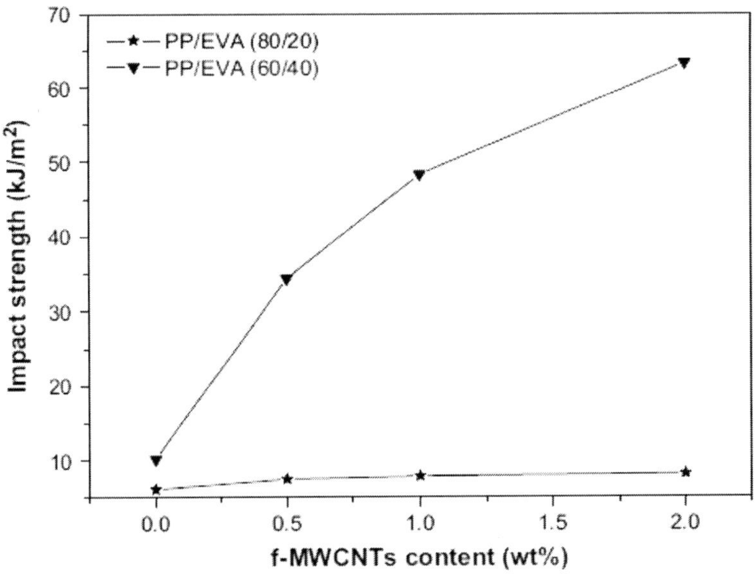

Figure 54. Notched Izod impact strength of PP/EVA blends with different contents of functionalized MWCNTs (f-MWCNTs). Reproduced with permission from ref. [122].

Young's modulus, tensile strength, ultimate strain and toughness of iPP can be improved since neat CNTs have higher mechanical properties than neat iPP.

CNTs/PP nanocomposites exhibit superior thermal stability compared to polyolefin materials.

Electrical and thermal conductivity are also enhanced by the addition of CNTs creating a conductive network in the polymer matrix.

It has been demonstrated that the incorporation of nanotubes affects the crystalline behavior and structure of the iPP matrix. In particular, CNTs accelerate the heterogeneous nucleation and crystal growth mechanisms of iPP, this effect being more noticeable at the lower filler content analyzed. From optical microscopy studies, it was found that the iPP spherulites decreased in size when CNTs was introduced into the polymer. The nonisothermal crystallization rate increases up to 2-4 wt % CNTs and then decreases slightly or remains almost constant at higher CNTs content. The decrease of the nucleation efficiency at high filler concentration is not due to a change in the crystallization mechanism, but to an aggregation of the filler particles at high concentration, which leads to a decrease of the number of heterogeneous nuclei.

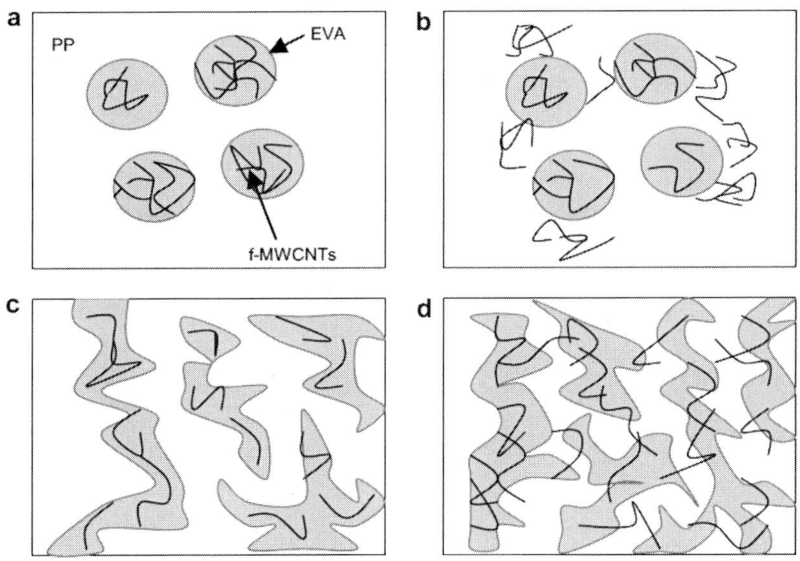

Figure 55. Schematic representations of the dispersion states of f-MWCNTs in PP/EVA blends. (a): f-MWCNTs tend to aggregate in dispersed EVA phase in PP/EVA (80/20) at low load; (b): f-MWCNTs migrate to PP phase and form a local "single-network structure" at high load; (c): f-MWCNTs exhibit good dispersion in continuous phase of EVA in PP/EVA (60/40) at low load and (d): f-MWCNTs form the network structure in the whole blend. Reproduced with permission from ref. [123].

However all these improvements depend on the amount of CNTs added in polymer matrix and mainly on the prepared microstructure. One of the advantages of CNTs as reinforcement agents is their large surface area, which can induce better adhesion with the polymeric matrix, which is an important factor for an effective improvement of the composite properties. However, CNTs tend to aggregate into bundles and hence they are difficult to be dispersed homogeneously in polymer matrices. For this reason mechanical properties are reducing after an added amount 2-3 wt % of CNTs.

Furthermore, the polymer/CNTs interfacial adhesion is weak, preventing an efficient load transfer from the polymer matrix to CNTs. As a result of poor dispersion and inefficient load transfer, the mechanical properties of polymer/CNTs composites are often not as good as anticipated. Appropriate surface treatment of nanotubes will result into improved fiber to matrix interaction. This in turn will enhance the dispersion and wetting properties of the nanotubes leading to further property improvement of composite materials. CNTs can chemically modified by grafting alkyl chains to improve compatibility between CNTs and iPP and to enhance dispersion of CNTs in iPP matrix. CNTs can also be grafted with functional groups to facilitate better dispersion of nanotubes within iPP.

Nanocomposites based on iPP and different types of CNTs could successfully be also prepared by *in situ* polymerization of propylene with a metallocene/methylaluminoxane catalyst system. In this case mechanical properties can be further enhanced due to increased matrix adhesion.

REFERENCES

1. Balow, M.J. Handbook of Polypropylene and Polypropylene Composites; Karian, H.G., Ed.; CRC Press: New York, NY, USA, 2003.

2. Ajayan, P.M. Nanotubes from Carbon. Chem. Rev. 1999, 99, 1787–1799.

3. Velasco-Santos, C.; Martinez-Hernandez, A.L.; Castano, V.M. Carbon nanotube-polymer nanocomposites: The role of interfaces. Comp. Interface 2005, 11, 567–586.

4. Moniruzzaman, M.; Winey, K.I. Polymer nanocomposites containing carbon nanotubes. Macromolecules 2006, 39, 5194–5205.

5. Ramanathan, H.; Liu, H.; Brinson, L.C. Functionalized SWNT/polymer nanocomposites for dramatic property improvement. J. Polym. Sci. B-Polym. Phys. 2005, 43, 2269–2279.

6. Meo, M.; Rossi, M. Prediction of Young's modulus of single wall carbon nanotubes by molecular-mechanics based finite element modelling. Comp. Sci. Technol. 2006, 66, 1597–1605.

7. Ritter, U.; Scharff, P.; Dmytrenko, O.P.; Kulish, N.P.; Prylutskyy, Y.I.; Grabovskiy, Y.E.; Belyi, N.M.; Pinchuk, T.N.; Alekseev, A.N.; Sementsov, Y.I.; Gavrylyuk, N.A.; Shlapatskaya, V.V. Strength improvement of iPP/MWCNT nanocomposites. Polym. Comp. 2010, 31, 179–184.

8. Dondero, W.E.; Gorga, R.E. Morphological and mechanical properties of carbon nanotube/polymer composites via melt compounding. J. Polym. Sci. B-Polym. Phys. 2006, 44, 864–878.

9. Thomassin, J.-M.; Huynen, I.; Jerome, R.; Detrembleur, C. Functionalized polypropylenes as efficient dispersing agents for carbon nanotubes in a polypropylene matrix; application to electromagnetic interference (EMI) absorber materials. Polymer 2010, 51, 115–121.

10. Zhou, Z.; Wang, S.F.; Zhang, Y.; Zhang, Y.X. Effect of different carbon fillers on the properties of PP composites: Comparison of carbon black with multiwalled carbon nanotubes. J. Appl. Polym. Sci. 2006, 102, 4823–4830.

11. Baddour, C.E.; Briens, C. Carbon nanotube synthesis: A review. Int. J. Chem. Reactor Eng. 2005, 3. Review R3.

12. Koval'chuk, A.A.; Shevchenko, V.G.; Shchegolikhin, A.N.; Nedorezova, P.M.; Klyamkina, A.N.; Aladyshev, A.M. Isotactic and syndiotactic polypropylene/multi-wall carbon nanotube composites: Synthesis and properties. J. Mater. Sci. 2008, 43, 7132–7140.

13. Logakis, E.; Pollatos, E.; Pandis, Ch.; Peoglos, V.; Zuburtikudis, I.; Delides, C.G.; Vatalis, A.; Gjoka, M.; Syskakis, E.; Viras, K.; Pissis, P.

Structure–property relationships in isotactic polypropylene/multi-walled carbon nanotubes nanocomposites. Comp. Sci. Technol. 2010, 70, 328–335.

14. Yang, B.-X.; Shi, J.-H.; Pramoda, K.P.; Goh, S.H. Enhancement of the mechanical properties of polypropylene using polypropylene-grafted multiwalled carbon nanotubes. Comp. Sci. Technol. 2008, 68, 2490–2497.

15. Masuda, J.; Torkelson, J.M. Dispersion and major property enhancements in polymer/multiwall carbon nanotube nanocomposites via solid-state shear pulverization followed by melt mixing. Macromolecules 2008, 41, 5974–5977.

16. Xia, H.; Wang, Q.; Li, K.; Hu, G.-H. Preparation of polypropylene/carbon nanotube composite powder with a solid-state mechanochemical pulverization process. J. Appl. Polym. Sci. 2004, 93, 378–386.

17. Tjong, S.C.; Liang, G.D.; Bao, S.P. Electrical behavior of polypropylene/multiwalled carbon nanotube nanocomposites with low percolation threshold. Scripta Mater. 2007, 57, 461–464.

18. Bikiaris, D.; Vassiliou, A.; Chrissafis, K.; Paraskevopoulos, K.M.; Jannakoudakis, A.; Docoslis, A. Effect of acid treated multi-walled carbon nanotubes on the mechanical, permeability, thermal properties and thermo-oxidative stability of isotactic polypropylene. Polym. Degrad. Stab. 2008, 93, 952–967.

19. Ganβ, M.; Satapathy, B.K.; Thunga, M.; Weidisch, R.; Pötschke, P.; Jehnichen, D. Structural interpretations of deformation and fracture behavior of polypropylene/multi-walled carbon nanotube composites. Acta Mater. 2008, 56, 2247–2261.

20. Duangphattra, N.; Aramphongphunb, C. A study of the effects of processing conditions on mechanical properties of polypropylene/multiwall carbon nanotube nanocomposites using design of experiments. Adv. Mater. Res. 2008, 55-57, 877–880.

21. Kearns, J.C.; Shambaugh, R.L. Polypropylene fibers reinforced with carbon nanotubes. J. Appl. Polym. Sci. 2002, 86, 2079–2084.

22. Li, C.; Chou, T.W. Multiscale modeling of compressive behavior of carbon nanotube/polymer composites. Comp. Sci. Technol. 2006, 66, 2409–2415.

23. Prashantha, K.; Soulestin, J.; Lacrampe, M.F.; Krawczak, P.; Dupin, G.; Claes, M. Masterbatch-based multi-walled carbon nanotube filled polypropylene nanocomposites: Assessment of rheological and mechanical properties. Comp. Sci. Technol. 2009, 69, 1756–1763.

24. Zhao, P.; Wang, K.; Yang, H.; Zhang, Q.; Du, R.; Fu, Q. Excellent tensile ductility in highly oriented injection-molded bars of polypropylene/carbon nanotubes composites. Polymer 2007, 48, 5688–5695.

25. Mičušík, M.; Omastová, M.; Krupa, I.; Prokeš, J.; Pissis, P.; Logakis, E.; Pandis, C.; Pötschke, P.; Pionteck, J. A comparative study on the Electrical and mechanical behaviour of multi-walled carbon nanotube composites prepared by diluting a masterbatch with various types of polypropylenes. J. Appl. Polym. Sci. 2009, 113, 2536–2551.

26. Hemmati, M.; Rahimi, G.H.; Kaganj, A.B.; Sepehri, S.; Rashidi., A.M. Rheological and mechanical characterization of multi-walled carbon nanotubes/polypropylene nanocomposites. J. Macromol. Sci. B Phys. 2008, 47, 1176–1187.

27. Chang, T.E.; Jensen, L.R.; Kisliuk, A.; Pipes, R.B.; Pyrz, R.; Sokolov, A.P. Microscopic mechanism of reinforcement in single-wall carbon nanotube/polypropylene nanocomposites. Polymer 2005, 46, 439–444.

28. Seidel, G.D.; Lagouda, D.C. Micromechanical analysis of the effective elastic properties of carbon nanotube reinforced composites. Mech. Mater. 2006, 38, 884–907.

29. Hammel, E.; Tang, X.; Trampert, M.; Schmitt, T.; Mauthner, K.; Eder, A.; Pötschke, P. Carbon nanofibers for composite applications. Carbon 2004, 42, 1153–1158.

30. Seo, M.-K.; Lee, J.-R.; Park, S.-J. Crystallization kinetics and interfacial behaviors of polypropylene composites reinforced with multi-walled carbon nanotubes. Mater. Sci. Eng. A 2005, 404, 79–84.

31. Wagner, H.D.; Lourie, O.; Feldman, Y.; Tenne, R. Stress-induced fragmentation of multiwall carbon nanotubes in a polymer matrix. Appl. Phys. Lett. 1998, 72, 188–190.

32. Satapathy, B.K.; Ganß, M.; Weidisch, R.; Pötschke, P.; Jehnichen, D.; Keller, T.; Jandt, K.D. Ductile-to-semiductile transition in PP-MWNT nanocomposites. Macromol. Rapid Commun. 2007, 28, 834–841.

33. Yang, J.; Zhang, Z.; Friedrich, K.; Schlarb, A.K. Creep resistant polymer nanocomposites reinforced with multiwalled carbon nanotubes. Macromol. Rapid Commun. 2007, 28, 955–961.

34. Hine, P.; Broome, V.; Ward, I. The incorporation of carbon nanofibres to enhance the properties of self reinforced, single polymer composites. Polymer 2005, 46, 10936–10944.

35. Manchado, M.A.L.; Valentini, L.; Biagiotti, J.; Kenny, J.M. Thermal and mechanical properties of single-walled carbon nanotubes-polypropylene composites prepared by melt processing. Carbon 2005, 43, 1499–1505.

36. Katerelos, D. T. G.; Joffe, R.; Labou, D.; Wallstrom, L. Alteration of the mechanical behaviour of polypropylene owing to successive introduction of multiwall carbon nanotubes and stretching. Mech. Comp. Mater. 2009, 45, 423–434.

37. Teng, C.-C.; Ma, C.-C.M.; Huang, Y.-W.; Yuen, S.-M.; Weng, C.-C.; Chen, C.-H.; Su, S.-F. Effect of MWCNT content on rheological and dynamic mechanical properties of multiwalled carbon nanotube/polypropylene composites. Comp. A Appl. Sci. Manuf. 2008, 39, 1869–1875.

38. Lee, G.-W.; Jagannathan, S.; Chae, H.G.; Minus, M.L.; Kumar, S. Carbon nanotube dispersion and exfoliation in polypropylene and structure and properties of the resulting composites. Polymer 2008, 49, 1831–1840.

39. Lee, S.H.; Kim, M.W.; Kim, S.H.; Youn, J.R. Rheological and electrical properties of polypropylene/MWCNT composites prepared with MWCNT masterbatch chips. Eur. Polym. J. 2008, 44, 1620–1630.

40. Rahmatpour, A.; Aalaie, J. Steady shear rheological behavior, mechanical properties, and morphology of the polypropylene/carbon nanotube nanocomposites. J. Macromol. Sci. B-Phys. 2008, 47, 929–941.

41. Pötschke, P.; Krause, B.; Stange, J.; Münstedt, H. Elongational viscosity and foaming behavior of PP modified by electron irradiation or nanotube addition. Macromol. Symp. 2007, 254, 400–408.

42. Xu, D.; Wang, Z. Role of multi-wall carbon nanotube network in composites to crystallization of isotactic polypropylene matrix. Polymer 2008, 49, 330–338.

43. Grady, B.P.; Pompeo, F.; Shambaugh, R.L.; Resasco, D.E. Nucleation of polypropylene crystallization by single-walled carbon nanotubes. J. Phys. Chem. B 2002, 106, 5852–5858.

44. Bhattacharyya, A.R.; Sreekumar, T.V.; Liu, T.; Kumar, S.; Ericson, L.M.; Hauge, R.H.; Smalley, R.E. Crystallization and orientation studies in polypropylene/single wall carbon nanotube composite. Polymer 2003, 44, 2373–2377.

45. Assouline, E.; Lustiger, A.; Barber, A.H.; Cooper, C.A.; Klein, E.; Wachtel, E.; Wagner, H.T. Nucleation ability of multiwall carbon nanotubes in polypropylene composites. J. Polym. Sci. B-Polym. Phys. 2003, 41, 520–527.

46. Valentini, L.; Biagiotti, J.; Kenny, J.M.; Santucci, S. Morphological characterization of single-walled carbon nanotubes-PP composites. Comp. Sci. Technol. 2003, 63, 1149–1153.

47. Leelapornpisit, W.; Ton-That, M.; Perrin-Sarazin, F.; Cole, K.C.; Denault, J.; Simard, B. Effect of carbon nanotubes on the crystallization and properties of polypropylene. J. Polym. Sci. B-Polym. Phys. 2005, 43, 2445–2453.

48. Wang, K.; Tang, C.; Zhao, P.; Yang, H.; Zhang, Q.; Du, R.; Fu, Q. Rheological investigations in understanding shear-enhanced crystallization of isotactic poly(propylene)/multi-walled carbon nanotube composites. Macromol. Rapid Commun. 2007, 28, 1257–1264.

49. Kaganj, A.B.; Rashidi, A.M.; Arasteh, R.; Taghipoor, S. Crystallisation behaviour and morphological characteristics of poly(propylene)/multi-walled carbon nanotube nanocomposites. J. Experim. Nanosci. 2009, 4, 21–34.

50. Reyes-de Vaaben, S.; Aguilar, A.; Avalos, F.; Ramos-de Valle, L.F. Carbon nanoparticles as effective nucleating agents for polypropylene. J. Therm. Anal. Cal. 2008, 93, 947–952.

51. Avila-Orta, C.A.; Medellín-Rodríguez, F.J.; Dávila-Rodríguez, M.V.; Aguirre-Figueroa, Y.A.; Yoon, K.; Hsiao, B.S. Morphological features and melting behavior of nanocomposites based on isotactic polypropylene and multiwalled carbon nanotubes. J. Appl. Polym. Sci. 2007, 106, 2640–2647.

52. Peneva, Y.; Valcheva, M.; Minkova, L.; Miušík, M.; Omastová, M. Nonisothermal crystallization kinetics and microhardness of PP/CNT composites. J. Macromol. Sci. B-Phys. 2008, 47, 1197–1210.

53. Hou, Z.; Wang, K.; Zhao, P.; Zhang, Q.; Yang, C.; Chen, D.; Du, R.; Fu, Q. Structural orientation and tensile behavior in the extrusion-stretched sheets of polypropylene/multi-walled carbon nanotubes' composite. Polymer 2008, 49, 3582–3589.

54. Jose, M.V.; Dean, D.; Tyner, J.; Price, G.; Nyairo, E. Polypropylene/carbon nanotube nanocomposite fibers: Process-morphology-property relationships. J. Appl. Polym. Sci. 2007, 103, 3844–3850.

55. Anand, K.A.; Agarwal, U.S.; Joseph, R. Carbon nanotubes induced crystallization of poly(ethylene terephthalate). Polymer 2006, 47, 3976–3980.

56. Causin, V.; Yang, B.X.; Marega, C.; Goh, S.H.; Marigo, A. Nucleation, structure and lamellar morphology of isotactic polypropylene filled with polypropylene-grafted multiwalled carbon nanotubes. Eur. Polym. J. 2009, 45, 2155–2163.

57. Fereidoon, A.; Ahangari, M.G.; Saedodin, S. A DSC study on the nonisothermal crystallization kinetics of polypropylene/single-walled carbon nanotube nanocomposite. Polym. Plast. Techn. Eng. 2009, 48, 579–586.

58. Lu, K.; Grossiord, N.; Koning, C.E.; Miltner, H.E.; van Mele, B.; Loos, J. Carbon nanotube/isotactic polypropylene composites prepared by latex technology: Morphology analysis of CNT-induced nucleation. Macromolecules 2008, 41, 8081–8085.

59. Zhang, S.; Minus, M.L.; Zhu, L.; Wong, C.-P.; Kumar, S. Polymer transcrystallinity induced by carbon nanotubes. Polymer 2008, 49, 1356–1364.

60. Valentini, L.; Biagiotti, J.; Kenny, J.M.; Santucci, S. Effects of single-walled carbon nanotubes on the crystallization behavior of polypropylene. J. Appl. Polym. Sci. 2003, 87, 708–713.

61. Miltner, H.E.; Grossiord, N.; Lu, K.; Loos, J.; Koning, C.E.; Van Mele, B. Isotactic polypropylene/carbon nanotube composites prepared by latex technology. Thermal analysis of camon nanotube induced nucleation. Macromolecules 2008, 41, 5753–5762.

62. Yin, C.L.; Liu, Z.Y.; Yang, W.; Yang, M.B.; Feng, J.M. Crystallization and morphology of iPP/MWCNT prepared by compounding iPP melt

with MWCNT aqueous suspension. Coll. Polym. Sci. 2009, 287, 615–620.

63. Wu, D.F.; Sun, Y.R.; Wu, L.; Zhang, M. Linear viscoelastic properties and crystallization behavior of multi-walled carbon nanotube/polypropylene composites. J. Appl. Polym. Sci. 2008, 108, 1506–1513.

64. Deng, H.; Zhang, R.; Bilotti, E.; Loos, J.; Peijs, T. Conductive polymer tape containing highly oriented carbon nanofillers. J. Appl. Polym. Sci. 2009, 113, 742–751.

65. Kharchenko, S.B.; Douglas, J.F.; Obrzut, J.; Grulke, E.A.; Migler, K.B. Flow-induced properties of nanotube-filled polymer materials. Nat. Mater. 2004, 3, 564–568.

66. Li, C.; Liang, T.; Lu, W.; Tang, C.; Hu, X.; Cao, M.; Liang, J. Improving the antistatic ability of polypropylene fibers by inner antistatic agent filled with carbon nanotubes. Comp. Sci. Technol. 2004, 64, 2089–2096.

67. Andrews, R.; Jacques, D.; Minot, M.; Rantell, T. Fabrication of carbon multiwall nanotube/polymer composites by shear mixing. Macromol. Mater. Eng. 2002, 287, 395–403.

68. Deng, H.; Skipa, T.; Zhang, R.; Lellinger, D.; Bilotti, E.; Alig, I.; Peijs, T. Effect of melting and crystallization on the conductive network in conductive polymer composites. Polymer 2009, 50, 3747–3754.

69. Seo, M.K.; Park, S.J. Electrical resistivity and rheological behaviors of carbon nanotubes-filled polypropylene composites. Chem. Phys. Lett. 2004, 395, 44–48.

70. Du, F.M.; Scogna, R.C.; Zhou, W.; Brand, S.; Winey, K.I. Nanotube networks in polymer nanocomposites: Rheology and electrical conductivity. Macromolecules 2004, 37, 9048–9055.

71. Alig, I.; Lellinger, D.; Dudkin, S.M.; Pötschke, P. Conductivity spectroscopy on melt processed polypropylene-multiwalled carbon

nanotube composites: Recovery after shear and crystallization. Polymer 2007, 48, 1020–1029.

72. Immonen, K.; Nättinen, K.; Sarlin, J.; Hartikainen, J. Conductive plastics with hybrid materials. J. Appl. Polym. Sci. 2009, 114, 1494–1502.

73. Liao, S.-H.; Yen, C.-Y.; Weng, C.-C.; YLin, F.; Ma, C.-C.M.; Yang, C.-H.; Tsai, M.-C.; Yen, M.-Y.; Hsiao, M.-C.; Lee, S.-J.; Xie, X.-F.; Hsiao, Y.-H. Preparation and properties of carbon nanotube/polypropylene nanocomposite bipolar plates for polymer electrolyte membrane fuel cells. J. Power Sources 2008, 185, 1225–1232.

74. Liang, G.D.; Bao, S.P.; Tjong, S.C. Mechanical behaviors of polypropylene/carbon nanotube nanocomposites: The effects of loading rate and temperature. Mater. Sci. Eng. B 2007, 142, 55–61.

75. Bao, H.-D.; Guo, Z.-X.; Yu, J. Effect of electrically inert particulate filler on electrical resistivity of polymer/multi-walled carbon nanotube composites. Polymer 2008, 49, 3826–3831.

76. Lozano, K.; Barrera, E.V. Nanofiber-reinforced thermoplastic composites. I. Thermoanalytical and mechanical analyses. J. Appl. Polym. Sci. 2001, 79, 125–133.

77. Yang, J.; Lin, Y.; Wang, J.; Lai, M.; Li, J.; Liu, J.; Tong, X.; Cheng, H. Morphology, thermal stability, and dynamic mechanical properties of atactic polypropylene/carbon nanotube composites. J. Appl. Polym. Sci. 2005, 98, 1087–1091.

78. Seo, M.K.; Park, S.J. A kinetic study on the thermal degradation of multi-walled carbon nanotubes-reinforced poly(propylene) composites. Macromol. Mater. Eng. 2004, 289, 368–374.

79. Marosföi, B.B.; Szabó, A.; Marosi, G.; Tabuani, D.; Camino, G.; Pagliari, S. Thermal and spectroscopic characterization of polypropylene-carbon nanotube composites. J. Therm. Anal. Cal. 2006, 86, 669–673.

80. Vladimirov, V.; Betchev, C.; Vassiliou, A.; Papageorgiou, G.; Bikiaris, D. Dynamic mechanical and morphological studies of isotactic

polypropylene/fumed silica nanocomposites with enhanced gas barrier properties. Comp. Sci. Technol. 2006, 66, 2935–2944.

81. Vassiliou, A.; Bikiaris, D.; Pavlidou, E. Optimizing melt-processing conditions for the preparation of iPP/fumed silica nanocomposites: Morphology, mechanical and gas permeability properties. Macromol. React. Eng. 2007, 1, 488–501.

82. Kashiwagi, T.; Grulke, E.; Hilding, J.; Harris, Jr R.H.; Awad, W.H.; Dougls, J. Thermal degradation and flammability properties of poly(propylene)/carbon nanotube composites. Macromol. Rapid Commun. 2002, 23, 761–765.

83. Kashiwagi, T.; Grulke, E.; Hilding, J.; Groth, K.; Harris, R.; Butler, K.; Shields, J.; Kharchenko, S.; Douglas, J. Thermal and flammability properties of polypropylene/carbon nanotube nanocomposites. Polymer 2004, 45, 4227–4239.

84. Bom, D.; Andrews, R.; Jacques, D.; Anthony, J.; Chen, B.; Meier, M.S.; Selegue, J.P. Thermogravimetric Analysis of the Oxidation of Multiwalled Carbon Nanotubes: Evidence for the Role of Defect Sites in Carbon Nanotube Chemistry. Nano Lett. 2002, 2, 615–619.

85. Andrews, R.; Jacques, D.; Qian, D.; Dickey, E.C. Purification and structural annealing of multiwalled carbon nanotubes at graphitization temperatures. Carbon 2001, 39, 1681–1687.

86. Moore, E.M.; Ortiz, D.L.; Marle, V.T.; Shambaugh, R.L.; Grady, B.P. Enhancing the strength of polypropylene fibers with carbon nanotubes. J. Appl. Polym. Sci. 2004, 93, 2926–2933.

87. Xiao, Y.; Zhang, X.; Cao, W.; Wang, K.; Tan, H.; Zhang, Q.; Du, R.; Fu, Q. Dispersion and mechanical properties of polypropylene/multiwall carbon nanotubes composites obtained via dynamic packing injection molding. J. Appl. Polym. Sci. 2007, 104, 1880–1886.

88. Liao, S.H.; Weng, C.C.; Yen, C.Y.; Hsiao, M.C.; Ma, C.C.M.; Tsai, M.C.; Su, A.; Yen, M.Y.; Lin, Y.F.; Liu, P.L. Preparation and properties of functionalized multiwalled carbon nanotubes/polypropylene

nanocomposite bipolar plates for polymer electrolyte membrane fuel cells. J. Power Sources 2010, 195, 263–270.

89. McIntosh, D.; Khabashesku, V.N.; Barrera, E.V. Nanocomposite fiber systems processed from fluorinated single-walled carbon nanotubes and a polypropylene matrix. Chem. Mater. 2006, 18, 4561–4569.

90. Zhou, Z.; Wang, S.; Lu, L.; Zhang, Y.; Zhang, Y. Isothermal crystallization kinetics of polypropylene with silane functionalized multi-walled carbon nanotubes. J. Polym. Sci. B-Polym. Phys. 2007, 45, 1616–1624.

91. Kelarakis, A.; Yoon, K.; Sics, I.; Somani, R.H.; Chen, X.M.; Hsiao, B.S.; Chu, B. Shear-induced orientation and structure development in isotactic polypropylene melt containing modified carbon nanofibers. J. Polym. Sci. B-Polym. Phys. 2006, 45, 247–261.

92. Ristolainen, N.; Vainio, U.; Paavola, S.; Torkkeli, M.; Serimaa, R.; Seppälä, J. Polypropylene/organoclay nanocomposites compatibilized with hydroxyl-functional polypropylenes. J. Polym. Sci. B- Polym. Phys. 2005, 43, 1892–1903.

93. Xu, D.-H.; Wang, Z.-G. Influence of carbon nanotube aspect ratio on normal stress differences in isotactic polypropylene nanocomposite melts. Macromolecules 2008, 41, 815–825.

94. Chen, X.H.; Hu, J.; Li, W.H.; Liu, Y.Q.; Chen, C.S.; Wang, Y.G. Preparation and crystallization of carbon nanotube/maleic anhydride-grafted polypropylene composites. J. Mater. Sci. Technol. 2008, 24, 279–284.

95. Li, W.-H.; Chen, X.-H.; Yang, Z.; Xu, L.-S. Structure and properties of polypropylene-wrapped carbon nanotubes composite. J. Appl. Polym. Sci. 2009, 113, 3809–3814.

96. Zhou, Z.; Wang, S.; Lu, L.; Zhang, Y.; Zhang, Y. Functionalization of multi-wall carbon nanotubes with silane and its reinforcement on polypropylene composites. Comp. Sci. Technol. 2008, 68, 1727–1733.

97. Koval'chuk, A.A.; Shevchenko, V.G.; Shchegolikhin, A.N.; Nedorezova, P.M.; Klyamkina, A.N.; Aladyshev, A.M. Effect of carbon nanotube functionalization on the structural and mechanical properties of polypropylene/MWCNT composites. Macromolecules 2008, 41, 7536–7542.

98. Vaisman, L.; Marom, G.; Wagner, H.D. The role of surfactants in dispersion of carbon nanotubes. Adv. Funct. Mater. 2006, 16, 357–363.

99. Garg, A.; Sinnott, S.B. Effect of chemical functionalization on the mechanical properties of carbon nanotubes. Chem. Phys. Lett. 1998, 295, 273–278.

100. Lee, S. H.; Cho, E. N.; Jeon, S. H.; Youn, J. R. Rheological and electrical properties of polypropylene composites containing functionalized multi-walled carbon nanotubes and compatibilizers. Carbon 2007, 45, 2810–2822.

101. Ahangari, M.G.; Fereidoon, A.; Saedodin, S. Mechanical and thermal properties of PP/compatibilized PP/acid treated SWCNTs nanocomposites: Effect of different acid treatment times. E-Polymers 2008, 153, 1–15.

102. Wu, D.; Sun, Y.; Zhang, M. Kinetics study on melt compounding of carbon nanotube/polypropylene nanocomposites. J. Polym. Sci. B-Polym. Phys. 2009, 47, 608–618.

103. Jin, S.H.; Kang, C.H.; Yoon, K.H.; Bang, D.S.; Park, Y.-B. Effect of compatibilizer on morphology, thermal, and rheological properties of polypropylene/functionalized multi-walled carbon nanotubes composite. J. Appl. Polym. Sci. 2009, 111, 1028–1033.

104. Prashantha, K.; Soulestin, J.; Lacrampe, M.F.; Claes, M.; Dupin, G.; Krawczak, P. Multi-walled carbon nanotube filled polypropylene nanocomposites based on masterbatch route: Improvement of dispersion and mechanical properties through PP-g-MA addition. eXPRESS Polym. Lett. 2008, 2, 735–745.

105. Velasco-Santos, C.; Martinez-Hernandez, A.L.; Fisher, F.T.; Ruoff, R.; Castano, V.M. Improvement of Thermal and Mechanical Properties of Carbon Nanotube Composites through Chemical Functionalization. Chem. Mater. 2003, 15, 4470–4475.

106. Philip, B.; Xie, J.; Abraham, J.K.; Varadan., V.K. Polyaniline/carbon nanotube composites: Starting with phenylamino functionalized carbon nanotubes. Polym. Bull. 2005, 53, 127–138.

107. Wiemann, K.; Kaminsky, W.; Gojny, F.H.; Schulte, K. Synthesis and properties of syndiotactic poly(propylene)/carbon nanofiber and nanotube composites prepared by in situ polymerization with metallocene/MAO catalysts. Macromol. Chem. Phys. 2005, 206, 1472–1478.

108. Kaminsky, W.; Wiemann, K. Polypropene nanocomposites by metallocene/MAO catalysts. Comp. Interf. 2006, 13, 365–375.

109. Kaminsky, W.; Funck, A.; Wiemann, K. Nanocomposites by in situ polymerization of olefins with metallocene catalysts. Macromol. Symp. 2006, 239, 1–6.

110. Funck, A.; Kaminsky, W. Polypropylene carbon nanotube composites by in situ polymerization. Comp. Sci. Technol. 2007, 67, 906–915.

111. Bonduel, D.; Bredeau, S.; Alexandre, M.; Monteverde, F.; Dubois, P. Supported metallocene catalysis as an efficient tool for the preparation of polyethylene/carbon nanotube nanocomposites: Effect of the catalytic system on the coating morphology. J. Mater. Chem. 2007, 17, 2359–2366.

112. Bredeau, S.; Boggioni, L.; Bertini, F.; Tritto, I.; Monteverde, F.; Alexandre, M.; Dubois, P. Ethylene-norbornene copolymerization by carbon nanotube-supported metallocene catalysis: Generation of high-performance polyolefinic nanocomposites. Macromol. Rapid Commun. 2007, 28, 822–827.

113. Kaminsky, W.; Funck, A. In situ polymerization of olefins with nanoparticles by metallocene-catalysis. Macromol. Symp. 2007, 260, 1–8.

114. Koval'chuk, A.A.; Shchegolikhin, A.N.; Shevchenko, V.G.; Nedorezova, P.M.; Klyamkina, A.N.; Aladyshev, A.M. Synthesis and properties of polypropylene/multiwall carbon nanotube composites. Macromolecules 2008, 41, 3149–4156.

115. Auriemma, F.; De Rosa, C. New Concepts in Thermoplastic Elastomers: The Case of Syndiotactic Polypropylene, an Unconventional Elastomer with High Crystallinity and Large Modulus. J. Am. Chem. Soc. 2003, 125, 13134–13147.

116. Gorrasi, G.; Romeo, V.; Sannino, D.; Sarno, M.; Ciambelli, P.; Vittoria, V.; De Vivo, B.; Tucci, V. Carbon nanotube induced structural and physical property transitions of syndiotactic polypropylene. Nanotechnology 2007, 18, 275703.

117. Sarno, M.; Gorrasi, G.; Sannino, D.; Sorrentino, A.; Ciambelli, P.; Vittoria, V. Polymorphism and thermal behaviour of syndiotactic poly(propylene)/carbon nanotube composites. Macromol. Rapid Commun. 2004, 25, 1963–1967.

118. Wu, M.; Shaw, L. Electrical and mechanical behaviors of carbon nanotube-filled polymer blends. J. Appl. Polym. Sci. 2006, 99, 477–488.

119. Zhang, L.; Wan, C.; Zhang, Y. Morphology and electrical properties of polyamide 6/polypropylene/multi-walled carbon nanotubes composites. Comp. Sci. Technol. 2009, 69, 2212–2217.

120. Zou, J.; Zhang, Y.; Juang, J.; Wu, H.; Qiu, Y.P. Preparation and properties of PP/PLA/multiwall carbon nanotube composites filaments obtained by melt compounding. Mater. Sci. Forum 2009, 620-622, 465–468.

121. Valentini, L.; Biagiotti, J.; J.; Kenny, M.; Manchado, M.A.L. Physical and mechanical behavior of single-walled carbon nanotube/

polypropylene/ethylene-propylene-diene rubber nanocomposites. J. Appl. Polym. Sci. 2003, 89, 2657–2663.

122. Liu, L.; Wang, Y.; Li, Y.; Wu, J.; Zhou, Z.; Jiang, C. Improved fracture toughness of immiscible polypropylene/ethylene-co-vinyl acetate blends with multiwalled carbon nanotubes. Polymer 2009, 50, 3072–3078.

123. Khare, R.A.; Bhattacharyya, A.R.; Kulkarni, A.R.; Saroop, M.; Biswas, A. Influence of multiwall carbon nanotubes on morphology and electrical conductivity of PP/ABS blends. J. Polym. Sci. B-Polym. Phys. 2008, 46, 2286–2295.

6

FUNCTIONALIZATION OF CARBON NANOTUBES

In-Yup Jeon[1], Dong Wook Chang[2], Nanjundan Ashok Kumar[1] and Jong-Beom Baek[1]

[1] Interdisciplinary School of Green Energy/Institute of Advanced Materials & Devices, Ulsan National Institute of Science and Technology (UNIST), Ulsan, South Korea

[2] Department of Chemical Engineering, Case Western Reserve University,Cleveland, Ohio, USA

1. INTRODUCTION

Albeit initially reported by a Russian group and reported again by Oberlin and co-workers (Oberlin et al., 1976), carbon nanotubes (CNTs) were first brought to the attention of greater scientific community by Ijima in 1991. It was not until 1991 that CNTs came to limelight unfolding its exotic properties. Following this report, academic interest in this one-dimensional nanomaterial has grown remarkably and phenomenal number of research are being carried out worldwide. Its outstanding properties such as high mechanical strength, flexibility and impressive electrical and thermal conductivity make itself attractive and potential candidate for various applications. Despite CNTs's exceptional properties, there are two main limitations that hinder its use. The surface energy of CNTs is significantly different from that of matrices such as common organic solvents or polymers and CNTs may not have chemical affinity to the organic matrices and thus, the dispersion of CNTs into matrices is the biggest obstacle in practice (Song et al., 2005). In addition, the seamless surface of CNTs cannot provide physical interaction between CNTs and matrix (Ajayan et al., 2000). The modification of CNTs with other materials, makes it the most attractive and ultimate candidate for a plethora of applications which includes nanodevices, to organic electronics. In terms of the practical challenges that must be addressed to expand CNTs application scope in various potential fields of nanotechnology; many protocols have been proposed and considerable amount of research work is being focused on the interfacial molecular engineering of CNTs, aiming at improving their state of interaction with macromolecules, mainly to optimize the final properties of the hybrid nanomaterials. These protocols to prepare such hybrid composites have triggered burgeoning interest and to realize most of these applications practically, CNTs are most frequently functionalized with various kinds of technologically important polymers to improve the properties of the end material. Such assemblies have already suggested that the functionalization of polymers to CNTs draws equal importance in materials development for future applications in constructing optoelectronic devices. This chapter will focus and discuss the frequently used methods to functionalize CNTs namely; covalent and non-covalent strategies. Of particular importance, this chapter will also shed light on the 'pros and cons' of the frequent methods and alternative methods that have been developed in the recent past which can open the path for their facile manipulation and processing in physiological environments and also to overcome difficulties in processing CNTs.

2. CARBON NANOTUBES

CNT is one of carbon allotropes such as diamond, graphite, graphene, fullerene and amorphous carbon. But, it is the one-dimensional carbon form which can have an aspect ratio greater than 1000 makes it interesting (Figure 1). The bonding in carbon nanotubes is sp^2, with each atom joined to three neighbours, as in graphite. CNTs are rolled-up graphene sheets (graphene is an individual graphite layer). This type of structural bonding, which is stronger than the sp^3 bonds found in diamond, provides the molecules with their unique strength. Under high pressure, nanotubes can merge together, trading some sp^2 bonds for sp^3 bonds, giving the possibility of producing strong, unlimited length wires through high-pressure nanotube linking.

The properties of nanotubes depend on atomic arrangement (on how the sheets of graphite are 'rolled'), the diameter and length of the tubes, and morphology, or nano structure. Single walled nanotubes (SWNTs) consist of a single sheet of graphene rolled seamlessly to form a cylinder with diameter of order of 1 nm and length of up to centimeters (Bethune et al., 1993; Ijima et al., 1993)·Multi-walled nanotubes (MWNTs) consist of an array of such cylinders formed concentrically and separated by 0.35 nm, similar to the basal plane separation in graphite (Ijima 1991). MWNTs can have diameters from 2 to 100 nm and lengths of tens of microns. Apart from these two main types, there was also the few-walled carbon nanotube (FWNTs) that was reported and found to have peculiar properties, often with much better structural integrity (Hutchinson et al. 2001).

Figure 1. Crystal structures of different allotropes of carbon. (Left to right) Three-dimensional diamond and graphite (3D); two-dimensional graphene (2D); one-dimensional nanotubes (1D); and zero-dimensional buckyballs (0D) (**Mikhail & Katsnelson, 2007**).

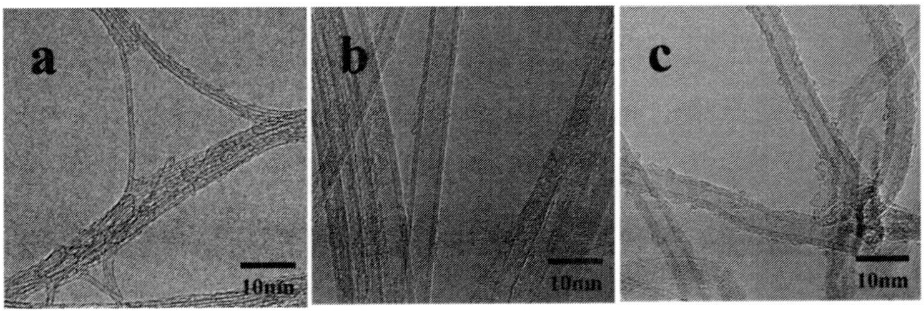

Figure 2. TEM images of different types of CNTs: (a) single-walled carbon nanotubes; (b) few-walled carbon nanotubes; (c) multi-walled carbon nanotubes (Huo et al. 2009).

SWNTs are formed by rolling a sheet of graphene into a cylinder along an (m, n) lattice vector in the graphene (Figure 3). The (m, n) indices determine the diameter and chirality. Three types of SWNTs are revealed with these values: when $n = m$, the nanotube is called "armchair" type (= 0º); when $m = 0$, then it is of the "zigzag" type (=

30º). Otherwise, when n ≠ *m*, it is a "chiral" tube and takes a value between 0º and 30º. The value of (*m, n*) determines the chirality of the nanotube and affects the optical, mechanical and electronic properties. For typical diameters, all armchair SWNTs and one-third of all zigzag nanotubes are metallic, the rest are semiconducting (Hamada et al.; 1992).

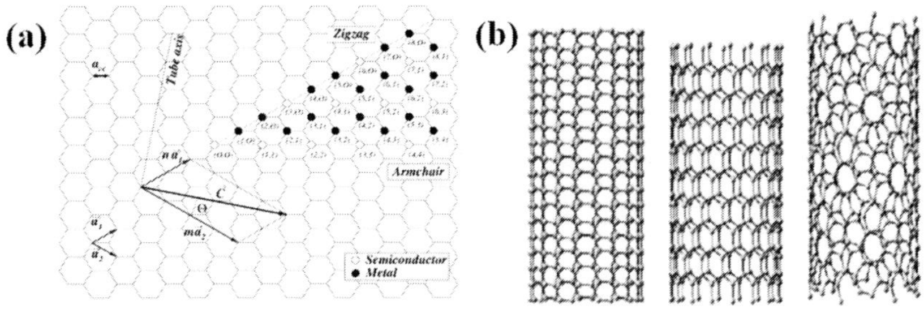

Figure 3. a) definition of the chiral vector describing the unit cell of a SWNT (**Belin & Epron 2005**); (b) CG images of armchair (*n, n*), zigzag (*n*, 0), and chiral nanotubes (**Popov 2004**).

It is clear that CNTs have many advantages over other materials in terms of mechanical, electrical and thermal properties although there is no consensus on the exact properties of CNTs (Table 1). Theoretical and experimental results have shown unusual mechanical properties with Young' modulus as high as 1.2 TPa and tensile strength of 50–200 GPa (Qian et al.; 2002). The relatively high values of modulus and strength values have been measured on high quality SWNT and arc discharge MWNT. However, CVD-MWNT was measured to have a modulus of 0.45 TPa and tensile strength of ~4 GPa (Xie et al.; 2000). The much larger variation in modulus for CVD-MWNT compared to arc-MWNT strongly suggests that the modulus is very sensitive to defect concentration and type.

Depending on the relationship between this axial direction and the unit vectors describing the hexagonal lattice, CNTs can either be metallic or semi-metallic. Semi-metallic nanotubes have band gaps that scale inversely with diameter, ranging from approximately 1.8 eV for very small diameter nanotubes to 0.18 eV for the widest possible stable SWNT (Elliot et al.; 2004). In addition, carrier mobility as high as 105 cm^2/Vs have been observed in semi-conducting nanotubes (Kim et al.;

2004). Superconductivity has also been observed in SWNTs at the transition temperatures of 5 K (Tang et al.; 2001). CNTs are also very conductive for phonons. Theory predicts a room temperature thermal conductivity of up to 6000 W/m K (Che et al.; 2000 & Osman et al.; 2001) but, the values around 200 W/m K have been measured (Hone et al.; 2002). These properties offer CNTs great potential for wide applications in field emission, conducting plastics, thermal conductors, energy storage, conductive adhesives, thermal interface materials, structural materials, fibers, catalyst supports, biological applications, air and water filtration, ceramics to name a few (Thostenson et al. 2001; Ajayan et al.; 2003; Coleman et al.; 2006).

Table 1. Theoretical and experimentally measured properties of carbon nanotubes (Schadler 2004).

Property	Carbon nanotubes	Graphite
Lattice structure	(Cylindrical) hexagonal lattice helicity Nanotubes: ropes, tubes arranged in triangular lattice with lattice parameters of a =1.7 nm, tube-tube distance = 0.314	Planar hexagonal, plane-to-plane distance c=0.335
Specific gravity	0.8-1.8 gcc^{-1} (theoretical)	2.26 gcc^{-1}
Elastic modulus	~1 TPa for SWNT ~0.3-1 TPa for MWNT	1 TPa (in-plane)
Strength	50-500 GPa for SWNT 10-50 GPa for MWNT	
Resistivity	~5-50 micro-ohm cm	50 (in-plane)
Thermal conductivity	3000 W $m^{-1}K^{-1}$ (theoretical)	3000 $Wm^{-1}K^{-1}$ (in-plane) 6 $Wm^{-1}K^{-1}$ (c axis)
Thermal expansion	Negligible (theoretical)	-1×10^{-6} K^{-1} (in-plane) $29 \times 10^{-6} K^{-1}$ (c axis)
Oxidation in air	>700 °C	450-650 °C

The synthesis methods for CNTs include arc-discharge (Journet et al. 1997), laser ablation (Rinzler et al. 1998), gas-phase catalytic growth from carbon monoxide (Nikolaev et al. 1999) and chemical vapor deposition (CVD) from hydrocarbons (Ren et al 1999& 2007). The arc-discharge and laser ablation are well established in producing high-quality and nearly perfect nanotube structures. But, due to their scale-up limitations the production cost is too hight hence limiting its access. During the synthesis process, impurities in the form of catalyst particles, amorphous carbon and non-tubular fullerenes are also produced. Thus, subsequent purification steps are required to separate the tubes. The gas-phase catalytic growth with carbon monoxide as the carbon source yielded in quantity high-purity single walled nanotubes at the highest accessible temperature and pressure (1200º C, 10 atm). However, this method is not economically practical due to high temperature and pressure of the synthesis conditions. The CVD process tended to tend nanotubes with fewer impurities and was more amenable to large-scale processing. Therefore, the CVD method has the highest potential to produce a large quantity and high purity of CNTs.

3. FUNCTIONALIZATION OF CARBON NANOTUBES

The main problem with the majority of popular synthetic methods is that they produce samples yielding a mixture of various diameters and chiralities of nanotubes that are normally contaminated with metallic and amorphous impurities. Thus, post-synthesis chemical processing protocols (Strano et al.; 2004 & Li et al.; 2002) that purify tubes that can also separate individual tubes according to diameter and chirality by taking advantage of their differential reactivities are often the only viable routes to rational and predictable manipulation of the favorable electronic and mechanical properties of these materials (Niyogi et al.; 2002& Tasis et al.; 2003).

The full potential of CNTs as reinforcements has been severely limited due to poor interfacial interaction, van der Waals interaction, between CNTs and polymer matrix. The nature of dispersion problem for CNTs is rather different from other conventional fillers, such as spherical particles and carbon fibers, because CNTs are characteristic of small diameter in nanometer scale with high aspect ratio (>1000) and thus possessing large surface area. In addition, the commercialized CNTs are supplied in the form of heavily entangled bundles, resulting in inherent difficulties in dispersion.

To resolve those problems, it has been directed towards developing methods to modify surface properties of CNTs. These approaches can be simply divided into chemical (covalent) and physical (noncovalent) functionalization as interactions between active materials and CNTs. The characterization corresponding to these methods are summarized in Figure 4.

Table 2. Comparison of the established techniques for CNT synthesis (See et al. 2007).

method	arc discharge	laser ablation	chemical vapor deposition
description	arc evaporation of graphite in the presence of inert gas; CNT formed on electrodes during quenching	vaporization of graphite target by laser; CNT formed on receiver during quenching	decomposition of hydrocarbons over transition metal catalyst to form CNT
operating temperature	>3000 °C	>3000 °C	<1200 °C
operating pressure	50–7600 Torr generally under vacuum	200–750 Torr generally under vacuum	760–7600 Torr
advantages	good quality CNTs	good quality CNTs; single conformation SWNT formed (10, 10)	easy scale up; synthesis on templates possible
disadvantages	difficult to scale up	difficult to scale up; expensive	quality of CNT not as good

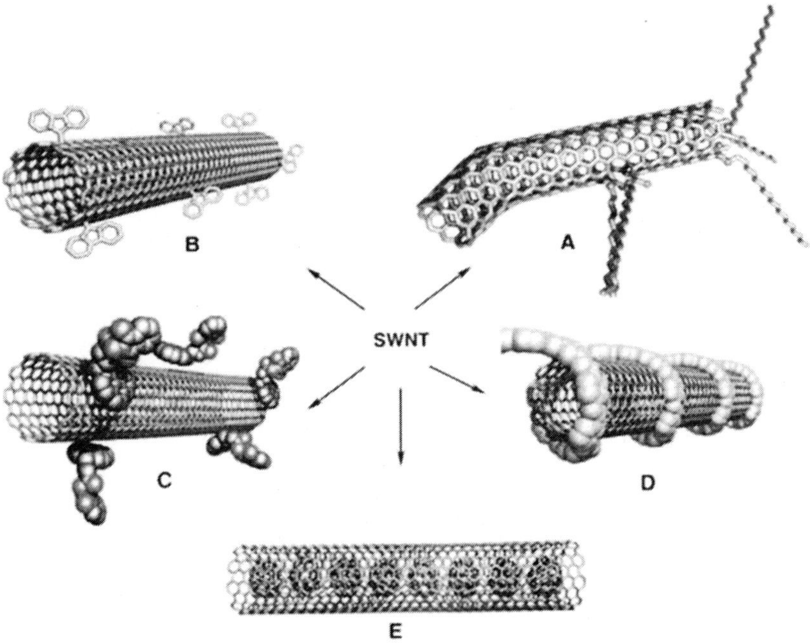

Figure 4. Functionalization possibilities for SWNTs: a) defect-group functionalization, b) covalent sidewall functionalization, c) non-covalent exohedral functionalization with surfactants, d) non-covalent exohedral functionalization with polymers, and e) endohedral functionalization with, for example, C_{60}. For methods B-E, the tubes are drawn in idealized fashion, but defects are found in real situations (**Hirsch 2002**).

3.1. Covalent Functionalization

The end caps of nanotubes tend to be composed of highly curved fullerene-like hemispheres, which are therefore highly reactive, as compared with the side walls (Hirsch 2002& Sinnott 2002). The sidewalls themselves contain defect sites such as pentagon-heptagon pairs called Stone-Walls defects, sp^3-hybrideized defects, and vacancies in the nanotube lattice (Figure 5) (Hirsch 2002).

Chemical functionalization is based on the covalent bond of functional groups onto carbon form of CNTs. It can be performed at the end caps of nanotubes or at their sidewalls which have many defects. Direct covalent sidewall functionalization is associated with a change of hybridization from sp^2 to $sp^3$and a simultaneous loss of *p*-conjugation system on graphene layer (Figure 6). This process can be made by reaction with some molecules of a high chemical reactivity. In the first approach, fluorination of CNTs has become popular for initial investigation of the covalent functionalization because the CNTs sidewalls are expected to be inert (Bianco et al.; 2008& Mickelson et al.; 1998). The fluorinated CNTs have C-F bonds that are weaker than those in alkyl fluorides (Kelly et al. 1993) and thus providing substitution sites for additional functionalization (Touhara et al.; 2002). Successful replacements of the fluorine atoms by amino, alkyl and hydroxyl groups have been achieved (Stevens et al.; 2003). Other methods, including cycloaddition, such as Diels-Alder reaction, carbene and nitrene addition (Hu et al. 2003), chlorination, bromination (Unger et al. 2002), hydrogenation (Kim et al.; 2002), azomethineylides (Tagmatarchis 2004) have also been successfully employed.

Another method is defect functionalization of CNTs. These intrinsic defects are supplemented by oxidative damage to the nanotube framework by strong acids which leave holes functionalized with oxygenated functional groups (Chen et al.; 1998). In particular, treatment of CNTs with strong acid such as HNO_3, H_2SO_4 or a mixture of them (Esumi et al.; 1996& Liu et al.; 1998), or with strong oxidants such as $KMnO_4$ (Yu et al.; 1998), ozone (Sham et al.; 2006), reactive plasma (Wang et al.; 2009) tend to open these tubes and to subsequently generate oxygenated functional groups such as carboxylic acid, ketone, alcohol and ester groups, that serve to tether many different types of chemical moieties onto the ends and defect sites of these tubes. These functional groups have rich chemistry and the CNTs can be used as precursors for further chemical reactions, such as silanation (Ma et al.; 2006), polymer grafting Sano et al. 2001& Kong et al. 2003), esterification (Hamon et al. 2002), thiolation (Liu et al.; Science 1998), and even some biomolecules (Coleman et al.; 2006). The CNTs functionalized by the covalent methods has good advantage which that was soluble in various organic solvents because the CNTs possess many functional groups such as polar or non-polar groups.

However these methods have two major drawbacks. First, during the functionalization reaction, especially along with damaging ultrasonication process (Figure 7), a large number of defects are inevitably created on the CNT sidewalls, and in some extreme cases, CNTs are fragmented into smaller pieces (Figure 8). Namely, the carbon hybridization of CNTs was changed from sp^2 to sp^3. These damaging effects result in severe degradation in mechanical properties of CNTs as well as disruption of π electron system in nanotubes. The disruption of π electrons is detrimental to transport properties of CNTs because defect sites scatter electrons and phonons that are responsible for the electrical and thermal conductions of CNTs, respectively. Secondly, concentrated acids or strong oxidants are often used for CNT functionalization, which are environmentally unfriendly. Therefore, many efforts have been put forward to developing methods that are convenient to use, of low cost and less damage to CNT structure.

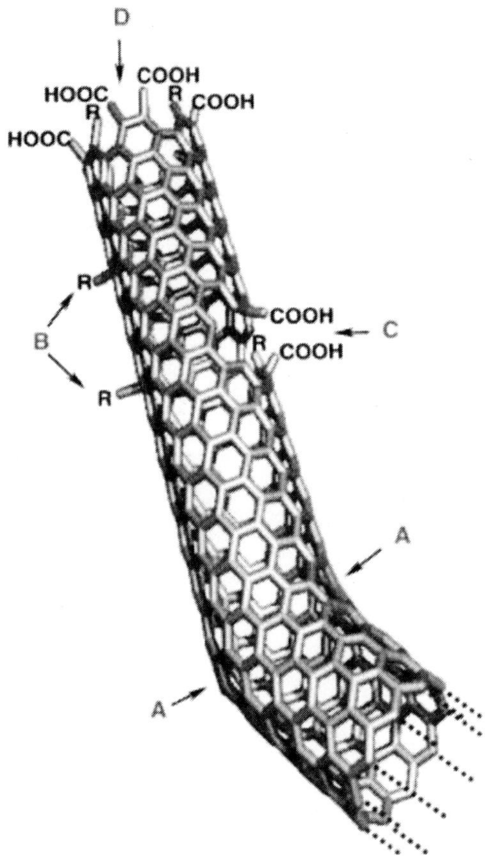

Figure 5. Typical defects in a SWNT. a) five-or seven-membered rings in the carbon framework, instead of the normal six-membered ring, leads to a bend in the tube. b) *sp* ^{3} -hybrideized defects (R=H and OH). c) carbon framework damaged by oxidative conditions, which leaves a hole lined with –COOH groups. d) open end of the SWNT, terminated with COOH groups. Besides carboxyl termini, the existence of which has been unambiguously demonstrated, other terminal groups such as -NO_2, -OH, -H, and =O are possible (**Hirsch 2002**).

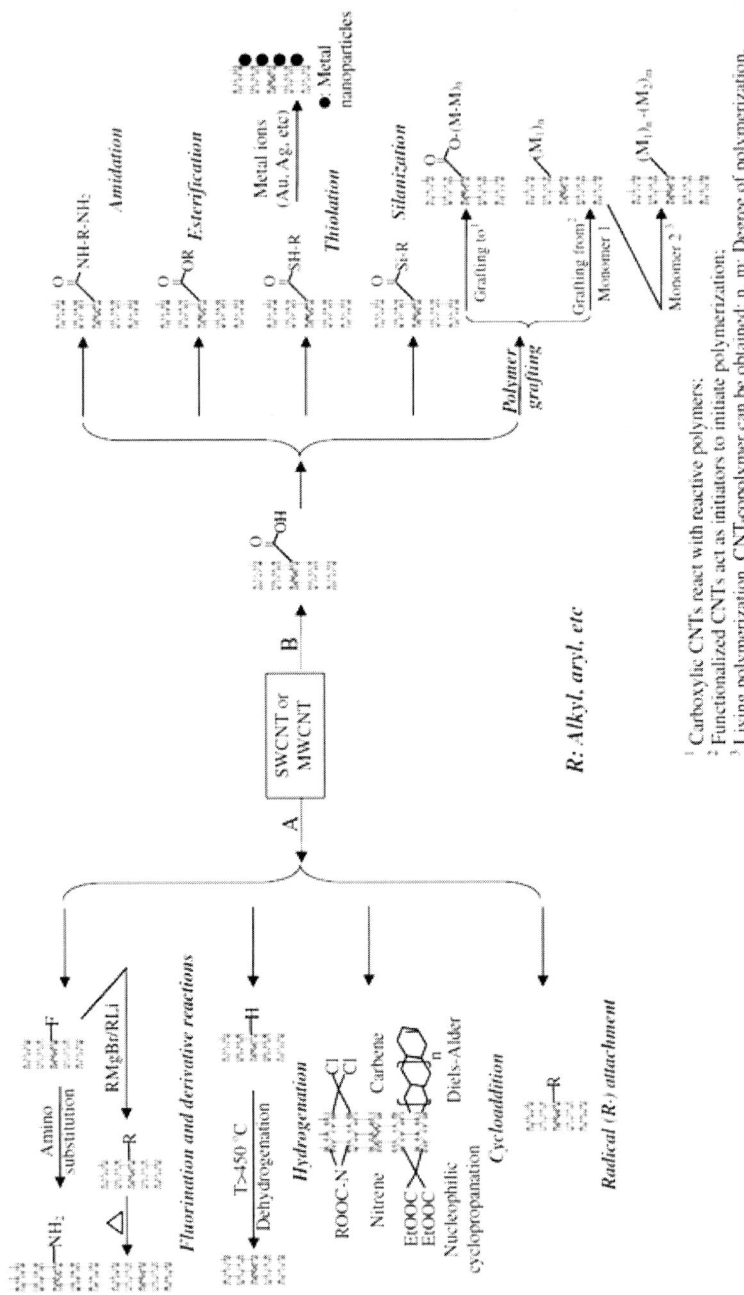

Figure 6. Strategies for covalent functionalization of CNTs (A: direct sidewall functionalization; B: defect functionalization) (Ma et al.; 2010).

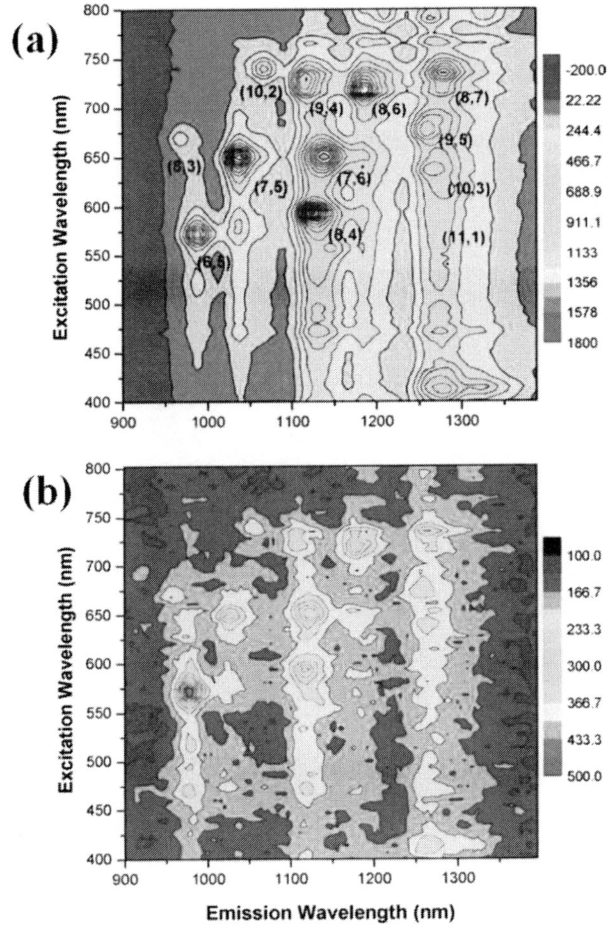

Figure 7. a) fluorescence excitation profile of sodium cholate-suspended SWNT in water, (b) profile of the same SWNT after 10min of probe-tip sonication conducted without controlling temperature (Heller et al.; 2005).

3.2. Non-Covalent Functionalization

The advantage of non-covalent functionalization is that it does not destroy the conjugated system of the CNTs sidewalls, and therefore it does not affect the final structural properties of the material. The non-covalent functionalization is an alternative method for tuning the interfacial properties of nanotubes. The CNTs are functionalized non-covalently by aromatic compounds, surfactants, and polymers, employing π-π stacking or hydrophobic interactions for the most part. In these approaches, the non-covalent modifications of CNTs can do much to preserve their desired properties, while improving their solubilities quite remarkably. It will summarize as followed: aromatic small molecule absorption, polymer wrapping, surfactants, biopolymers and endohedral method.

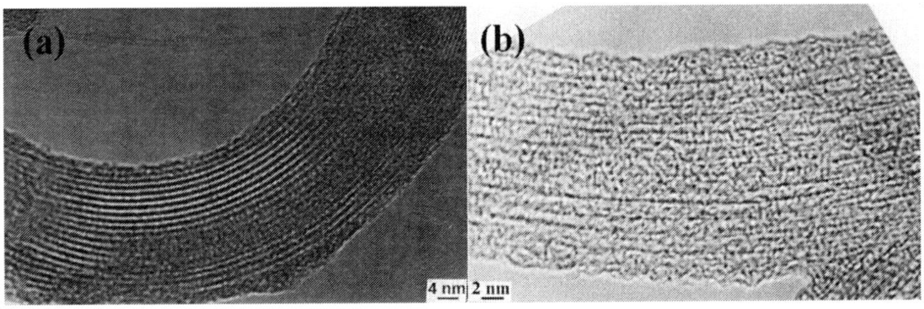

Figure 8. TEM images: (a) as prepared SWNTs rope; (b) acid treated SWNTs rope (**Monthioux et al.; 2001**).

Aromatic molecules, such as pyrene, porphyrin, and their derivatives, can and do interact with the sidewalls of CNTs by means of π-π stacking interactions, thus opening up the way for the non-covalent functionalization of CNTs (Figure 9). Dai and co-workers have reported a general and attractive approach to the non-covalent functionalization of CNTs sidewalls and the subsequent immobilization of biological molecules onto CNTs with a high degree of control and specificity (Dai et al.; 2001). Hecht et al. fabricated CNTs/FET devices functionalized non-covalently with a zinc porphyrin derivative, was used to detect directly a photo induced electron transferring within the zinc porphyrin derivative-CNTs system (Hecht et al.; 2006). (Hu et al.; 2008) prepared CdSe-CNTs hybrids by self-assembling the pyrene-functionalized CdSe (pyrene/CdSe) nanoparticles onto the surfaces of the CNTs.

Polymers, especially conjugated polymers, have been shown to serve as excellent wrapping materials for the non-covalent functionalization of CNTs as a result of π-π stacking and van der Waals interactions between the conjugated polymer chains containing aromatic rings and the surfaces of CNTs (Figure 10) (Star et al.; 2001, 2002, 2003 Cheng et al.; 2006& Yi et al.; 2008). Those has reported some organic-soluble conjugated poly(*m*-phenylenevinylene)-co-(2,5-dioctoxy-*p*phenylene) vinylene (PmPV) (Star et al.; 2001) poly(2,6-pyridinlenevinylene)- co-(2,5-dioctoxy-*p*-phenylene)vinylene (PPyPV) (Steuerman et al.; 2002), poly-(5-alkoxy-*m*-phenylenevinylene)-co-(2,5-dioctoxy-*p*-phenylene)-vinylene (PAmPV) (Star et al.; 2003), and stilbene-like dendrimers(Star et al.; 2002), to investigate their non-covalent functionalization for CNTs.

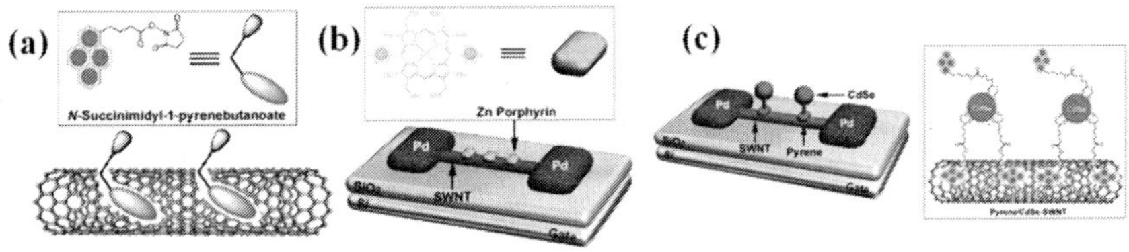

Figure 9. Aromatic small-molecule based non-covalent functionalization: (a) *N*-succinimidyl-1-pyrenebutanoate coated CNTs; (b) zinc porphyrin-coated CNTs; (c) pyrene/CdSe coated CNTs (**Zhao et al. 2009**).

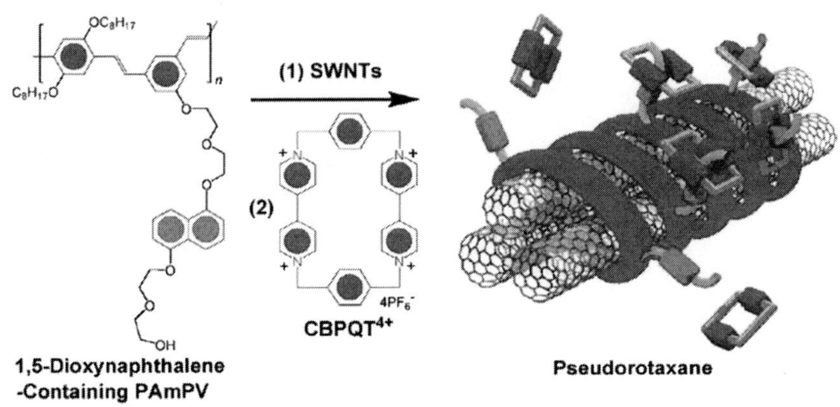

Figure 10. The side arms of the 1,5-dioxynaphthalene containing PAmPV-decorated CNTs hybrids associate with cyclobis(paraquat-p-phenylene) ($CBPQT^{4+}$) rings (Zhao et al.; 2006).

In addition, surfactants polymers have also been employed to functionalize CNTs (Figure 11). The physical adsorption of surfactant on the CNTs surface lowered the surface tension of CNTs that effectively prevented the aggregation of CNTs. Furthermore, the surfactant-treated CNTs overcame the van der Waals attraction by electrostatic/steric repulsive forces. The efficiency of this method depended strongly on the properties of surfactants, medium chemistry and polymer matrix. The relation between surfactants and CNTs studied previously as followed: (i) non-ionic surfactants, such as polyoxyethylene 8 lauryl or $C_{12}EO_8$ (Gong et al.; 2000), polyoxyethylene octylphenylether (Triton X-100) (Vaisman et al.; 2006), (ii) anionic surfactants, such as sodium dodecylsulfate (SDS), sodium dodecylbenzenesulfonate (NaDDBS), poly(styrene sulfate) (PSS) (Islam et al.; 2003& Yu et al.; 2007) (iii) cationic surfactants, such as dodecyl tri-methyl ammoniumbromide (DTAB) (Whitsitt et al.; 2003), cetyltrimethylammounium 4-vinylbenzoate (Kim et al.; 2007). Although surfactants may be efficient in the solubilization of CNTs, they are known to be permeable plasma membranes. They are toxic for biological applications. Therefore, the use of surfactant-stabilized CNTs complexes is potentially limited for biomedical applications (Klumpp et al.; 2006).

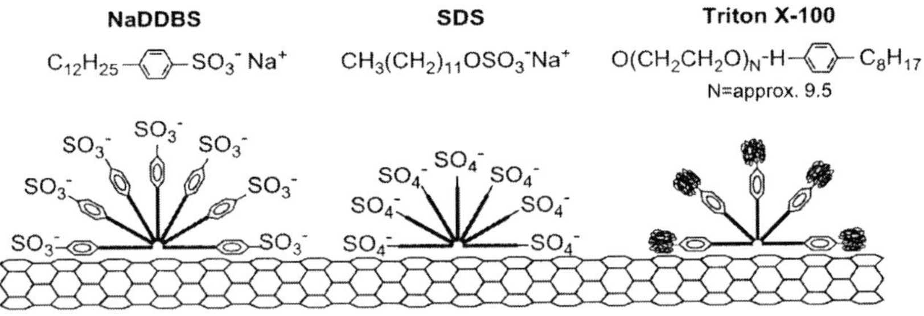

Figure 11. The schematic representation of how surfactants may adsorb onto the CNTs surface (**Islam et al 2003**).

The solubilization of CNTs with biological components is certainly more appropriate towards integration of this new type of material with living systems (Figure 12). The biomacromolecules for non-covalent functionalization of CNTs has included simple saccharides and polysaccharides (Barone et al.; 2006; Star et al.; 2002; Chambers et al.; 2003& Ikeda et al.; 2007), proteins (Chen et al.; 2001), enzymes, DNA, etc. Various biomaterials such as *n*-decyl-*β*-Dmaltoside (Ishibashi et al.; 2006), γ-cyclodextrin (Chambers et al.; 2003), *η*-cyclodextrin (Dodzuik et al.; 2003), chitosan (Yan et al.; 2008), pullulan (Kim et al.; 2003) and phospholipid-dextran (Goodwin et al.; 2009) have been used for the non-covalent functionalization of CNTs. They have been aided due to the fact that such saccharides and polysaccharides have almost no light absorption in UV-vis wavelength region, so that the CNTs hybrids can be characterized by photochemical experiments and also the saccharide- and polysaccharide-coated CNTs hybrids are usually biocompatible and may be applicable for many medicinal purposes.

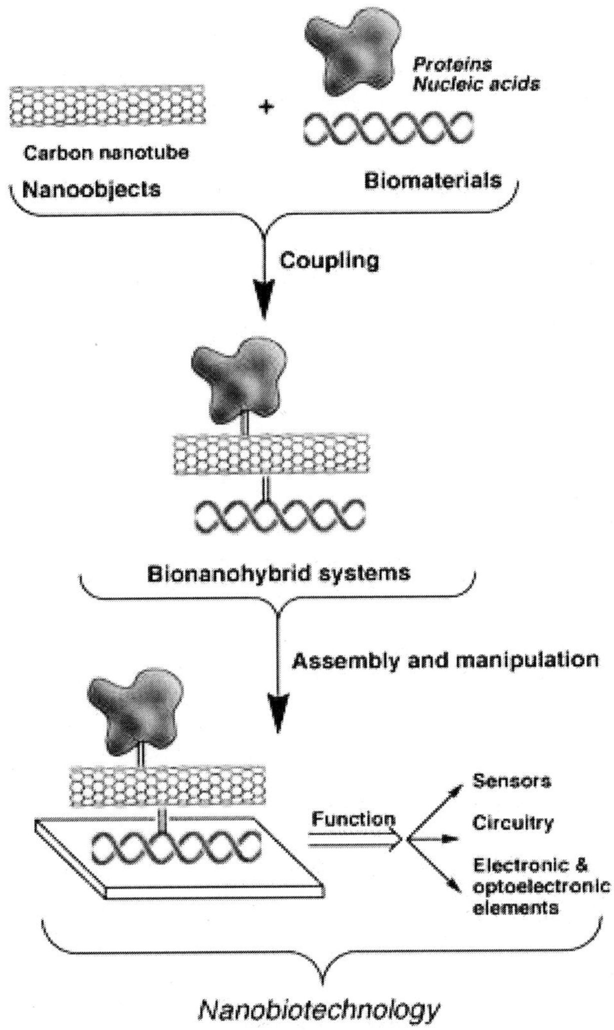

Figure 12. The conceptual generation of biomolecules-CNTs conjugates and their assembly to yield functional devices.

3.3. Alternative Routes For The Functionalization Of Carbon Nanotubes

Though various methods for enhancing the interaction between CNTs and polymers were adopted in the recent past, the two main routes; namely covalent and non-covalent functionalization generally provided. As discussed earlier, both methods had their own merits depending on the platform. Additionally, the traditional covalent functionalization strategy of CNT is most frequently initiated by chemical acid oxidation acid treatment. However, dramatic amounts of induced defects during functionalization hinder the intrinsic mobility of carriers along CNTs, which is not preferred in any case. This method not only functionalizes the nanotube surfaces with carboxylic acid groups but leaves behind detrimental structures, hence hampering their potential for practical applications and can also compromise the mechanical properties of the nanotubes. Therefore, as a common rule, and a now widespread approach to alleviate these problems is to find alternative routes such as an effective functionalization method that can not only introduce high density and homogenous surface functional groups, which enhance the compatibility between CNTs and the foreign matrix, but allow direct grafting and has little or no structural damage to the CNTs, thus, optimizing their properties for various applications.

To overcome this challenge, Baek et al. (2005; 2004 ; 2006; 2007; 2010) have reported an efficient route to covalently functionalize CNTs via direct Friedel-Crafts acylation technique (Figure 13). This kind of covalent grafting of the nanotubes is a promising strategy to not only improve nanotube dispersion but also provide a means for creating microscopic interlinks. On the whole, this kind of surface functionalization not only enhances the reactivity, but also improves the specificity and provides an avenue for further chemical modification of CNTs. Considerable achievements have been made in enhancing the various functionalities of CNT-polymer nanocomposites, generally not achievable for each of the components individually. The approach is conceptualized on the basis of our foray into the CNT chemistry using 'direct' Friedel-Crafts acylation technique which has superior operational simplicity. Not only a mild and an alternative route to functionalize CNTs, this strategy was previously shown to be a less-destructive and/or nondestructive reaction condition for the efficient dispersion and functionalization of carbon nanomaterials. As a result, CNT damage from severe chemical treatments including oxidation and sonication can be avoided to a larger extent. Thus, maximum enhanced properties can be expected from improved dispersion stability as well as chemical affinity with matrices.

Table 3. Advantage and disadvantage of various CNT functionalization methods (Ma et al.; 2010).

Method		Principle	Possible damage to CNTs	Easy to use	Interaction with polymer matrix[a]	Re-agglomeration of CNTs in matrix
Chemical method	Side wall	Hybridization of C atoms from sp^2 to sp^3	√	×	S	√
	Defect	Defect transformation	√	√	S	√
Physical method	Polymer wrapping	Van der Waals force, π–π stacking	×	√	V	×
	Surfactant adsorption	Physical adsorption	×	√	W	×
	Endohedral method	Capillary effect	×	×	W	√

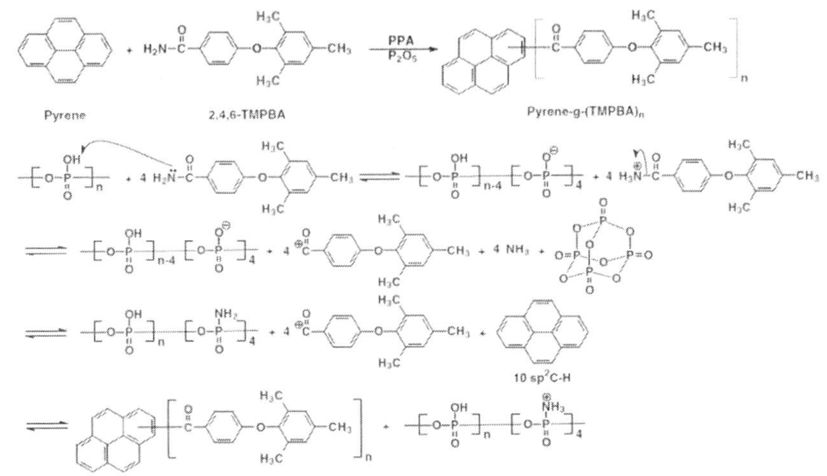

Figure 13. Friedel-Crafts acylation reaction of pyrene as a miniature graphene and organic material in poly(phosphoric acid)/phosphorous pentoxide medium (Jeon et al.; 2010).

4. CONCLUSION AND OUTLOOK

In light of the continuous progress of nanotechnology and material science over the last two decades, CNT based materials have opened new pathways for developing novel functional materials. In summary, CNTs were always been regarded as new and interesting type of materials, especially for energy and electronic applications. All types of CNTs are being investigated with equal importance. While few strategies have been developed so that the handling and manipulation of CNTs be relatively easier, on the whole there is much room for investigations in areas such as supercapacitors. Very recent reports show CNT based supercapacitor devices with better capacitor performance. Precisely, the combination of CNTs with particular macromolecules that offer to enhance the conductivity of the material is one interesting research. Particularly, the large electrochemical window and the environmental stability draw critical importance on such materials. Furthermore, the extension of these functional methods to the 2D forms of carbon namely graphene based materials are also now a fast growing area. While the quest for new materials are always on, the research on CNT based materials in special fields such as doping are still open for investigations and discussion.

ACKNOWLEDGEMENTS

This research work was supported by World Class University (WCU), US-Korea NBIT, and Mid-Career Researcher (MCR) programs through the National Research Foundation (NRF) of Korea funded by the Ministry of Education, Science and Technology (MEST) and the U.S. Air Force Office of Scientific Research (AFOSR).

REFERENCES

1. Radushkevich O.L.V. & Lukyanovich V.M. 1952 Zurn. Fisic. Chim. 28 88 95 .

2. A. Oberlin, M. Endo, T. Koyama, 1976 Filamentous growth of carbon through benzene decomposition. J. Cryst. Growth 32 335 349 .

3. S. Iijima, 1991 Helical microtubules of graphitic carbon. Nature 354 56 58 .

4. Y. S. Song, J. R. Youn, 2005 Influence of dispersion states of carbon nanotubes on physical properties of epoxy nanocomposites. Carbon 43 1378 1385 .

5. P. M. Ajayan, L. S. Schadler, C. Giannaris, A. Rubio, 2000 Single-Walled Carbon Nanotube-Polymer Composites: Strength and Weakness. Adv. Mater. 12 750 753 .

6. I. Mikhail, Katsnelson, 2007 Graphene: carbon in two dimensions. Materials Today 10 20 27 .

7. D. S. Bethune, C. H. Kiang, M. S. Devries, G. Gorman, R. Savoy, J. Vazquez, et al. 1993 Cobalt-catalyzed growth of carbon nanotubes with singleatomic-layerwalls. Nature 363 (6430), 605-607.

8. S. Iijima, T. Ichihashi, 1993 Single-shell carbon nanotubes of 1 -nm diameter. Nature 363, (6430), 603-605.

9. J. L. Hutchison, N. A. Kiselev, E. P. Krinichnaya, A. V. Krestinin, R. O. Loutfy, A. P. Morawsky, V. E. Muradyan, E. D. Obraztsova, J. Sloan, S. V. Terekhov, D. N. Zakharov, 2001 Double-walled carbon nanotubes fabricated by a hydrogen arc discharge method. Carbon 39 761 770 .

10. Y. Hou, J. Tang, H. Zhang, C. Qian, Y. Feng, J. Liu, 2009 Functionalized Few-Walled Carbon Nanotubes for Mechanical Reinforcement of Polymeric Composites. ACS Nano 3 1057 1062 .

11. N. Hamada, S. Sawada, A. Oshiyama, 1992 New one-dimensional conductors: Graphitic microtubules Phys. Rev. Lett. 68 1579 1581 .

12. T. Belin, F. Epron, 2005 Characterization methods of carbon nanotubes: A review. Mater. Sci. Eng., B 119, (2), 105-118.

13. V. N. Popov, 2004 Carbon nanotubes: Properties and application. Mater. Sci. Eng., R 43 (3), 61-102.

14. D. Qian, G. J. Wagner, W. K. Liu, M. F. Yu, R. S. Ruoff, 2002 Mechanics of carbon nanotubes. Appl. Mech. Rev. 55 495 533 .

15. S. Xie, W. Li, Z. Pan, B. Chang, L. Sun, 2000 Mechanical and physical properties on carbon nanotube. J. Phys. Chem. Solids 61, (7), 1153-1158.

16. L. S. Schadler, 2004 Polymer-Based and Polymer-Filled Nanocomposites, in Nanocomposite Science and Technology (eds P. M. Ajayan, L. S. Schadler and P. V. Braun), Wiley-VCH Verlag GmbH & Co. KGaA, Weinheim.

17. J. A. Elliott, J. K. W. Sandler, A. H. Windle, R. J. Young, M. S. P. Shaffer, 2004 Collapse of single-wall carbon nanotubes is diameter dependent. Phys. Rev. Lett. 92(9), 1-4.

18. B. M. Kim, A. M. S. Fuhrer, 2004 Properties and applications of high-mobility semiconducting nanotubes. J. Phys: Condens. Matter 16(18) R553 -R580.

19. Z. K. Tang, L. Y. Zhang, N. Wang, X. X. Zhang, G. H. Wen, G. D. Li, et al. 2001 Superconductivity in 4 Angstrom single-walled carbon nanotubes. Science 292, (5526), 2462-2465.

20. J. W. , T. Cagin, W. A. Goddard, 2000 Thermal conductivity of carbon nanotubes. Nanotechnology 11, (2), 65-69.

21. M. A. Osman, D. Srivastava, 2001 Temperature dependence of the thermal conductivity of single-wall carbon nanotubes. Nanotechnology 12, (1), 21-24.

22. J. Hone, M. C. Llaguno, M. J. Biercuk, A. T. Johnson, B. Batlogg, Z. Benes, et al. 2002 Thermal properties of carbon nanotubes and nanotube-based materials. Appl. Phys. A-Mater. Sci. Process 74, (3), 339-343.

23. E. T. Thostenson, Z. F. Ren, T. W. Chou, 2001 Advances in the science and technology of CNTs and their composites: a review. Compos. Sci. Technol. 61 1899 1912 .

24. P. M. Ajayan, L. S. Schadler, P. V. Braun, Nanocomposite science and technology. Weinheim: Wiley-VCH; 2003 77 80 .

25. J. N. Coleman, U. Khan, Y. K. Gunko, 2006 Mechanical reinforcement of polymers using carbon nanotubes. Adv. Mater. 18 689 706 .

26. C. Journet, W. K. Maser, P. Bernier, A. Loiseau, M. L. de la Chapelle, S. Lefrant, et al. 1997 Large-scale production of single-walled carbon nanotubes by the electric-arc technique Nature 388 756 758 .

27. A. G. Rinzler, J. Liu, H. Dai, P. Nikolaev, C. B. Huffman, Macias. F. J. Rodriguez, et al. 1998 Large-scale purification of single-wall carbon nanotubes: process, product, and characterization. Appl.Phys. A 1998 67 29 37 .

28. P. Nikolaev, M. J. Bronikowski, R. K. Bradley, F. Fohmund, D. T. Colbert, K. A. Smith, et al. 1999 Gas-phase catalytic growth of single-walled carbon nanotubes from carbon monoxide. Chem. Phys. Lett. 313 91 97 .

29. Z. F. Ren, Z. P. Huang, J. W. Xu, D. Z. Wang, J. G. Wen, J. H. Wang, et al. 1999 Growth of a single freestanding multiwall carbon nanotube on each nanonickel dot. Appl. Phys. Lett. 75 1086 1088 .

30. Z. F. Ren, Z. P. Huang, J. W. Xu, J. H. Wang, P. Bush, M. P. Siegal, et al. 1998 Synthesis of Large Arrays of Well-Aligned Carbon Nanotubes on Glass. Science. 282 1105 1107 .

31. C. H. See, A. T. Harris, 2007 A Review of Carbon Nanotube Synthesis via Fluidized-Bed Chemical Vapor Deposition. Ing. Eng. Chem. Res. 46 997 1012 .

32. M. S. Strano, C. A. Kyke, M. L. Usrey, P. W. Barone, M. J. Allen, H. Shan, C. Kittrell, R. H. Hauge, J. M. Tour, R. E. Smalley, 2003 Electronic

Structure Control of Single-Walled Carbon Nanotube Functionalization Science 301 1519 1522 .

33. H. Li, B. Zhou, Y. Lin, L. Gu, W. Wang, K. A. S. Fernando, S. Kumar, L. F. Allard, Y. Sun, P. , 2004 Selective Interactions of Porphyrins with Semiconducting Single-Walled Carbon Nanotube. J. Am. Chem. Soc. 126 1014 1015 .

34. S. Niyogi, M. A. Hamon, H. Hu, B. Zhao, P. Bhowmik, R. Sen, M. E. Itkis, R. C. Haddon, 2002 Chemistry of Single-Walled Carbon Nanotubes. Acc. Chem. Res. 35, (12), 1105-1113.

35. D. Tasis, N. Tagmatarchis, V. Georgakilas, M. Prato, 2003 Soluble Carbon Nanotubes. Chem. Eur. J. 9, (17), 4000-4008.

36. A. Hirsch, 2002 Functionalization of Single-Walled Carbon Nanotubes. Angew, Chem, Int. Ed. 41, (11), 1853-1859.

37. S. B. Sinnott, 2002 Chemical Functionalization of Carbon Nanotubes. Journal of Nanoscience and Nanotechnology J. Nanosc. Nanotechnol. 2 113 123 .

38. A. Bianco, M. Prato, K. Kostarelos, A. Bianco, 2008 Functionalized Carbon Nanotubes in Drug Design and Discovery. Acc. Chem. Res. 41, (1), 60-68.

39. E. T. Mickelson, C. B. Huffman, A. G. Rinzler, R. E. Smalley, R. H. Hauge, J. L. Margrave, 1998 Fluorination of single-wall carbon nanotubes. Chem. Phys. Lett. 296, (1-2), 188-194.

40. K. F. Kelly, I. W. Chiang, E. T. Mickelson, et al. 1993 Insight into the mechanism of sidewall functionalization of single-walled nanotubes: an STM study. Chem. Phys. Lett. 313 445 450

41. H. Touhara, J. Inahara, T. Mizuno, et al. 2002 Fluorination of cup-stacked carbon nanotubes, structure and properties. Fluorine Chem.114 181 188 .

42. J. L. Stevens, A. Y. Huang, H. Peng, et al. 2003 Sidewall amino-functionalization of SWNTs through fluorination and subsequent reactions with terminal diamines. Nano Lett. 3 331 336 .

43. H. Hu, B. Zhao, M. A. Hamon, et al. 2003 Sidewall functionalization of single-walled carbon nanotubes by addition of dichlorocarbene. J. Am. Chem. Soc.125 14893 14900 .

44. E. Unger, A. Graham, F. Kreupl, M. Liebau, W. Hoenlein, 2002 Electrochemical functionalization of multi-walled carbon nanotubes for solvation and purification. Curr. Appl. Phys. 2 107 111 .

45. K. S. Kim, D. J. Bae, J. R. Kim, K. A. Park, S. C. Lim, J. J. Kim, W. B. Choi, C. Y. Park, Y. H. Lee, 2002 Modification of Electronic Structures of a Carbon Nanotube by Hydrogen Functionalization. Adv. Mater. 14, (24), 1818-1821

46. N. Tagmatarchis, M. J. Prato, 2004 Functionalization of carbon nanotubes via 1 3 -dipolar cycloadditions. J. Mater. Chem. 14, 437-439.

47. J. Chen, M. A. Hamon, H. Hu, Y. Chen, A. M. Rao, P. C. Eklund, R. C. Haddon, 1998 Solution Properties of Single-Walled Carbon Nanotubes. Science 282, (5386), 95-98.

48. K. Esumi, M. Ishigami, A. Nakajima, K. Sawada, H. Honda, 1996 Chemical treatment of carbon nanotubes. Carbon 34 279 281 .

49. J. Liu, A. G. Rinzler, H. Dai, J. H. Hafner, R. K. Bradley, P. J. Boul, et al. 1998 Fullerene Pipes. Science 280 1253 1256 .

50. R. Yu, L. Chen, Q. Liu, J. Lin, K. Tan, L. , S. C. Ng, H. S. O. Chan, G. Xu, Q. , T. S. A. Hor, 1998 Platinum Deposition on Carbon Nanotubes via Chemical Modification. Chem. Mater. 10 718 722 .

51. M. Sham, L. , J. Kim, K. , 2006 Surface functionalities of multi-wall carbon nanotubes after UV/Ozone and TETA treatments. Carbon 44, (4), 768-777.

52. S. C. Wang, K. S. Chang, C. J. Yuan, 2009 Enhancement of electrochemical properties of screen-printed carbon electrodes by oxygen plasma treatment. Electrochimica Acta 54, (21), 4937-4943.

53. P. C. , J. K. Kim, B. Z. Tang, 2006 Functionalization of carbon nanotubes using a silane coupling agent. Carbon 44 3232 3238 .

54. M. Sano, A. Kamino, J. Okamura, S. Shinkai, 2001 Self-Organization of PEO-graft-Single-Walled Carbon Nanotubes in Solutions and Langmuir–Blodgett Films. Langmuir 17, (17), 5125-5128.

55. H. Kong, C. Gao, D. Yan, 2003 Controlled Functionalization of Multiwalled Carbon Nanotubes by in Situ Atom Transfer Radical Polymerization. J. Am. Chem. Soc. 126 412 413 .

56. M. A. Hamon, H. Hui, P. Bhowmik, 2002 Ester-functionalized soluble single-walled carbon nanotubes. Appl. Phys. A. 74 333 338 .

57. J. N. Coleman, U. Khan, Y. K. Gun'ko, 2006 Mechanical Reinforcement of Polymers Using Carbon Nanotubes. Adv. Mater. 18 689 706 .

58. P. , C. , N. A. Siddiqui, G. Marom, J. Kim, K. , 2010 Dispersion and functionalization of carbon nanotubes for polymer-based nanocomposites: A review. Composites: Part A 41 1345 1364 .

59. D. A. Heller, P. W. Barone, M. S. Strano, 2004 Sonication-induced changes in chiral distribution: A complication in the use of single-walled carbon nanotube fluorescence for determining species distribution. Carbon 43 651 673 .

60. M. Monthioux, B. W. Smith, B. Burteaux, A. Claye, J. E. Fischer, D. E. Luzzi, 2001 Sensitivity of single-wall carbon nanotubes to chemical processing: an electron microscopy investigation. Carbon 39, (8), 1251-1272.

61. R. J. Chen, Y. Zhang, D. Wang, H. Dai, 2001 Noncovalent Sidewall Functionalization of Single-walled carbon nanotubes for protein immobilization. J. Am. Chem. Soc. 123 3838 3839 .

62. D. S. Hecht, R. J. A. Ramirez, M. Briman, E. Artukovic, K. S. Chichak, J. F. Stoddart, G. Gruner, 2006 Bioinspired detection of light using a porphyrin-sensitized singlewall nanotube field effect transistor. Nano Lett. 6 2031 2036 .

63. L. Hu, Y. Zhao, L. , K. Ryu, C. Zhou, J. F. Stoddart, G. Gruner, Light-induced charge transfer in pyrene/CdSe-SWNT hybrids. Adv. Mater. 2008 2008 20 939 946 .

64. Y. Zhao, L. , J. F. Stoddart, 2009 Noncovalent Functionalization of Single-Walled Carbon Nanotubes. Acc. Chem. Res. 42, (8), 1161-1171

65. A. Star, J. F. Stoddart, D. Steuerman, M. Diehl, A. Boukai, E. W. Wong, X. Yang, S. Chung, W. , H. Choi, J. R. Heath, 2001 Preparation and Properties of Polymer-Wrapped Single-Walled Carbon Nanotubes. Angew. Chem., Int. Ed. 2001 40 1721 1725 .

66. D. W. Steuerman, A. Star, R. Narizaano, H. Choi, R. S. Ries, C. Nicolini, J. F. Stoddart, J. R. Heath, 2002 Interactions between conjugated polymers and singlewalled carbon nanotubes. J. Phys. Chem. B 106 3124 3130 .

67. A. Star, Y. Liu, K. Grant, L. Ridvan, J. F. Stoddart, D. W. Steuerman, M. R. Diehl, A. Boukai, J. R. Heath, 2003 Noncovalent side-wall functionalization of singlewalled carbon nanotubes. Macromolecules 36 553 560 .

68. A. Star, J. F. Stoddart, 2002 Dispersion and solubilization of single-walled carbon nanotubes with a hyperbranched polymer. Macromolecules 35 7516 7520 .

69. F. Cheng, A. Adronov, 2006 Noncovalent functionalization and solubilization of carbon nanotubes by using a conjugated Zn-porphyrin polymer. Chem. Eur. J. 12 5053 5059 .

70. W. Yi, A. Malkovskiy, Q. Chu, A. P. Sokolov, M. L. Colon, M. Meador, Y. Pang, 2008 Wrapping of single-walled carbon nanotubes by a π-conjugated polymer: The role of polymer conformation-controlled size selectivity. J. Phys. Chem. B 112 12263 12269 .

71. X. Gong, J. Liu, S. Baskaran, R. D. Voise, J. S. Young, 2000 Surfactant-Assisted Processing of Carbon Nanotube/Polymer Composites. Chem. Mater. 12 (4), 1049-1052.

72. L. Vaisman, G. Marom, H. D. Wagner, 2006 Dispersions of Surface-Modified Carbon Nanotubes in Water-Soluble and Water-Insoluble Polymers. Adv. Funct. Mater. 16 (3), 357-363

73. M. F. Islam, E. Rojas, D. M. Bergey, A. T. Johnson, A. G. Yodh, 2003 High Weight Fraction Surfactant Solubilization of Single-Wall Carbon Nanotubes in Water. Nano Lett. 3 (2), 269-273

74. J. Yu, N. Grossiord, C. E. Koning, J. Loos, 2007 Controlling the dispersion of multi-wall carbon nanotubes in aqueous surfactant solution. Carbon 45 (3), 618-623.

75. E. A. Whitsitt, A. R. Barron, 2003 Silica Coated Single Walled Carbon Nanotubes. Nano Lett. 3, (6), 775-778.

76. T. H. Kim, C. Doe, S. R. Kline, S. M. Choi, 2007 Water-Redispersible Isolated Single-Walled Carbon Nanotubes Fabricated by In Situ Polymerization of Micelles. Adv. Mater. 19, (7), 929-933.

77. C. Klumpp, K. Kostarelos, M. Prato, A. Bianco, 2006 Functionalized carbon nanotubes as emerging nanovectors for the delivery of therapeutics. Biochim. Biophys. Acta, (BBA)- Biomembranes 1758, (3), 404-412.

78. P. W. Barone, M. S. Strano, 2006 Reversible Control of Carbon Nanotube Aggregation for a Glucose Affinity Sensor. Angew. Chem., Int. Ed. 45 8138 8141 .

79. A. Star, D. W. Steuerman, J. R. Heath, J. F. Stoddart, (200, carbon. Starched, Angew. nanotubes, Chem, Int, 41 41 2508 2512 .

80. G. Chambers, C. Carroll, G. F. Farrell, A. B. Dalton, M. Mc Namara, het. in, M. Panhuis, H. J. Byrne, 2003 Characterization of the interaction of gamma cyclodextrin with single-walled carbon nanotubes. Nano Lett. 3 843 846 .

81. M. Ikeda, T. Hasegawa, M. Numata, K. Sugikawa, K. Sakurai, M. Fujiki, S. Shinkai, 2007 Instantaneous inclusion of a polynucleotide and hydrophobic guest molecules into a helical core of cationic 1 3 - glucan polysaccharide. J. Am. Chem. Soc. 129, 3979-3988.

82. R. J. Chen, Y. Zhang, D. Wang, H. Dai, 2001 Single-walled carbon nanotubes for protein immobilization. J. Am. Chem. Soc. 123 3838 3839 .

83. A. Ishibashi, N. Nakashima, 2006 Individual dissolution of single-walled carbon nanotubes in aqueous solution of steroid or sugar compounds and their Raman and Near-IR spectral properties. Chem. Eur. J. 12 7595 7602 .

84. H. Dodziuk, A. Ejchart, W. Anczewski, H. Ueda, E. Krinichnaya, G. Dolgonos, W. Kutner, 2003 Water solubilization, determination of the number of different types of single-wall carbon nanotubes and their partial separation with respect to diameters by complexation with η-cyclodextrin. Chem. Commun. 986 987 .

85. L. Y. Yan, Y. F. Poon, M. B. Chan-Park, Y. Chen, Q. Zhang, 2008 Individually dispersing single-walled carbon nanotubes with novel neutral pH water-soluble chitosan derivatives. J. Phys. Chem. C 112 7579 7587 .

86. O. Kim, K. , J. Je, J. W. Baldwin, S. Kooi, P. E. Pehrsson, L. J. Buckley, 2003 Solubilization of single-walled carbon nanotubes by supramolecular encapsulation of helical amylase. J. Am. Chem. Soc. 125 4426 4427 .

87. A. P. Goodwin, S. M. Tabakman, K. Welsher, S. P. Sherlock, G. Prencipe, H. Dai, 2009 Phospholipid-dextran with a single coupling point: A useful amphiphile for functionalization of nanomaterials. J. Am. Chem. Soc. 131 289 296 .

88. J. Baek, B. , C. B. Lyons, L. Tan, S. , 2004 Covalent modification of vapour-grown carbon nanofibers via direct Friedel-Crafts acylation in polyphosphoric acid. J. Mater, Chem. 14 (13), 2052-2056.

89. J. Baek, B. , C. B. Lyons, L. Tan, S. , 2004 Grafting of Vapor-Grown Carbon Nanofibers via in-Situ Polycondensation of 3 - Phenoxybenzoic Acid in Poly(phosphoric acid). Macromolecules 37 (22), 8278-8285.

90. H. Lee, J. , S. Oh, J. , J. Choi, Y. , J. W. Kim, J. Han, L. Tan, S. , J. Baek, B. , 2005 In Situ Synthesis of Poly(ethylene terephthalate) (PET) in Ethylene Glycol Containing Terephthalic Acid and Functionalized Multiwalled Carbon Nanotubes (MWNTs) as an Approach to MWNT/PET Nanocomposites. Chem. Mater. 17, (20), 5057-5064.

91. S. Oh, J. , H. Lee, J. , D. Keum, K. , S. Lee, W. , D. H. Wang, S. Park, Y. , L. Tan, S. , J. Baek, B. , 2006 Multiwalled carbon nanotubes and nanofibers grafted with polyetherketones in mild and viscous polymeric acid. Polymer 47, (4), 1132-1140.

92. J. Choi, Y. , S. Oh, J. , H. Lee, J. , D. H. Wang, L. Tan, S. , J. Baek, B. , 2007 In-Situ Grafting of Hyperbranched Poly(ether ketone)s onto Multiwalled Carbon Nanotubes via the A3 + B2 Approach. Macromolecules 40, (13), 4474-4480.

93. J. Choi, Y. , S. Han, W. , W. Huh, S. , L. Tan, S. , J. Baek, B. , 2007 In situ grafting of carboxylic acid-terminated hyperbranched poly(ether-ketone) to the surface of carbon nanotubes. Polymer 48, (14), 4034-4040.

94. D. H. Wang, M. J. Arlen, J. Baek, B. , L. Tan, S. , 2007 Nanocomposites Derived from a Low-Color Aromatic Polyimide (CP2) and Amine-Functionalized Vapor-Grown Carbon Nanofibers: In Situ Polymerization and Characterization. Macromolecules 40, (17), 6100-6111.

95. I. Jeon, Y. , E. Choi, K. , S. Bae, Y. , J. Baek, B. , 2010 Edge-Functionalization of Pyrene as a Miniature Graphene via Friedel-Crafts Acylation Reaction in Poly(Phosphoric Acid). Nanoscale Res. Lett. 5, (10), 1686-1691.

7

INSIGHT INTO THE BROAD FIELD OF POLYMER NANOCOMPOSITES: FROM CARBON NANOTUBES TO CLAY NANOPLATELETS, VIA METAL NANOPARTICLES

Eduard A. Stefanescu [1,*], Codrin Daranga [2] and Cristina Stefanescu [3,*]

[1]Department of Chemical & Life Science Engineering, Virginia Commonwealth University, Richmond, VA 23284, USA

[2]Department of Civil & Environmental Engineering, University of Wisconsin, Madison, WI 53706, USA

[3]Department of Chemistry, Louisiana State University, Baton Rouge, LA 70803, USA

ABSTRACT

Highly ordered polymer nanocomposites are complex materials that display a rich morphological behavior owing to variations in composition, structure, and properties on a nanometer length scale. Metal-polymer nanocomposite materials are becoming more popular for applications requiring low cost, high metal surface areas. Catalytic systems seem to be the most prevalent application for a wide range of metals used in polymer nanocomposites, particularly for metals like Pt, Ni, Co, and Au, with known catalytic activities. On the other hand, among the most frequently utilized techniques to prepare polymer/CNT and/or polymer/clay nanocomposites are approaches like melt mixing, solution casting, electrospinning and solid-state shear pulverization.

Additionally, some of the current and potential applications of polymer/CNT and/or polymer/clay nanocomposites include photovoltaic devices, optical switches, electromagnetic interference (EMI) shielding, aerospace and automotive materials, packaging, adhesives and coatings. This extensive review covers a broad range of articles, typically from high impact-factor journals, on most of the polymer-nanocomposites known to date: polymer/carbon nanotubes, polymer/metal nanospheres, and polymer/clay nanoplatelets composites. The various types of nanocomposites are described form the preparation stages to performance and applications. Comparisons of the various types of nanocomposites are conducted and conclusions are formulated.

KEYWORDS

carbon nanotube; laponite; montmorillonite; metal; nanocomposite

1. INTRODUCTION TO POLYMER NANOCOMPOSITES

Highly ordered polymer nanocomposites are complex materials that display a rich morphological behavior because of variations in composition, structure, and properties on a nanometer length scale [1,2,3,4]. Novel physical properties of soft and bulk polymer nanocomposite materials are also dependent on the supramolecular organization of the nanostructures [5]. The presence of the nanoparticle and the interaction of the polymer with the particle, as well as the particle orientation in a dispersed fluid composition may lead to a variety of ordered composite materials in the bulk or film [6,7,8]. Incorporation of metals into polymer matrixes has been shown to produce unique optical, magnetic and dielectric properties at the nano- and macro-scale in nanocomposites, apparently due to the surface and confinement effects of the nanoparticles [9,10,11,12,13,14,15]. The preparation of magnetic nanocomposites that comprise polymer shells and magnetic nanoparticles constitutes a very attractive approach since the modular nature of polymeric materials facilitates the design of a wide range of hybrid nanocomposites of various compositions [10]. Additionally, the inherent dipole moment in ferromagnetic colloids enables the one- and two-dimensional assembly of such materials into novel mesostructures. The development of magnetic assemblies from ferromagnetic and superparamagnetic nanoparticles on supporting surfaces has been demonstrated in the past. Some of the structures that have been shown to occur as a result of magnetic assembly include 1-D chains [16], flux closure rings [17], 2-D superlattices of closed packed nanocrystals [18], and 3-D labyrinth-like suprastructures [19].

During the past decade carbon nanotubes (CNTs) have attracted much attention for their ability to increase electrical conductivity and thermo-mechanical properties of polymers at relatively low content [20,21,22,23,24,25,26,27,28]. For example, much of the literature describing conductive polymer/CNT nanocomposites is mainly focused on decreasing the percolation threshold in order to reduce the cost and improve the physical properties and processability of composites [29,30]. CNTs are characterized by very high aspect ratios, having diameters in the order of a few nanometers and lengths in the range of hundreds nanometers. It is known that, despite their light weight characteristics, CNTs are roughly 10 to 100 times stronger than the strongest steel [21]. One of the main challenges in creating high performance polymer/CNT nanocomposites consists in one's ability to obtain an optimum dispersion and a proper adhesion of CNTs within the polymer matrix [8,30,31,32,33]. A proper dispersion is often times impossible to achieve with pristine carbon nanotubes because of their small diameter and completely hydrophobic surface [34]. These practical difficulties are most times overcome by functionalizing the surface of CNTs with small polar groups or with pre-polymers that are either identical or structurally similar to the matrix material that needs to be reinforced [35,36]. Various processing methods have been designed to produce polymer/CNT nanocomposites. For example, solution processing can be used if the polymer is soluble in a solvent that is capable of suspending the CNTs [28,35,37,38]. Melt processing is also available for thermoplastic polymers, where the nanotubes are mechanically sheared and compounded in the molten polymer matrix [27,34,39]. Some of the current and potential applications of polymer/CNT nanocomposites include photovoltaic devices, optical switches, electromagnetic interference (EMI) shielding, aerospace and automotive materials, bicycle and tennis racquet frames, packaging, adhesive and coatings [21].

On the other hand, a large number of metals have been utilized in the past to produce polymer composites with distinct characteristics. For example, polyaniline (PANI)/platinum (Pt) nanocomposites are extensively studied for the extraordinary ability of the metal to catalyze redox reactions [40]. Gold (Au) nanoparticles have attracted considerable attention for their catalytic properties towards carbon monoxide and methanol oxidation [41]. Palladium (Pd) and its polymer-based composites represent an important catalytic class for the formation of C-C bonds, while exhibiting tolerance for several functional groups such as carbonyl and hydroxyl [42]. Cobalt (Co) is a versatile ferromagnetic metal used in the fabrication of magnetic, wear-resistant alloys and composites [10,12,43,44]. Furthermore, nickel (Ni) is a magneto-selective material characterized by negative magneto-striction [45,46]. Nanocrystalline Ni/polyvinylidene fluoride (PVDF) composites exhibit a high dielectric constant at very small metal concentrations [14]. In addition, silver (Ag) particles characterized by spherical shapes exhibit a very strong absorption band in the visible region due to their surface plasmon resonance [47]. Owing to their extraordinary characteristics, polymer/metal nanocomposites can be utilized for the design of several advanced devices that include optical limiters and polarizers, optical sensors, magneto-optical modulators, electro-optical filters, superparamagnetic materials, electro-catalytic electrodes, chemical absorbers, and so on [20].

In addition to metal- and CNT-composites, in recent years a broad literature has emerged that examines the fundamental relationships between network structure, chain dynamics, ionic conductivity and dimensional stability, in cross-linked nanosized polymer-clay networks [5,48,49,50,51,52,53,54]. Perhaps the two most discussed clays in literature are laponite and montmorillonite. Laponite clay is an inexpensive and environmentally benign disc-shaped silicate with a plate diameter of 25–30 nm and a thickness of approximately 1nm [54,55]. In aqueous solutions and gels hydrophilic polymers strongly adsorb to the charged laponite nanoparticles leading to the formation of transparent systems. Smectites such as montmorillonite are 2:1 charged phyllosilicates that contain exchangeable interlayer cations and show the ability to intercalate various polymers [6]. Montmorillonite clay produces an opaque suspension of predominantly exfoliated platelets that range on average in size from ca. 70 to 100 nm across and are approximately 1 nm thick [5]. When dispersed alone in water such clays exhibit a Newtonian behavior, but in the presence of polymers the interaction between the polymer chains and the particles causes a major change in the rheological behavior of dispersions [56]. Flow birefringence studies demonstrated that upon shear the clay platelets orient along the flow direction [3,4]. In solution these clay particles can only adsorb a maximum amount of polymer until all the clay surfaces are covered [1]. The polymer and clay build a network-like structure which is interpenetrated by a sub-network of interconnecting pores containing excess polymer and water. During the layer by layer film preparation method the exfoliated solution-structure collapses, reorders and re-intercalates into blob-like chains and layers [5].

The aim of the present paper is to provide an extended review of the recent scientific methodologies utilized to prepare and characterize polymer nanocomposites containing metals, carbon nanotubes and clay nanoplatelets as reinforcing agents. Keeping in mind the importance of nanotechnology, our efforts were concentrated on gathering information from some of the best journals in the area of polymer composites, with the purpose of helping the reader stay up-to-date with the current cutting-edge developments in the field. The reader is cautioned, however, that although the review contains a great deal of information, it is far from being an exhaustive source of information in this matter.

2. POLYMER/CARBON NANOTUBES NANOCOMPOSITES

Following the recent developments in the fields of nanotechnology, carbon nanotube (CNT) reinforced polymer composites received tremendous attention in the past decades, and currently occupies an elite place among the most studied polymer composites. When properly dispersed within polymer matrixes, CNTs have the ability to improve the properties of the resultant materials several orders of magnitude relative to the unfilled polymers. The enhanced properties may include tensile behavior, strength, toughness, stiffness, electrical and thermal conductivity and crystallization kinetics. The focus of the present section is to bring together, in a coherent way, various aspects of the recently published research in the field of polymer/CNT nanocomposites. The polymer matrixes that are covered in this section are among the most researched and utilized in academia and industry, and include polystyrene, poly(ethylene oxide), poly(ε-caprolactone), polypropylene, various nylons andpoly(ethylene terephthalate).

2.1. Polystyrene/Carbon Nanotubes Composites

Xu *et al.* used an electrostatic assembly to synthesize polystyrene (PS)-carbon nanofibers (CNF) via a bottom-up method [57]. The authors mixed a home-made PS latex with an aqueous suspension of oxidized

CNF, and obtained the nanocomposites through the electrostatic interaction of the cationic PS with the anionic CNF. They claim that the molding temperature has an important effect on the morphology and electrical conductivity of PS-CNF composites, where the optimal temperature and pressure of 185 °C and 25 MPA, respectively, result in a percolation threshold below 2 wt%. Thermal analysis measurements showed that the presence of the CNF increases the onset and the thermo-oxidative temperature by 60 °C. However, the CNF do not alter the glass-transition temperature (T_g) of the polymer. Similar results were obtained by Wang *et al.*, who prepared a composite of syndiotactic PS (sPS) filled with carbon nanocapsules (CNC) using a solution-blending method coupled with ultrasonication [58]. Just like Xu *et al.*, they observed that the filler addition had no impact on the Tg of the polymer, but that it led to an increase in the melt-crystallization temperature of the β–form sPS. In addition, the presence of the filler triggered a fast crystallization of the sPS, induced either by melt-quenching or by slow cooling of the molten state, which was further attributed to the existence of a profound primary nucleation density. The authors concluded that CNCs are excellent nucleating agents that improve the crystallinity and thermal stability of sPS. Likewise, Shen *et al.* used CNFs as nucleating agents to produce PS microcellular nanocomposite foams with uniform size distributions [59]. They found that small amounts of CNFs could drastically reduce the composite's cell dimensions and increase the cellular density. The favorable surface and geometrical characteristics of CNFs were found to be the main reasons for the excellent nucleation efficiency of the filler.

On the other hand, combining polystyrene with single-wall carbon nanotubes (SWCNT) Mu *et al.* produced electrically conductive composites with good rheological and mechanical properties [60]. In short, they created a contiguous SWCNT cellular-structure by coating PS pellets with SWCNTs and pressing them afterwards at elevated temperatures. When compared to nanocomposites with well-dispersed SWCNTs, the novel SWCNT cellular-structures proved to be better electric conductors, showing a lower electrical percolation threshold. Such PS composites having the SWCNTs well dispersed in the polymer matrix are described by Chang *et al.* [61]. The authors varied the concentration of SWCNTs and analyzed the morphologies, as well as the electrical and mechanical properties, of the resultant nanocomposites. They observed that composites that were subjected to an annealing treatment exhibited higher electrical conductivities than the raw composites. However, even after annealing, the SWCNT/PS composites described by Chang *et al.* displayed lower electrical conductivities than the PS/SWCNT cellular-structures described by Mu *et al.* It has been suggested that, in general, it is difficult to disperse SWCNTs in polymers since the energy of mixing is endothermic up to nanotube diameters of 2.2 nm [62]. Furthermore, using a "grafting to" technique Xie *et al.* prepared composites of SWCNTs with styrene copolymers [63]. In their method the reaction occurred at relatively mild conditions between the alkyne groups present on the SWCNTs and the benzyl chloride groups from the styrene copolymers. This coupling reaction significantly improved the bonding density at the polymer/SWCNT interface, with an overall efficiency of 81 wt% as revealed by thermogravimetric analysis (TGA). The authors claimed that the steric hindrance created by the polymer, which was physically adsorbed at the surface of SWCNTs, limited the number of cross-links between individual SWCNTs. At last, Putz *et al.* studied the effect of interfacial interactions in functionalized SWCNT/PS nanocomposites and found that for concentrations less than 0.1% of SWCNTs the elasticity of the composites remained comparable to the one of pure PS [64]. As the concentration of SWCNTs increased, however, a more pronounced elastic behavior of nanocomposites was observed. The authors correlated their findings with theoretical calculations and found significant inconsistencies in the resultant values.

Often times highly conductive polymer/CNT composites lose their conductivity after melt processing. Using multi-wall carbon nanotubes (MWCNT) and PS, Cipriano *et al.* designed an efficient route to obtain highly conductive extruded plastics, of which conductivity could be recovered after processing through annealing at temperatures higher than Tg of the polymer [65]. The authors suggested that the conductivity recovery involved a transition of the unconnected particles, which were aligned in the flow direction during extrusion-processing, to an interconnected network after annealing. In line with this idea, they performed a series of electrical measurements on commercial polydisperse polystyrene PS/MWCNT composites and observed very high conductivities (ca. 1 S/m) at a filler loading of 8 vol% [66]. The same group studied the impact of the aspect ratio of MWCNTs on the rheology and flammability of extruded PS/MWCNT nanocomposites and found that MWCNTs with larger aspect-ratios triggered higher storage moduli and complex viscosities in nanocomposites [67]. Additionally, the larger aspect-ratio nanocomposites were more efficient at reducing flammability, since they resulted in a reduced mass-loss of the composites subjected to combustion.

Cui *et al.* used reversible addition fragmentation chain-transfer (RAFT) to synthesize polymer-connected-MWCNT nanocomposites by grafting a

RAFT agent, and subsequently PS chains, on the surface of MWCNTs [68]. Figure 1 describes the steps used by the authors to prepare the nanotube-attached RAFT agents. After being synthesized, the nanotube-attached RAFT agents were further reacted with the styrene monomers in the presence of AIBN and THF to obtain the final polymer composites. The grafting of the PS chains to the surface of the MWCNTs led to the formation of core-shell structures. The authors claim that their method can be used to synthesize other polymer/CNT systems, including composites based on copolymers of acrylate and styrene-type monomers. In fact two years after their initial report the same group synthesized a composite of MWCNT and poly(methylmethacrylate)-block-polystyrene (PMMA-b-PS) following a similar route [69].

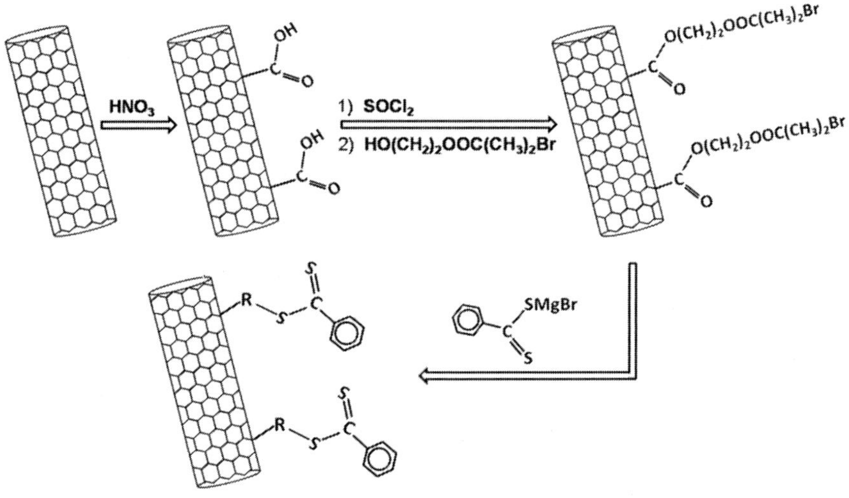

Figure 1. Schematic showing the steps used by Cui *et al.* to synthesize nanotube-attached RAFT agents.

2.2. Poly(Ethylene Oxide)/Carbon Nanotubes Composites

Chatterjee *et al.* have prepared and characterized poly(ethylene oxide) (PEO) nanocomposites containing SWCNTs [33]. The authors used a lithium based surfactant to improve the dispersion of SWCNTs in the polymeric matrix. Their composites exhibited an electrical percolation of 0.03 wt% and a geometrical percolation of 0.09 wt%. Inside the polymer matrix the SWCNTs were found to be either under tension or strongly interacting with the PEO-surfactant complex. At very small loadings (ca. 0.2 wt%) the filler was capable of significantly lowering the PEO crystallinity and crystallization temperature. Furthermore, the same group studied later, via rheological measurements, the response of fractal networks in SWCNT/PEO nanocomposites [32]. They found that the composites exhibited a strong shear-thinning behavior. Additionally, for systems corresponding to the semidilute regime the authors observed that the constant rate shear triggered a stress response that went through a maxima and decayed to steady values at long times. They attributed the stress maxima to the SWCNT network superstructure, since the molecular weight (Mw) of the PEO used was too low to trigger a chain entanglement that would result in a stress overshoot. Single-wall carbon nanotubes have been also used as fillers in amphiphilic cross-linked PEO-fluoropolymer networks [70]. Prior to the dispersion of filler in the polymer matrix the surface of SWCNTs was functionalized with diamine-terminated polyethylene glycol (PEG) chains. As anticipated, the surface functionalization led to an improved dispersion of the treated nanotubes in the polymer matrix, compared to the non-treated ones, which was attributed to an enhancement of the non-covalent interactions between the filler and the polymer. In turn, the improved dispersion resulted in a greater reinforcing effect, where the tensile modulus (E) of the composite was increased by 430% relative to the unfilled material.

Similar to the case of SWCNTs, various surface treatments have been also applied to MWCNTs to improve their reinforcing effects in the PEO matrix [26,71,72]. Song observed that acid-treated MWCNTs had a significantly higher effective volume fraction than the real volume of the untreated MWCNTs [26]. This effect triggered an improvement in the rheological response of the composites filled with acid-treated MWCNTs. In contrast, the nanotubes treated with long, amine-terminated, alkyl chains (e.g. octadecylamine) did not make a significant impact on the viscosity and modulus of resultant composites, as observed through comparison with unfilled PEO [26]. Jin *et al.* studied the crystallization behavior of PEO nanocomposites, containing modified and unmodified MWCNTs, by means of DSC measurements [72]. The chemically modified MWCNTs contained either carboxylic or hydroxyl functional groups. The results suggested that incorporating either the treated or the untreated MWCNTs generated a decrease in the polymer crystallinity, similar to what has been observed by Chatterjee *et al.* for SWCNTs [33]. In short, the filler decreased the nucleation sites in PEO nanocomposites, but led to an increase in the spherical crystal size compared to neat PEO [26]. Moreover, to improve the properties of PEO/MWCNT nanocomposites, Yang *et al.* have functionalized the nanotubes by reacting the carboxylic acid groups from the filler's surface with the glycidyl terminal groups of poly (bisphenol-A-co-epichlorohydrin) [71]. The grafting of phenoxy chains onto MWCNTs improved the dispersion of the filler in the polymer matrix owing to the high affinity of PEO and phenoxy chains for each other. In turn, the improved dispersion caused the enhancement of several properties of the composites, such as stiffness, strength, ductility, and toughness. As expected, an increase in the amount of MWCNTs in the composites resulted in a decrease in the size of the PEO spherulites [71].

At this point it should be noted that the technique known as *"solution-sonication"* represents a very common way to prepare PEO/MWCNT nanocomposites, and that the method is often used by various scientific research-groups [24,25,26,72,73,74]. Using this technique, Narh *et al.* studied the impact of agglomerated versus non-agglomerated MWCNTs on the thermo-mechanical properties of PEO/MWCNTs [25]. The authors found that the breaking-strain values, during mechanical measurements, were much lower for the nanocomposites than for the unfilled polymer. The resultant breaking-strain values were slightly higher for the composites containing non-agglomerated MWCNTs than for the agglomerated counterparts. The tensile modulus of nanocomposites, however, was observed to be 110 times higher than the one of the unfilled PEO. Furthermore, Song studied the rheological properties of PEO/MWCNT nanocomposites as a function of filler loading, and found that the shear viscosity of composites increased when the filler content was increased [74]. Complimentary, oscillatory shear experiments revealed that an increase in the filler loading generated a more pronounced solid-like behavior of the composite dispersion, owing to stronger polymer-particle and particle-particle interactions. Similarly, Abraham *et al.* observed an increase in the rheological properties with increasing the amount of MWCNTs [24]. The electric conductivity was also observed to improve with the filler addition. However, the increase in MWCNT loading induced a prominent decrease in the crystallization temperature of the PEO in the solid composites.

Finally, Wang *et al.* proposed a method for the preparation of PEO nanocomposites containing multiwall carbon nanotubes/montmorillonite (MWCNT/MMT) hybrids [73]. The authors developed their technique relying on the formation of stable MWCNT/MMT aqueous dispersions, which are typically caused by the strong interactions between MWCNTs and MMT platelets. As indicated in

Figure 2, in water the negatively charged surface of MWCNTs can interact with the Na cations from the surface of MMT platelets. This interaction was the key for the formation of stable MWCNT/MMT aqueous dispersions. The PEO composites containing the MWCNT/MMT hybrids exhibited high mechanical properties, owing to the formation of an extended network structure in the ternary system [73].

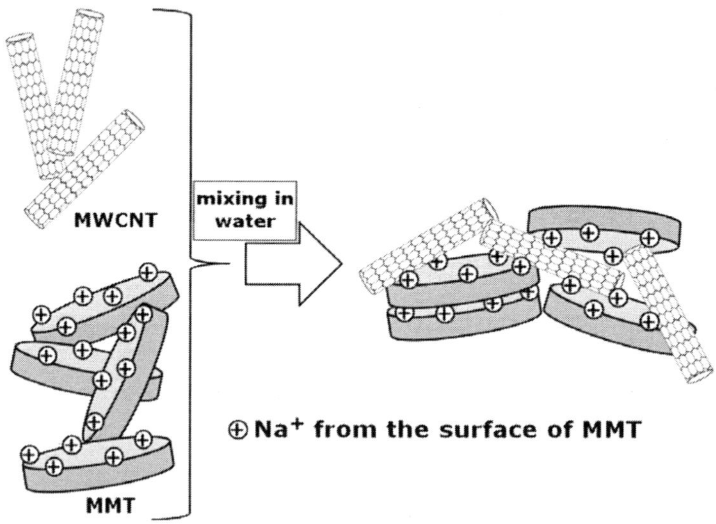

Figure 2. Schematic showing the interaction between MWCNTS and MMT nanoplatelets in water.

2.3. Poly(ε-Caprolactone)/Carbon Nanotubes Composites

Poly(ε-caprolactone) (PCL) is a biodegradable polymer that has been intensively studied over the past decades for its potential applications as implantable biomaterial. There are several reports in literature describing addition of either SWCNTs [8,31,75] or MWCNTs [22,23,76,77,78,79,80] to PCL to improve its physical properties. PCL/SWCNT composites have the unique advantage of combining the lightweight of these materials with their enhanced mechanical, thermal and electrical properties. Mitchell *et al.* prepared PCL/SWCNT nanocomposites with the aid of a zwitterionic surfactant (12-aminododecanoic acid) (ADA) [31]. Owing to the ability of surfactant to interact with both SWCNTs and PCL, an excellent dispersion of the filler within the polymer matrix was achieved. As a result, the nanocomposites exhibited geometrical and electrical percolations at SWCNT loadings of only 0.08 wt%. Furthermore, the same research group also reported on the ability to control the nano- and meso-scale arrangement of polymer chains in PCL/SWCNT composites containing the same zwitterionic surfactant (ADA) [8]. It was found that the SWCNTs, which exhibited an effective aspect ratio of *ca.* 750, were able to act as nucleators for the crystallization of PCL. In consequence, the polymer's rate of crystallization increased a hundred fold in the presence of filler. Additionally, the authors subjected the nanocomposites to uniaxial extension and observed, using Raman spectroscopy, a strong alignment of SWCNTs within the polymer matrix. At last, Priftis *et al.*reported a novel approach in which they functionalized SWCNTs (along with MWCNTs) with a titanium alkoxide catalyst and used the modified filler to surface-initiate the coordination polymerization of ε-caprolactone monomer [75]. Their results proved that the surface grafted catalyst was able to effectively promote the polymerization of monomer from SWCNTs, leading to the formation of PCL/SWCNT nanocomposites.

Ring opening polymerization (ROP) of ε-caprolactone monomer represents a very utilized method to fabricate PCL/MWCNT nanocomposites [22,36,81]. Chrissafis *et al.* used *in situ* ROP to fabricate PCL composites containing MWCNTs, along with other fillers such as fumed silica and montmorillonite [22]. The tensile strength of nanocomposites was observed in all cases to be higher than the one of neat PCL. Thermal analysis revealed that MWCNTs could decelerate the thermal degradation of PCL owing to a shielding effect introduced by the filler. In contrast, fumed silica and modified montmorillonite were found to accelerate the polymer's degradation due to chemical reactions that these fillers could induce in the PCL matrix. All fillers were observed to affect the activation energy of decomposition, but none of them could alter the decomposition mechanism of PCL [22]. Furthermore, Castro *et al.* grafted PCL chains on the surface of carbon nanotubes (CNT), via a ROP approach, and tested the vapor sensing properties of the resultant nanocomposites [81]. For that, the authors prepared conductive polymer composite transducers utilizing a layer-by-layer approach. The resultant PCL/CNT transducers were able to provide quantitative and discriminating responses after exposure to different polar and non-polar chemical vapors, such as water, methanol, toluene, tetrahydrofuran and chloroform. Finally, Wang *et al.* functionalized carbon nanofibers (CNF) with PCL chains by *in situ* ROP in the presence of stannous octoate [36]. The authors claim that the content of polymer grafted to the filer's surface could be controlled by adjusting the feed ratio of monomer to hydroxyl-functionalized CNFs. The highest polymer amount that they were able to graft was around 64 wt%.

Another effective way to produce PCL/MWCNT nanocomposites is the melt mixing of constituents [23,78,80]. Using such a method, Thomassin *et al.* fabricated PCL composites containing MWCNTs of two different diameters, aiming to improve the electromagnetic interference (EMI) shielding characteristics of the polyester [23]. Although the thin MWCNTs were found less prone to break down during melt mixing than thick MWCNTs, the use of thick MWCNTs triggered a more significant improvement of mechanical and electrical properties in the resultant composites. As a result of breaking-down, thick MWCNTs generated electrical conductivities higher than 1 S/m at fairly low loadings (0.7 wt%), inducing very good EMI shielding properties in the material [23]. Also using a melt compounding method Xu *et al.* prepared a series of PCL/MWCNT nanocomposites, and studied their crystallization kinetics and mechanical behavior [80]. The authors observed that the presence of MWCNTs, even in very small amounts, triggers a significant increase of the shear moduli of the composite compared to neat PCL. Since the incorporation of MWCNTs accelerated the nucleation and crystal growth of PCL, the filler loading was found to critically affect the degree of crystallinity and the time of crystallization in composites. The authors claim that the activation energy of crystallization in composites was higher for small MWCNT loadings and it decreased gradually as the filler loading was increased [80]. Moreover, Wu *et al.* blended PCL with polylactide (PLA) and coupled the resultant blend with functionalized MWCNTs via melt mixing [78]. The authors observed that the carboxylic MWCNTs were predominantly dispersed either in the PCL phase or at the interface between the two polymeric phases, but not too much in the PLA phase. The positioning of the filler at the interface resulted in an enhancement of the phase morphology of the ternary system, owing to a superior MWCNT-mediated interfacial adhesion. In consequence, rheological and electrical percolation threshold values of the ternary composites resulted lower than the ones corresponding to the PCL/PLA systems.

In addition to incorporating carbon nanotubes in polymers, some effort has been recently devoted to the addition of carbon nanospheres, also known as fullerenes or buckyballs, to polymers [82,83,84]. Kai *et al.* synthesized, via ring opening polymerization of ε-caprolactone, star shaped PCL composites containing a fullerene core [84]. As presented in Figure 3, the fullerenes' surface was functionalized with hydroxyl groups prior to the polymerization of ε-caprolactone. Conducting bulk polymerizations the authors obtained fullerene cores having 2 and 3 arms, and they estimated the ratio of the two structures to be 1:1.

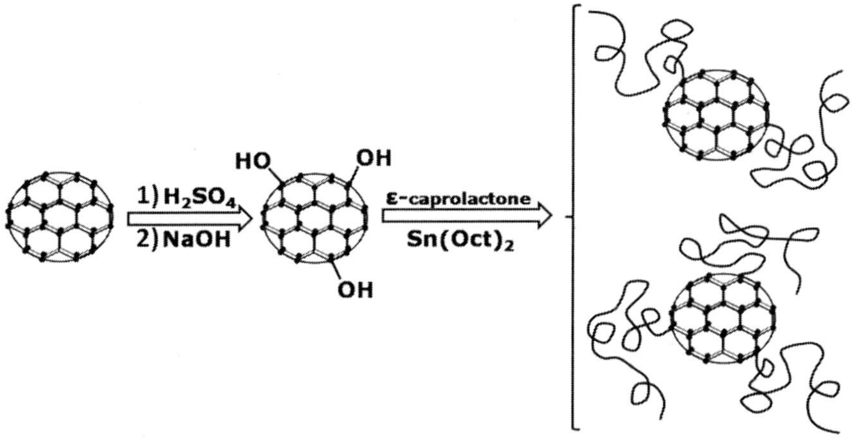

Figure 3. Schematic displaying the steps needed to obtain 2 and 3 arm PCL-grafted-fullerene.

As opposed to this case, polymerization of ε-caprolactone in toluene resulted in fullerene cores having predominantly three arms attached. The same group later developed a method to produce PCL/fullerene nanocomposite films [82]. Within such a composite film the fullerene domains were found to aggregate and form microscopic fullerene particles with an average diameter of ca. 2 μm. Furthermore, the authors argued that the microscopic aggregates impacted the polymer crystallinity via two major effects: a confinement effect and a nucleating effect. Owing to these two effects the crystallization temperature of the composites was found to be higher than that of the neat polymer. Additionally, the final crystallinity of the PCL in the composite was lower than the one of the unfilled polymer. Finally, the authors found that in fullerene-capped PCL composites the mechanical properties were strongly dependent on the fullerene loading and on the end-capping type (*i.e.,* single *vs.* double) [83].

2.4. Polypropylene/Carbon Nanotubes Composites

Owing to its low price and excellent physical properties, polypropylene (PP), and more specifically isotactic polypropylene (iPP), is one of the most extensively studied thermoplastics, with widespread applications such as home appliances and automotive construction [85,86]. Probably the most convenient and efficient way to produce PP/CNT composites is represented by melt compounding of components [87,88,89]. Using such a method, Pujari *et al.* prepared and characterized a set of MWCNT dispersions in PP [88]. Additionally they prepared a second set of dispersions employing a solid-state shear pulverization method, and compared the composites with the ones obtained through melt mixing. The authors observed that the nanocomposites prepared via melt compounding exhibited slower crystallization kinetics and lower mechanical stiffness than the shear pulverized samples. However, the melt mixed samples showed higher electrical conductivity. In addition, as opposed to the melt mixed samples, the pulverized composites showed no measurable low elastic plateau in melt rheology [88]. Furthermore, Hou *et al.* investigated the structural orientation and tensile properties of extrusion elongated PP/MWCNT melts [87]. The authors adjusted the orientation level in the resultant composite sheets by altering the draw-speeds after extrusion. By utilizing fast draw-speeds on nanocomposites with high MWCNT loadings, they were able to optimize simultaneously the strengthening and the toughening effects in the samples. The authors also noticed that although the MWCNTs acted as nucleators for the polymeric crystalline superstructure, the filler did not modify the orientation degree of the crystalline lamellae.

Wu *et al.* employed a maleic anhydride (MA) grafted polypropylene to prepare PP/MWCNT nanocomposites via melt mixing [89]. The MA grafts played the role of compatibilizers. The authors showed that while in the absence of MA the filler was only dispersed at the microscale within the PP matrix, addition of MA triggered the nanoscale dispersion of MWCNTs. It was further observed that the chain motion of the bulk polymer was not strongly affected by the presence of MWCNTs, and that the effective diffusivity of the composites was very similar to the one of the neat PP. Moreover, Xu and Wang investigated the impact of MWCNT aspect-ratio on the normal stress differences in iPP/MWCNT melts [90]. The aspect ratio of the nanotubes utilized was in the range of 22 to 45. At high MWCNT loadings the authors observed the occurrence of a non-equilibrium MWCNT network structure. A strong impact of the filler aspect ratio on the composites rheology was detected by means of die-swell measurements. While the composites containing large aspect-ratio MWCNTs exhibited die-shrinkage, the composites containing short MWCNTs showed a rather strong die-swelling effect [90].

Besides studying the rheological properties of composites, researchers often investigate the ability of fillers to act as nucleating agents for the polymer crystallization [85,86,87]. Liu *et al.* studied the nucleating ability of halloysite nanotubes (HNTs) towards crystallization kinetics of iPP [85]. Apparently, HNTs trigger predominantly the formation of the trigonal β-crystal iPP in the detriment of the monoclinic (α) and orthorhombic (γ) crystal geometries. The authors stated that the occurrence of the β-crystal iPP was strongly influenced by the HNT loading, and showed that composites with 20 phr (parts per hundred rubber) HNT had the highest β-crystal content. Additionally, the authors claimed that a decrease in the cooling rate during non-isothermal crystallization could lead to an increase of the β-crystal iPP content, with a maximum obtained at a cooling rate of 2.5 °C/min. When isothermal crystallization was employed, the occurrence of β-crystal iPP was detected between 115 – 140 °C. The authors attributed the formation of the β-crystal iPP to the unique surface characteristics of HNTs [85]. Finally, Lu *et al.* investigated the influence of SWCNTs and MWCNTs on the crystallization behavior of iPP [86]. The authors observed that when iPP was crystallized from quiescent melt both SWCNTs and MWCNTs exhibited a strong nucleating effect that led to the formation of a transcrystalline layer of ordered iPP lamellar crystals in the vicinity of the fillers. Contrary to what was detected for the halloysite nanotubes [85], CNTs were observed to induce the formation of monoclinic α-crystal iPP exclusively [86]. The authors attributed the strong alignment of the transcrystalline layer around the nanotubes surface to the fact that iPP macromolecules started wrapping around the CNTs in the melt, which triggered a melt confinement that was maintained during the nucleation process.

2.5. Nylon/Carbon Nanotubes Composites

Nylon 6, or polyamide 6, is typically obtained through the ring opening polymerization of the cyclic caprolactam monomer, but it can also be synthesized through the step polymerization of aminocapronic acid. Several reports in literature talk about the incorporation of carbon nanotubes in nylon 6 to improve its physical properties [21,29,30]. Jose *et al.*synthesized through an electrospinning method nanofibrous composites of nylon 6 and MWCNTs [21]. Carboxylic acid functional groups were attached to the surface of MWCNTs prior to nanocomposite preparation. The diameter of the electrospun nanocomposite fibers was observed to decrease at higher MWCNT loadings. The authors also noticed a transformation of the γ-phase crystals of nylon 6 into a mixture of α- and γ-phase crystals as the collector speed increased. Additionally, a significant alignment of the MWCNTs parallel to the main axis of fibers was detected [21]. Furthermore, Brosse *et al.* investigated the effect of MWCNTs on the lamellae morphology of nylon 6 [29]. The authors observed that when MWCNTs were well dispersed in the polymer matrix, through melt compounding, the crystalline lamellae of nylon 6 grew with a perpendicular alignment to the surface of nanotubes. Although the authors claim that the MWCNT-induced epitaxial growth is particular to polyamide 6, this feature was also noticed by Lu *et al.* for PP/MWCNT composites [86]. As the dispersion of the filler within the nylon 6 matrix was further improved, an increase in the amount of polymeric transcrystalline epitaxial crystallites could be detected. Li and Shimizu prepared hierarchically structured nanocomposites by dispersing MWCNTs in a poly(vinylidene fluoride) (PVDF)/nylon 6 matrix via high shear processing [30]. The authors observed an unexpected positioning of the filler exclusively in the nylon 6 phase, forming domains with sizes of 10 to 150 nm that were dispersed within the PVDF phase. They further attributed the high electrical conductivity and ductility characteristics of composites to the existence of nylon 6 nanodomains uniformly distributed in the PVDF matrix.

Besides nylon 6, another highly valued polyamide is nylon 6-6, which is typically obtained through the condensation polymerization of hexamethylene diamine with adipic acid. Incorporation of CNTs in nylon 6-6 has also been reported in recent years [91,92]. Using SWCNTs, Haggenmuller *et al.* prepared nylon 6-6 composites via an *in situ* interfacial polymerization [91]. Prior to nanocomposite preparation, SWCNTs were subjected to either purification or to alkyl-chain functionalization. In addition to preparing the two distinct sets of composites containing either the purified or the functionalized nanotubes, the authors fabricated a third type in which a surfactant (sodium salt of dodecylbenzenesulfonic acid) was added along with the SWCNTs. Although both the functionalization and the surfactant stabilization enhanced the filler dispersion in the solvent, only the functionalized nanotubes achieved a uniform dispersion within the polymer matrix. The other two types of composites, containing purified and surfactant-stabilized fillers, exhibited nanotubes agglomeration, and in consequence, an unsatisfactory dispersion degree. When nanocomposites were subjected to low shear forces nanotubes agglomeration was detected in all cases [91]. Furthermore, Li *et al.* studied the morphology and crystallization behavior of nylon 6-6/MWCNT nanocomposites using nanotubes that were functionalized with poly(hexamethylene adipamide) prior to composite fabrication [92]. The authors detected a very good dispersion of the filler within the polymer matrix, through optical and electron microscopy measurements. Additionally, several melting peaks were observed in DSC measurements, which were attributed to changes in the lamellar thickness of crystallites upon heating. The authors concluded that, on one hand the MWCNTs provided heterogeneous nucleation sites for the polyamide crystallization, but on the other hand the filler network hindered the growth of large crystals.

Other polyamide types were also used to fabricate CNT-based composites, such as Nylon 6-10 [93], nylon 10-10 [94], or nylon 12 [95] following various approaches. One very simple method to prepare nylon/CNT composites is exemplified by Kang *et al.* for polyamide 6-10 (see Figure 4) [93].

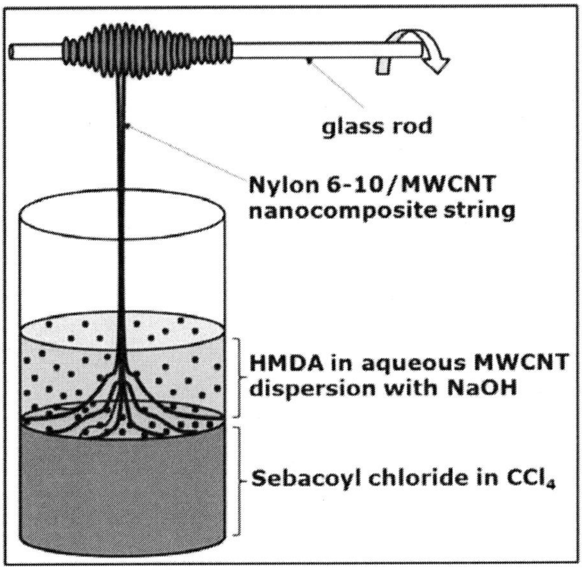

Figure 4. Schematic showing the preparation of nylon 6–10/MWCNT composites via *in situ* interfacial polycondensation of co-monomers.

The method consisted in the *in situ* interfacial polycondensation of co-monomers (hexamethylene diamine, HMDA, and sebacoyl chloride) dissolved in two immiscible liquid phases (one of which contained also the MWCNTs), followed by the separation of the polymeric string with the help of a glass rod. Theoretically, the method can be applied for the synthesis any other nylon/CNT composites in which the starting co-monomers dissolve in immiscible liquids. Regardless of the type of nylon used by various researchers, all nanocomposites exhibit physical properties superior to the ones of the neat polymers. Generally, an excellent dispersion of the filler leads to good mechanical properties of composites, fact that is attributed in part to the ability of carbon nanotubes to act as nucleation sites in the polymer crystallization process [93,94,95].

2.6. Poly(Ethylene Terephthalate)/Carbon Nanotubes Composites

Poly(ethylene terephthalate) (PET) is a widely known semicrystalline thermoplastic polymer that is characterized by excellent physico-mechanical properties, such as high stiffness and strength, and by very good chemical and heat resistance [27,35,39,96,97]. One effective way to prepare PET/CNT composites is represented by *in situ* polymerization of co-monomers in the presence of MWCNTs [96,97]. Using such an approach Antoniadis *et al.* prepared a series of PET/MWCNT composites through a two stage melt-polycondensation [96]. The authors observed that an increase in the filler content triggered a decrease of the intrinsic viscosity of composites. They attributed this behavior to branching and crosslinking of macromolecules owing to specific chemical reactions of nanotubes with co-monomers and/or polymer chains. In addition, the polymer crystallinity in the nanocomposites was observed to decrease with increasing the filler loading. While concentrations up to 1 wt% MWCNTs were found to improve the polymer crystallization rate, loadings of 2 wt% decreased the rate on the basis of chain crosslinking. The authors also found that the MWCNTs acted as nucleating agents for PET crystallization, in agreement with what was observed by other researchers for other polymer/CNT systems [96]. Furthermore, Mun *et al.* used MWCNTs functionalized with (2-hydroxyethyl) triphenyl phosphonium bromide to prepare PET nanocomposites through an *in situ* polymerization approach [97]. After polymerization, nanocomposite fibers were fabricated via a melt spinning approach. A significant improvement in the tensile strength of the polymer was detected after addition of only a small portion of functionalized MWCNTs (0.5 wt%). For MWCNT concentrations between 0 and 1.5 wt%, a direct proportionality was proved to exist between the composites modulus and the filler loading. In addition to the improvement in mechanical characteristics, the thermal properties of nanocomposites were superior to the ones of neat PET [97].

Another valuable method to produce PET/CNT composites is the melt compounding of components, approach that has been also used extensively to prepare other types of polymer composites too [27,34,39]. Yoo *et al.* utilized the melt extrusion method to fabricate PET composites containing MWCNTs functionalized with either benzyl isocyanate or phenyl isocyanate [34]. The mechanical properties (tensile strength and modulus) of composites containing MWCNTs functionalized with isocyanate groups were found superior than in composites containing pristine or acid treated fillers. The authors attributed this behavior to the $\pi - \pi$ interactions between the isocyanate groups on the surface of nanotubes and the six member aromatic rings from the PET structure. Additionally, the π–π interactions triggered an improved dispersion of the functionalized MWCNTs in the polymer matrix relative to the unmodified fillers. Interestingly, higher crystallinities were detected for the composites containing the well dispersed functionalized MWCNTs than for the counterparts containing the pristine fillers [34]. Moreover, Averett *et al.* investigated the effects of fatigue and residual strain on the mechanical behavior of PET microfibers reinforced with vapor grown carbon nanofibers (VGCNF) [27]. The authors prepared the composite fibers, with diameters of ca. 25 ± 2 μm through a melt mixing procedure, and compared their resultant properties with the ones of non-reinforced PET fibers (diameters 24 ± 3 μm). The PET composite fibers contained roughly 5 wt% VGCNFs. To induce the fatigue effects the authors subjected the fibers to an uniaxial stress that was ca. 60% of the fracture stress, at an elongation rate of 10 mm/min. Then, to determine the fatigue induced consequences they tested the non-fractured samples under uniaxial stress conditions. It was determined that the residual fatigue strains in the uniaxial tension measurements (subsequent to fatigue) affected most of the mechanical characteristics, such as elastic and hardening moduli, fracture strength and yield strain, of reinforced as well as non-reinforced fibers. The authors concluded that the degradation of mechanical properties as a function of residual strain due to fatigue was greater for the PET/VGCNF composite fibers than for the neat PET fibers [27].

Solution casting is another efficient way to produce PET/CNT composites [28,37]. Wakamatsu *et al.* used a novel approach in which they reacted anionic SWCNTs with a cationic ammonium lipid in several organic solvents (methylene chloride, chloroform, benzene and toluene) [37]. The resultant complex was cast on pretreated transparent PET films to form honeycomb structures. The authors observed that changes in the concentration of SWCNTs, or in the solvent used to prepare the solutions, led to variations in the cell sizes of the honeycomb networks. After removal of the cationic lipid through a simple ion exchange method, the nanotube honeycomb network exhibited thinner skeletons. Additionally, the nanotube network displayed a spectacular change in the resistivity characteristics going from insulating to conducting [37]. Furthermore, Steinert *et al.* prepared a solution blend containing PET and SWCNTs, and cast it onto glass substrates to produce flexible nanocomposite films [28]. During processing the authors used two magnetic resonance imaging (MRI) scanners to produce magnetic fields utilized to induce alignment of the nanotubes in the composites, as indicated in Figure 5. Following the same method they prepared films containing three distinct nanotube loadings (0.5, 1, and 3 wt%). The nanotubes' degree of alignment could be significantly altered by changing the filler concentration and the strength of the magnetic field, where more anisotropic networks were obtained at elevated SWCNT concentrations (3 wt% vs. 0.5 or 1 wt%) and field strengths (9.4T vs. 3T). The authors concluded that the greatest impact on the conductivity of nanocomposite films was brought mainly by the concentration and degree of dispersion of SWCNTs in the polymer matrix. On the other hand, the alignment of SWCNTs was observed to play a secondary, less important role [28].

Yet another useful approach, besides solution casting, is represented by solution electrospinning [35,38]. Ahn *et al.* used H_2SO_4-functionalized MWCNTs dispersed in a PET solution, in trifluoro acetic acid, to prepare nanocomposites via electrospinning [35]. The authors noticed that by increasing the filler concentration they were able to obtain thinner composite nanofibers, and attributed this observation to the shear thinning and charge carrier characteristics of the MWCNTs. Mechanical properties, like tensile strength and modulus, and thermal stability of the resultant nanowebs, as well as the degree of PET crystallinity, were found to increase as a result of an elevated filler loading. Opposite trends were observed for properties like crystallization temperature and elongation at break [35]. Moreover, Chen *et al.* investigated the chain confinement in electrospun PET/MWCNT nanofibers [38]. The authors claim that addition of MWCNTs triggered a more pronounced confinement of PET chains in the electrospun fibers, which in turn led to an increase of the rigid amorphous fraction of the polymer. For example, while in the neat PET fibers the rigid amorphous fraction had a value of 0.23, addition of 2 wt% MWCNTs increased the PET amorphous fraction to 0.64. The authors attributed the decrease in polymer crystallinity to spatial constrains developed during filler addition, which hindered the folding of polymer chains. It was also noticed that while for entirely amorphous composites the filler did not affect the PET chain conformation, for cold crystallized composites the nanotubes favored the development of trans conformers in the electrospun fibers [38].

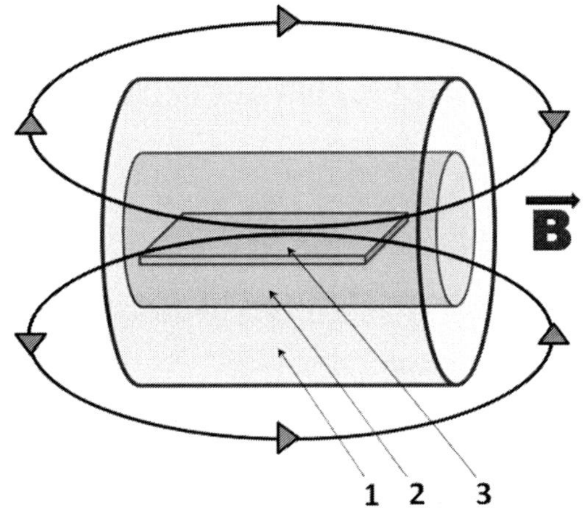

Figure 5. Schematic showing the placement of the samples with respect to the MRI instrument and to the magnetic field (B). The components are: 1 - magnet, 2 - MRI chamber, 3 – PET-SWCNT nanocomposite.

3. POLYMER/METAL NANOCOMPOSITES

As advanced technologies keep developing every day, there is a constant need for functional materials with special properties and/or combinations of unique properties. Materials based on nano-sized metal particles may help meet many of these needs, and with this thought in mind the central goal of this section is to present and highlight the work recently published on the synthesis and characterization of polymer/metal nanocomposites. In the literature covered here most of the metal-containing materials exhibit excellent mechanical properties and/or exceptional electrical or electro-magnetic characteristics. Additionally, while some of the covered reports emphasize the synthetic approaches taken to obtain metal-based nanocomposites, some others discuss materials properties that allow them to be considered for various applications, such as replacing parts of living tissues. Among the preparative methods presented in the scientific literature only a few are covered in this report. While some sections focus on broader and more general methods, some others present more tailored approaches, targeting either stabilization of the resultant materials, or their magnetic properties, or even other desirable features in the final product.

3.1. Polymer/Platinum Nanocomposites

A widely popular matrix for building polymer nanocomposites with Pt is polyaniline (PANI) as exemplified by Liu *et al.*, Nyczyk *et al.*, and Palmero *et al.* [40,98,99]. These papers describe applications containing PANI-Pt nanocomposites used to enhance the catalytic properties of Pt (as verified by catalytic alcohol conversion reactions) by reducing the metal particle size and increasing surface area. Key to the success of the Pt nanocomposites as catalytic systems is a good dispersion of the metal throughout the polymer matrix and a high surface area of the platinum nanoparticles. Nyczyk *et al.* describe polymer nanocomposites of Pt with PANI and its derivatives such as poly(o-methoxyaniline) and poly(o-toluidine) (POT) [40]. The Pt nanoparticles contained in these materials have dimensions between 2 and 4 nm. These are some of the smallest reported Pt particle sizes successfully incorporated into PANI polymer matrices or any of its derivatives. The metal is introduced in the matrices either from an aqueous Pt sol [40], or via electrodeposition [98,99]. A novel composite material based on the electrochemical generation of a layer-by-layer structure of PANI and Pt particles has been prepared by Palmero *et al.* [99]. Additionally, films of PANI synthesized independently by potentiostatic and galvanostatic method, were used by Liu *et al.*as the supporting matrix for loading Pt nanoparticles [98]. The porous network structure of PANI promotes the effective dispersion of Pt particles (about 10–20 nm) and facilitates an unrestricted access of methanol to the catalytic sites, as reported by the same authors [98]. An important aspect in the synthesis and use of these polymer nanocomposites with Pt is prevention from agglomeration of the Pt particles. It is believed that charge transfer interactions between polymer chains and Pt nanoparticles enable the polymer chains to act as stabilizers and to prohibit Pt-nanoparticle agglomeration. Among the polymers studied by Nyczyk *et al.*, POT shows the weakest stabilizing effect [40].

The synthesis of platinum nanoparticles organized inside rod or wire shaped superstructures, with diameters between 1.6–2.3 nm, is described by Tristany *et al.* [100]. Furthermore, carbon nanotubes (CNT) are used as support for polymer nanocomposites with Pt, as described by Sapurina *et al.* and Okamoto *et al.* [101,102]. The great advantage of the carbon nanotubes as support materials is that they facilitate charge transport throughout the assembly, thus greatly increasing the catalytic activity of the polymer nanocomposites with Pt [101]. A schematic showing the increased ability of these systems to distribute charges and their respective pathways is presented in Figure 6. Through their method, Sapurina *et al.* developed novel multipurpose electrode materials of potentially valuable catalytic applications. The materials are composites based on multiwalled carbon nanotubes and electroconducting polymer PANI, organized as "fiber in a jacket". Nanoparticles of metallic Pt and compounds of transition metals are additionally immobilized in the polymer layer. Experimental results demonstrate that the materials possess high electronic and protonic conductance, thermal stability, hydrophilicity, large specific surface area, and considerable porosity [102]. The polymer nanocomposites with Pt deposited on multiwalled carbon nanotubes may contain various matrixes, such as PANI [102], or poly(benzimidazole) (PBI) [101]. Moreover, nanocomposite materials of PBI incorporating multiwalled carbon nanotubes (MWCNTs) and Pt nanoparticles have been prepared by Okamoto *et al.* [101]. The nanocomposites fabrication involves the preparation of PBI-wrapped MWCNTs (MWCNT/PBI), followed by Pt loading onto the MWCNT/PBI. The average diameter of the Pt particles was found to be 4.8 ± 0.4 nm. Cyclic voltammogram measurements revealed that the composites with Pt nanoparticles deposited on the MWCNT/PBI show higher utilization efficiency (74%) for electrocatalysts than the ones with pristine MWCNT (39%) [101].

Either with the metal supported on CNT, or layered in a conductive polymer-matrix, polymer nanocomposites with small Pt particles show an increased catalytic activity versus materials containing larger Pt particles, as verified by catalytic alcohol conversion reactions [40,98,99,100,101,102,103,104]. Pt can also be used in combination with other metals in nanocomposite-based catalytic systems, such as Fe [104], or Ag [103]. The iron-paired Pt nanocomposites systems exhibit magnetic properties. Magnetic ordered mesoporous carbon with superparamagnetic FePt nanoparticles (average diameters 3–4 nm) has been successfully prepared by Zhu *et al.* via a simple nanocasting route

[104]. The magnetic properties are useful in simplifying the separation and regeneration of the catalysts at the end of a process. Furthermore, Compagnini employed two strategies to obtain noble metal particles for nanostructured and polymer-based nanocomposite thin films [103]. On one hand, the reduction of metal-organic compounds, generally used in liquids to obtain metal colloids, has been used directly in the solid state to obtain polymer/metal nanocomposites in a single stage (annealing) after the co-deposition of the two organic species. On the other hand, laser ablation of metallic targets in liquids, has been shown as a new and powerful method to obtain colloids in a wide range of polymer solvents [103].

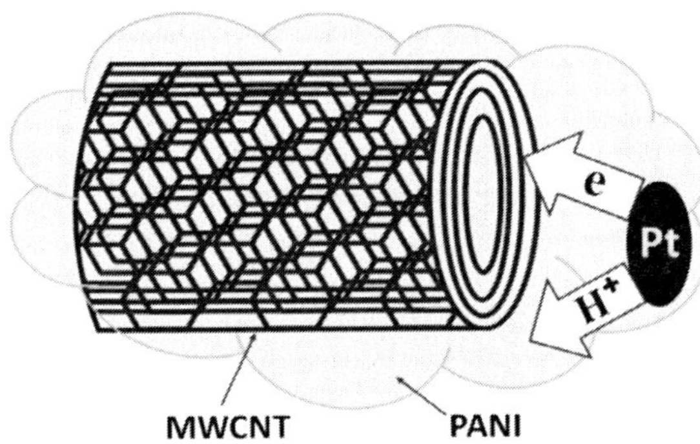

Figure 6. Schematic describing the pathways for the flow of charges in a PANI-Pt nanocomposite deposited on a multiwalled carbon nanotube, as imagined by Sapurina*et al.*

3.2. Polymer/Cobalt Nanocomposites

One of the most interesting applications of Co-polymer nanocomposites materials comes from the biomedical field. Fuhrer*et al.* report the use of a hydrogel as a matrix for incorporation of Co nanoparticles [105]. The mean diameter of Ca particles was found to be ca. 25 nm. The authors observed that the resultant material possesses very high flexibility coupled with surprising strength. They used covalent bonding of the particles to control loss of metal from the nanocomposites due to leaching. As a general desirable feature, it is the authors' opinion that these materials should contain metals rather than metal oxides. This increases the force to volume ratio for the Co-polymer nanocomposites, making them more suitable for use as magnetic actuators in a muscle-like application [105]. The researchers reported stable hydrogels, loaded up to a value as high as 60 wt%, with good flexibility and shape memory. To increase even further the stability of these systems, Fuhrer *et al.* also reported a *cross-linking of the polymer matrix* approach as a way to reduce Co particle migration. Such a device is represented schematically in Figure 7. The amazing properties of these materials may render them invaluable for use as soft actuators in medical applications [105].

A popular application of biopolymers is related to their ability to control the growth of inorganic particles. Due to differences in the chemical properties and structural variety, biopolymers exhibit various templating efficiencies. Brayner *et al.* used several sources of alginate to control the formation of Co, Ni, and Co/Ni nanoparticles [106]. The authors observed different particle sizes (mostly ranging from 10 nm to 26 nm), structures and magnetic properties in the resulting Co-polymer nanocomposites, all depending on the alginate's composition. Some of the gels obtained during this work are reported to be stable enough to allow formation of nanoalloys over the entire solid solution domain. The network like structure of the gels is enhanced by the cross-linking effect of metal cations found in close proximity to each other. The gels form cavities that trap the metal and, after reduction, play an active role in controlling the growth of the Co nanoparticles. The anisotropic properties of the cavities reported by the authors can lead to unusual hexagonal compact phase structures in the resulting Co-polymer nanocomposites [106].

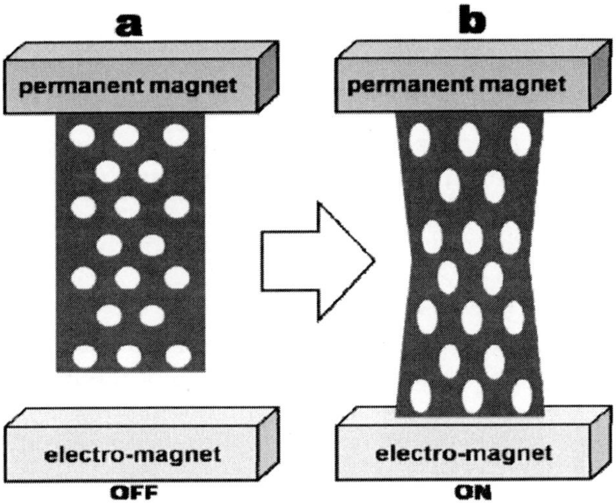

Figure 7. Schematic of a muscle-like Co polymer nanocomposites in a magnetic field.

Chemical methods for metal deposition represent good choices for creating Co-polymer nanocomposites. One such example was published by Carotenuto *et al.* [20]. The authors describe a method where polymer-embedded Co or Co sulfide clusters are prepared by annealing alkanethiolates of Co dissolved in polymers [20]. Nanoparticles obtained by this approach range in size from 7 to 15 nm. This method proves to be very flexible for obtaining metal polymer nanocomposites, since it allows for the incorporation of various transition metals into polymer matrices. The properties of the resultant materials can be accurately tuned by varying the amount of alkanethiolates introduced and their corresponding annealing times and temperatures [20]. Hybrid materials combining properties of metals, oxides, and polymers can be obtained by mixing these most different materials at the nanometer level, but their success is being strongly dependent on the ability to stabilize and control the interactions at the newly created interfaces [43,107,108]. Owing to such interfacial issues, the incorporation of nanoparticles into engineering thermoplastics by mechanical means (e.g., by melt blending techniques) has always proved to be difficult, especially when it came to the dispersal of inorganic nanoparticles within hydrophobic polymer matrices [109,110,111,112,113,114]. Luechinger *et al.* investigated the possibility to overcome this dispersion problem by mechanically entrapping metal nanoparticles in a grapheme-like carbon shell [115]. The Co core-nanoparticles, which ranged in size from 10 to 50 nm, were covered with a 1–2 nm thick grapheme coating. With this approach the authors aimed at successfully producing metal polymer composites in general, and Co-polymer nanocomposites in particular [115]. One major achievement of the work done by Luechinger *et al.* is the mediation of the surface energy difference between inorganic and organic materials, which was performed by means other than the typical incorporation of chemical additives, such as surfactants. In addition to this advantage, the metal loadings achieved in this study are very high, up to 90%wt, while the resultant materials exhibit an excellent metal dispersion throughout the polymer matrix, all done via a low cost process [115].

Similarly in scope to the research presented by Luechinger *et al.* [115], Pirkkalainen *et al.* [12] investigated the use of a porous cellulose matrix as support for Co nanoparticles, both in amorphous and crystalline form. Making use of a chemical reduction approach, the authors produced Co-polymer nanoparticles within a microcrystalline cellulose matrix [12]. Two kinds of Co containing nanoparticles were successfully obtained: amorphous Co–B or Co oxide composites, prepared through $NaBH_4$ reduction and exhibiting poorer ferromagnetic properties, and highly-ordered ferromagnetic hexagonal compact phase (hcp) Co nanocrystals, prepared via a NaH_2PO_2 reduction methodology. The authors noticed that the well ordered Co nanoparticles exhibit nanocrystals with average crystallite sizes ranging from 5.3 to 6.3 nm, along with some small amounts of larger micrometric aggregates [12]. Furthermore, Co particles (less than 50 nm diameter) embedded in polytetrafluoroethylene (PTFE) polymer matrix were described by Kwong *et al.* [11]. In their work the research team demonstrated the successful fabrication of Co-PTFE granular composite films by employing a pulse laser deposition (PLD) method [11]. They detected magneto-resistance (MR) values of 4% at room temperature in an applied magnetic field of 8 kOe, and observed that the MR values increase to 6% for measurements performed at a temperature of 20 K [11]. This temperature dependency of the MR was attributed to the spin-dependent tunneling of electrons and the superparamagnetic relaxation behavior of the Co nanoparticles embedded in the PTFE matrix [11]. Magnetic properties of Co polymer nanocomposites were also the subject of the work published by Yurkov *et al.* [44]. The authors present the synthesis of Co nanoparticles with a narrow size distribution (average diameter 8.3 nm), embedded in a polyethylene (PE) matrix. The resultant powder Co-polymer nanocomposite materials were processed in thick films and bulk materials [44]. An increased dielectric permittivity and absorption coefficient is reported by the researchers with the increase in Co nanoparticles concentration, thus leading us to conclude that magnetic properties of the synthesized Co-polymer nanoparticles depend directly on the concentration of metal particles in the composite. Co containing polymer nanocomposites were found to exhibit a higher magnetization (including the remnant one) than that of the iron-containing samples [44].

3.3. Polymer/Nickel Nanocomposites

In the work presented by Ghose *et al.*, the goal of the research team was to increase the thermal conductivity of a commercially available ethylene vinyl acetate copolymer (Elvax™ 260) by compounding the copolymer with carbon based nanofillers and including Ni nanostrands (400–800 nm) [116]. The tested material was obtained by extruding the polymer so that the Ni strands were aligned in one direction. A major improvement of thermal conductivity (up to 24 fold) in the direction of alignment was achieved, without any significant loss of flexibility [116]. The authors attributed this improvement in thermal conductivity to a more efficient phonon transfer through the nanotube axis. One interesting application of Ni is represented by its use as a bonding agent in polymer nanocomposites to improve the mechanical properties of the finite material. Such an example is presented in the work by Sun *et al.* [46]. Single-walled carbon nanotubes (SWCNT) are used to reinforce Ni-polymer nanocomposites by bonding the metal to the SWCNT. The authors claim that their method prevents separation of the carbon nanotube from the matrix [46]. An increased strength, up to 3-fold, was noticed in the nanocomposites and the improvement was attributed to the good interfacial bonding between Ni and SWCNT, as well as to the stiffer nature of the matrices [46]. A schematic of the composite material under tensile pressure is shown in Figure 8. Due to the significant differences in the "stretchability" of the CNT and the polymer matrix, separation can occur under load. By loading the polymer matrix with metal nanoparticles, the Young modulus and the stretchability of the matrix is brought closer to that of the CNT, thus reducing the risk of separation under loading [46].

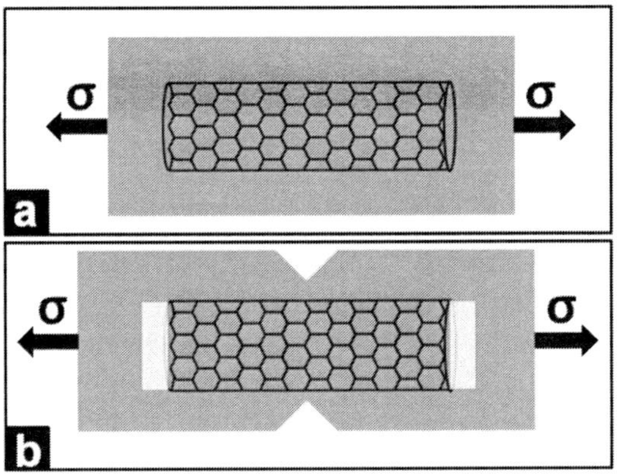

Figure 8. Sketch of a carbon nanotube under tensile pressure (σ), without (a) and with (b) potential separation between the carbon nanotube and matrix. In (b) separation is visible at the two ends of the carbon nanotube, while in (a) the presence of Ni in the polymer matrix prevents the matrix-CNT separation under tensile pressure.

The use of hydrogel networks for growth control of metal nanoparticles is a popular in-situ synthesis technique that can be utilized in conjunction with Ni-based materials [117]. With this technique in mind, Nurettin Sahiner focused his work on preparing metal particles of various sizes inside a three dimensional polymer matrix through the absorption of a $NiCl_2$aqueous solution followed by subsequent reduction to form Ni nanoparticles [117]. The author investigated the impact of cross-linking on the morphology of the hydrogels and the Ni particle size, and found it to have no significant effect on either property. On the other hand, the ion exchange capacity of the matrix material, in this case 2-acrylamido-2-methyl-1-propansulfonic acid (AMPS), was observed to play a major role on the size and loading capacity of the Ni particles [117]. Ni-polymer nanocomposites obtained by this method have potential applications in the biomedical field as catalysts, sensors, and/or signal-triggering delivery systems. The metal particles obtained in this study ranged in size from 40 to 100 nm.

Furthermore, Co and Ni polymer nanocomposites were also the focus of the work introduced by Sarkar *et al.* [45]. The authors present Ni and Co nanoparticles, with sizes between 3 to 5 nm, stable within a polyacrylamide matrix in the presence of air. The investigation of the magnetic properties of these particles showed a low superparamagnetic blocking temperature, as expected for very small Co and Ni crystallites [45]. The work is among the first to present Co-polymer nanocomposites stable in air, at room temperature. Additionally, the advantage of stabilizing magnetic Ni nanoparticles in a gel matrix is underlined in the report [45]. Moreover, Chelebaeva *et al.* present a method of obtaining soluble Ni-polymer nanocomposites by using cyano-ligands [118]. Organic amines were used as stabilizing agents. Particle sizes obtained during this work ranged from 2 to 5 nm, depending on the nature of the metal ion used [118]. According to the authors, these particles were easily precipitated in alcohols and redispersed in organic solvents (ex: hexane). Rather than being the final product, the Ni-polymer nanocomposites served as a precursor component in an interesting work described by Zhao *et al.*, where authors prepared hollow Ni nanospheres from Ni-polymer nanocomposite materials, [119]. The polystyrene (PS) nanospheres used in the study were nickel-plated. Through the use of this approach, the authors were able to obtain uniform Ni particles filled with PS. Subsequently, thermal decomposition of the organic polymer led to formations of the hollow Ni nanospheres, characterized by a high surface area to weight ratio [119]. Concisely, the research team demonstrated an easy, low-cost method of obtaining hollow particles via a Ni-polymer nanocomposite precursor.

On the other hand, focusing on the characterization of nanocomposites rather than synthesis, Mohanraj *et al.* studied the effect of temperature, pressure, and composition on electrical properties of conductive styrene-butadiene rubber-particulate metal alloys [13]. Nanosized particles (smaller than 70 nm) of Cu-Ni alloys were used as fillers, with loading values varied from zero to 40 phr, as reported. While an increase in the temperature was shown to improve the AC conductivity of the Ni-polymer nanocomposites, pressure was observed to result in an exponential decrease in resistivity [13]. It is apparent that the higher the metal loading of the polymer, the better the electrical properties of the nanocomposites are. The improvement can be attributed to the formation of some continuous conductive networks in the rubber matrix [13]. Finally, surface analysis and interfacial effects of filler (Ni) particles on polymer nanocomposite properties were the main goal of the work published by Panda *et al.* [14]. The polymer matrix of choice in this study was polyvinyledene fluoride. The size of the Ni particles, between 20–30 nm, was reported to be directly linked to the enhancement of the low frequency dielectric constant for these materials [14]. This observation has been attributed to the increase in surface area of the smaller size Ni nanoparticles. At the percolation threshold, a higher dielectric constant was observed due to increased leakage current resulting from smaller distances between particles [14].

3.4. Polymer/Silver Nanocomposites

Synthesis of hybrid organic/inorganic films by a single step process has been reported by Compton *et al.* [120]. Different complexes (a silver based complex and two palladium based complexes) were used by the authors as precursors in the process of obtaining polymer nanocomposites. Depending on the precursor used and the processing conditions employed, different resulting particle sizes, ranging from 5 to 35 nm, and size distributions were obtained. While a uniform dispersion of nanometer sized spheres was obtained by thermal curing applied to the initial homogeneous (silver complex/polyimide) solution, an increase in the final curing step time at 300 °C was found to lead to a more complex morphology. This morphology change induced small variations in the glass transition temperature (Tg) and the gas transport properties of the Ag-polymer nanocomposite films discussed [120]. In addition to this procedure, a one step photopolymerization process was employed by Balan *et al.* to fabricate Ag-polymer nanocomposites in a poly(ethylene glycol) matrix [47]. In short, the work shows how the radical mechanism that governs the process can be accelerated by the addition of an amine co-initiator, allowing the control of the resultant silver nanoparticles' sizes (3 to 10 nm). Another approach for the preparation of Ag-polymer nanocomposites, described by Wada *et al.*, involves successive polymerization of silver containing methacrylate monomers [121]. The silver nanoparticles with a mean diameter of 2.8 nm were initially dispersed in the monomer by microwave irradiation, without polymerization. In the authors' opinion a significant advantage of this process is that the reaction proceeds evenly throughout the reaction vessel, as opposed to thermally driven reactions/processes. The Ag-polymer nanocomposite is later obtained by the ultraviolet (UV) initiated polymerization of the previously metal loaded monomer [121].

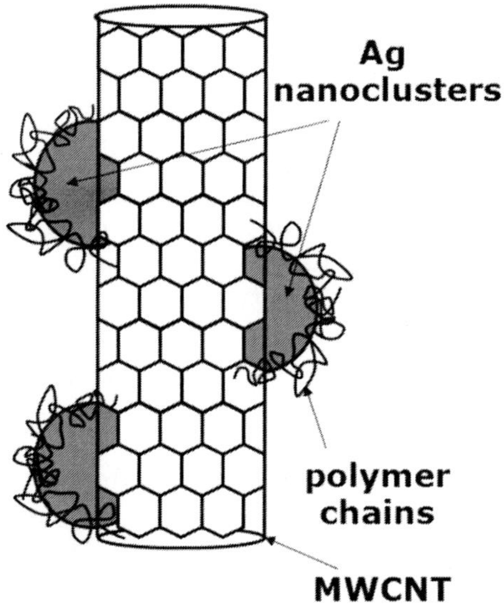

Figure 9. Carbon nanotube/Ag nanohybrids.

An interesting approach to Ag-polymer nanocomposites is depicted in the work published by Gao *et al.* [122]. The process starts with multiwalled carbon nanotubes (MWCNT) which are functionalized in order to become macroinitiators suitable for atom transfer radical polymerization (ATRP) processes. The authors have chosen glycerol monomethacrylate (GMA) as the monomer for the ATRP reaction [122]. Moreover, the researchers used the functionalized MWCNTs to entrap metal ions such as Ag^{+}, Co^{2+}, Ni^{2+}, Au^{3+}, La^{3+}, and Y^{3+}, forming MWCNT-polymer/metal hybrid nanocomposites [122]. The authors reported different shapes, including nanowires and/or necklace-like nanostructures, all depending on the content of the grafted polymer, and the type of the metal loaded. One surprising and unexpected result is the presence of polymer covered-silver nanoclusters, with iameters between 3–10 nm, attached on the convex surface of MWCNTs, as depicted schematically in Figure 9 [122]. Furthermore, the thermal decomposition of (1,1,1,5,5,5-hexafluoroacetylacetonato) silver(I) (AgHFA) in a polymer matrix is the process described by Deshmukh *et al.* to obtain Ag loaded nanocomposite films [123]. In the example described by the authors, AgHFA is decomposed in a poly(methyl methacrylate) film with a silver loading of up to 20 wt%. It is reported that most of the silver is evenly distributed throughout the entire film's thickness, with only a few percents being situated near the surface. Upon annealing, strong segregation in nanoparticles sizes is observed between the surface, the mid region, and the substrate of these materials. The silver beads have dimensions of 20 to 75 nm towards the surface, 6 nm in the mid region of the films, and 2 to 20 nm at the substrate region [123]. The researchers hope to achieve future control of the conductivity and reflectivity of Ag-polymer nanocomposites films through a rigorous understanding of the phenomenon behind this segregation. In a similar effort, conductivity enhancement in Ag-polymer nanohybrid films was examined by Jiang *et al.* [9]. Starting from silver nanoparticles (average diameter 16 nm) obtained through a chemical vapor condensation method, the authors describe the incorporation of functionalized particles into an epoxy resin (diglycidyl ether of bisphenol A) matrix. Through the use of a diacid surfactant for the functionalization of the silver particles, the researchers were able to obtain polymer nanocomposites with dramatically reduced resistivity. The authors advise that these results were possible due to the sintering of silver nanoparticles, a process that is continued during use, which induces stable properties in the Ag-polymer nanocomposites obtained by this method [9].

Moreover, improvement of the thermal and mechanical properties of poly(vinyl alcohol) (PVA) through silver addition was the goal of the work published by Mbhele *et al.* [124]. Silver was introduced in the polymer matrix by evaporation from water based PVA solutions and silver colloidal mixtures. By loading the polymer matrix with silver nanoparticles, the researchers were able to show an increase in the Tg and thermal stability, and an improvement in the Young's modulus for the obtained materials [124]. In spite of these promising results, experiments involving high metal loadings revealed a decrease in the stability of the nanocomposites. In consequence, the changes in properties are believed to be proportional with the degrees of metal loading. At last, Mukherjee *et al.* [125] studied the effects of nanoparticles size and density in Ag-polymer nanocomposite thin films. The choice polymer matrix for this study is poly(*o*-methoxyaniline)

(POMA). The authors found that the larger particle size silver nanobeads (21.9 ± 1.7 nm) exhibit switching behavior, as shown by the electronic bistability observed in I-V experiments, while the smaller sized particles (8.9 ± 0.8 nm) show rectification properties. They were able to control the electronic properties of the Ag-polymer nanocomposite films by tuning the metal particle size. This choice of polymer helped the authors conclude that the directionality detected in current flow experiments performed on the Ag-polymer composites is due to the asymmetry of the p/n junction assemblies arising from POMA donor (converts to p-type semiconductor) and Ag nanoparticles as acceptor (converts to n-type semiconductor) [125].

3.5. Polymer/Palladium Nanocomposites

The catalytic properties of Pd are well known to researchers; however, its inhibitor-like behavior is not so common. Lee *et al.* report both types of behaviors in their Pd-polymer nanocomposites [126]. For example, when combined with polymers like polystyrenes, polypropylene and methacrylates, Pd is observed to retard the thermal decomposition of nanocomposites. This behavior was attributed by the authors to the metal particles hindering the motion of the polymer chains, and thus acting as inhibitors. However, the presence of Pd was found to accelerate the thermal decomposition process of polymers such as polyamide 6 and poly(ethylene terephthalate), effect attributed by the researchers to the catalytic activity exhibited by the Pd nanoparticles. The catalytic properties of Pd are also the main focus of the work presented by Huang *et al.* [42]. A three-step process was used for synthesizing the Pd-polymer nanoparticles. The polymer matrices chosen were polyaniline and polypyrrole. Prior to nanocomposite preparation these matrixes were reduced to their lowest oxidation state in order to obtain a significant uptake of palladium from solution [42]. The catalytic properties of the resultant Pd-containing nanocomposites were investigated by reduction of oxygen dissolved in water, and reduction of nitrobenzene to aniline. The authors present a successful method for obtaining electro-active Pd-containing polymeric nanocomposites, with demonstrated catalytic abilities [42]. Continuing the trend of research underlying the catalytic activity of Pd-polymer nanocomposites materials, Wang *et al.* [127] report a facile way of preparing conductive polymer-supported Pd nanoparticles. The authors used a Pd salt, which was later reduced on the surface of a polyaniline film or membrane, to form a metal-polymer nanocomposite. Cluster sizes observed by the researchers were in the range of 200 to 500 nm, depending on weather a polyaniline membrane or film was used, respectively [127]. It was further demonstrated that those clusters were in fact comprised of much smaller Pd nanoparticles with an average size distribution of 13 nm. The variations in size of the metal particles typically translate into variations in surface area of the Pd particles, and they were further reported to be directly related to the catalytic properties of the resultant compounds [127]. Pd-polymer nanocomposites synthesized via the method described by Wang *et al.* exhibit efficient catalytic properties toward hydrogenation of alkynes and cinnamaldehyde, with good selectivity controlled by a kinetic mechanism. Due to the low cost of materials and ease of synthesis, the authors view such composites as perfect candidates for future development of fuel cell type applications [127].

Two synthetic routes are investigated by Valmikanathan *et al.* [15] in an attempt to obtain Pd-polymer nanocomposite materials: an *in-situ* method and an *ex-situ* method. Both methods are reported to produce materials in which the polymer matrix (polycarbonate) exhibits a glass transition temperature that is 16 °C lower than the one of the pure polymer [15]. One notable difference between the materials resulting from the two methods of synthesis is that the *ex-situ* method yields more dispersed Pd nanoparticles (diameter ca. 15 nm) while the *in-situ* path produces more agglomerated nanoparticles. As a direct result, it is suggested that one should use the *in-situ* synthesis if one wants to obtain electrically conducting Pd-polymer nanocomposites. The *ex-situ* route is more suited for producing a material that was found by the authors [15] to transmit more light than the *in-situ* produced material, for similar metal loadings. Stability of the Pd nanoclusters obtained by either synthetic route in the absence of a capping agent is believed to be a possible issue. On the other hand, the thermal degradation onset temperature of the Pd-polymer nanocomposites materials is shown to be 20 °C to 40 °C higher than the one of the pure polymer matrix [15]. Furthermore, a study on the catalytic activity of Pd nanoparticles encapsulated in spherical polyelectrolyte brushes (SPB) and microgels is presented by Mei *et al.* [128]. The catalytic properties of the resultant Pd-polymer nanocomposites were investigated photometrically, monitoring the reduction of *p*-nitrophenol by $NaBH_4$ [128]. In short, the data presented in this work suggests that the activity of the Pd nanospheres (2.4 to 3.8 nm diameter) is highly dependent on the carrier system used, although the authors could not see any clear indication for specific interactions between the polymer chains and the Pd nanoparticles [128].

In the work described by French *et al.*, palladium polymer nanocomposites are used as precursors to Pd-surface-metalized polyimide films [129]. Some of the investigation techniques used to characterize the resultant films include specular and diffuse reflectivity and conductivity measurements as a function of curing time and temperature. The authors observed that during the curing step, Pd in the 2+ oxidation state undergoes reduction to Pd metal [129]. At around 250 °C very small particles (ca. 3 to 10 nm) of Pd metal are formed at the surface, and they increase in size and agglomeration with increasing curing temperatures. Based on these observations, the researchers speculate that only a relatively narrow processing window exists during the thermal curing protocol where the optimum film, with regard to specular reflectivity, conductivity, and mechanical integrity, can be produced [129]. Because Pd is a very versatile catalyst, utilized in synthesis of various chemical compounds, including metal-polymer nanocomposites, Krebs *et al.* [130] focused their studies on the effects of the remnant Pd catalyst particles in polymer nanocomposites. The authors chose two synthetic routes for obtaining electrically conductive, thin polymer films. While one route required the use of Pd catalyst, the second route was based on a condensation reaction, with no need for Pd [130]. The authors showed that palladium nanoparticles (20 nm diameter) can act as very efficient charge carriers, influencing the properties of thin polymer films. The inability to completely remove all Pd catalyst traces severely impairs the electrical properties of the thin polymer films, limiting their application for electroactive devices, for which reason caution is advised when Pd is used as a catalyst in the polymerization processes (particularly when electrically active materials are sought) [130].

3.6. Polymer/Gold Nanocomposites

The preparation of regular and mechanically robust electric components based on Au-(conducting polymer) nanostructured materials is described by Zotti *et al.* [41]. The authors produced regular films, full monolayers and multilayers via solution processing, through the use of gold nanoparticles (average diameter 5 nm) and soluble conductive polymers, such as polypyrrole or poly(3,4-ethylenedioxy-thiophene). The resulting materials were reported to have good conductivities, in the order of $3–6 \times 10^{-2}$ Scm^{-1} [41]. In addition, the conductive polymers used in the study were found to strongly act as aggregating agents for the gold nanoparticles. The fact that addition of polyconjugated electron linkers did not modify the electron transport led the authors to conclude that the electron transport properties of the Au-polymer nanocomposites described in this work are mainly a result of gold-polymer contacts [41]. Furthermore, two acrylate monomers with different functionalities and gold nanoparticles were the starting materials for the Au-polymer nanocomposites synthesized by Goldberg *et al.* [131]. The authors utilized an original synthetic approach that includes a holographic photopolymerization methodology, schematically represented in Figure 10. To ensure the stability and dispersbility of the Au particles (1.5–3 nm core diameter) in the acrylic monomers, the particles were coated with an ester terminated alkanethiol [131]. The researchers showed that increasing the Au loading of the polymer matrix decreases significantly the rate of polymerization. Their preliminary investigations proved that the polymer-Au nanoparticle gratings (periodically ordered metal nanostructures) exhibit nonlinear optical properties [131].

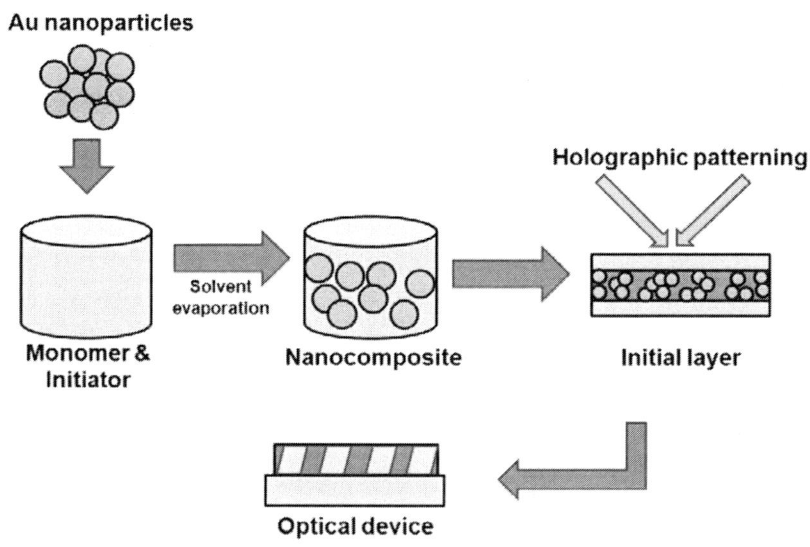

Figure 10. Schematic of the nanocomposite preparation and optical active element fabrication.

Moreover, Hata *et al.* proved that the intercalation of a basic polyelectrolyte poly(allylamine) (PAA) into a synthetic Na-fluoromica produces Au-polymer nanocomposites with very different properties under different pH conditions [132]. At high pH values the saturation loading was found to be lower, but the free energy of intercalation was observed to be more favorable. At low pH, the polymer chains are completely protonated, and at high loadings, chloride ions (from the addition of $HAuCl_4$) are also intercalated to screen the repulsive interactions between adjacent polycation chains [132]. The optimum pH for intercalating Au nanoparticles (smaller than 10 nm diameter) was reported to be 11.0 where typically the matrix has relatively little charge. At a high loading of Au nanoparticles, the authors observed some unusual features in the long-wavelength region (570–730 nm) of the plasmon resonance spectrum, features reportedly arising from interlayer particle interactions [132]. The work illustrates a convenient route for obtaining high-quality dispersions of noble metal nanoparticles intercalated into layered materials, that offers a rather attractive potential for new optical applications based on plasmon resonance. Li *et al.* obtained novel nanocomposites of gold nanoparticles and poly(4-vinylpyridine) by using surface initiated atom-transfer radical polymerization (ATRP) at ambient conditions [133]. The Au nanoparticles, with an average diameter of 20 nm, were stabilized using citrates, and then modified with a disulfide initiator in order to be active in the ATRP initiation. Authors claim that the polymerization reaction takes place at the surface of the gold particles and that due to the presence of protonated pyridine groups, the Au-polymer nanocomposite materials are pH responsive. The polymer chains attached to the gold nanoparticles are expanded at low pH and collapsed at high pH values [133]. The researchers suggests that materials exhibiting this type of behavior can act as smart catalysts [133].

Furthermore, Uhlenhaut *et al.* were able to obtain nanocomposites of poly(dimethylsiloxane) (PDMS) and thiol-modified gold particles (2–3 nm average diameter) that have the ability of offering a versatile reversible-color system, as a result of kinetically controlled particle dispersion states [134]. The authors claim that due to the elastic properties of the polymer matrix, dichroic states can be generated and can be transformed again to optically isotropic states by swelling. However, upon heating to 220–240 °C for a few minutes, irreversible thermochromic effects were detected as a result of desorption of alkanethiol molecules from the gold surfaces, which leads to the irreversible particle aggregate formation [134]. In addition to this work, several methods for synthesizing arrays of Au-conductive polymer nanocomposites and solid conductive polymer nanoparticles were studied by Marinakos *et al.* [135]. Although the authors believe that such Au-conductive polymer nanocomposite materials are ideally suited for studying electron hopping in nanoscale solid-state systems, the main goal of their research was the use of metal nanoparticles as templates for forming hollow polymeric nanocapsules [135]. It appears that the possibility of using Au-thiolate chemistry to introduce a desired molecule into the interior capsule core was the driving force in the study [135]. For this work the authors employed suspensions of citrate-stabilized Au nanoparticles with diameters of either 8 nm or 30 nm.

Moreover, Park *et al.* reported on the preparation of blue polymer light-emitting diodes (PLED) with improved luminescent stability [136]. The PLEDs were obtained by incorporating 5–10 nm gold nanoparticles as the quenchers of the triplet states of blue emitting polymer. The presence of the Au nanoparticles conferred several advantages to these devices: for example the gold-doped PLEDs are protected from oxidation by the quenching of the triplet state of the polymer. Additionally, the authors noticed that the Au nanoparticles appear to modify the interfacial morphology of the nanocomposites in manner that facilitates electron injection while blocking hole migration [136]. In short, the materials doped with Au nanoparticles showed greater external quantum efficiencies than the pristine PLEDs. Moreover, a polythiophene-gold nanoparticle alternate network film was prepared by Tanaka *et al.* [137]. The described synthesis involves mixing a thiol group-terminated polythiophene and gold nanoparticles (diameter 3.3 ± 0.8 nm) that result in nanocomposite materials exhibiting excellent conductive characteristics, and relatively high electric conductivity. These properties coupled with an extremely low activation energy led the researchers to conclude that the intrinsic conductive characteristic of the polythiophene backbone can be evaluated by connecting the polythiophene with Au nanoparticles through the linkage of conjugated S-Au bonds [137].

Gold-fluoropolymer nanocomposites can be utilized as vapor sensing devices, as indicated by Cioffi *et al.* who studied the composition and structure of such materials [138]. The purpose of their work was to better understand the swelling mechanism behind the material's vapor sensing capability. Although transmission electron microscopy results showed a uniform horizontal distribution of the Au nanoparticles (diameters between 2–8 nm) in the polymer layer, when analyzed by angle resolved X-ray photoelectron spectroscopy, a vertical concentration gradient of the Au nanoparticles was identified [138]. It was also proven that the extent of cross-linking in the polymer matrix increases with the gold particle concentration, and that the interface between the metal and polymer is comprised mostly from Au oxides and fluorides. The authors expect the Au oxides and fluorides species to play a critical role in the gas swelling phenomenon that occurs during the exposure of the materials to polar solvents [138]. At last, Laicer *et al.* studied the mechanism of morphological seeding in block-copolymer nanocomposites [139]. In their work the authors combined cylinder phases of polystyrene-block-polyisoprene diblock (as a solution in dibutylphthalate) and poly(styrene-blockisoprene-block-styrene) triblock (as a blend with homopolystyrene) copolymers with gold nanorods of different diameters (30, 100, 180 and 400 nm) and surface treatments. The overall goal of the work is to illustrate a general mechanism for the kinetic templating of block-copolymer domains from Au nanoparticle seeds. The researchers suggest that both kinetic and thermodynamic factors need to be considered when looking at the design rules for templating nanocomposite block copolymers [139]. It is also stated that it is possible to actively orient the block copolymers via kinetic templating throughout the bulk composite if electric, magnetic, or shear fields are applied to align nanorods suspended in an isotropic polymer melt prior to domain nucleation [139].

3.7. Other Metals/Polymer Nanocomposites

Iron-containing nanocomposites represent an important class of materials. A simple approach for fabricating a robust vinyl ester resin nanocomposite, reinforced with iron nanoparticles, is demonstrated by Guo *et al.* [140]. This approach eliminates the need for any surfactant or coupling agent. The authors show that the resin is chemically bound onto the nanoparticle surface and protects the iron nanoparticles (average diameter 20 nm) from agglomeration and oxidation. The mechanical properties of the resultant material are enhanced, as proven by tensile strength and Young's modulus values larger than those of cured pure resin. The resulting Fe-polymer nanocomposite material is also reported to be magnetically harder, with an increased thermal stability, showing ferromagnetic properties at room temperature [140]. The authors believe that these materials have potential applications in the marine systems, magnetoresistive sensors, and microwave absorption systems [140]. In addition, Baker *et al.* used iron along with its oxides to produce Fe-oxide/Fe core shell nanoparticles (average size 13 ± 2 nm) dispersed in polymethylmethacrylate (PMMA) films in varying concentrations [141]. The authors observed that the magnetic properties of the composite materials depend on the amount of Fe-oxide coating the iron particles, and hypothesized that this behavior is a result of a heightened anisotropy barrier that results from the ferrimagnetic oxide shell and the exchange interaction with the Fe core. Reduced dipolar interactions between particles are reported, which are believed to be triggered by the particles being separated from each other. As a result, the researchers observed an increase in coercivity, and a slow relaxation rate in thermoremanent magnetization experiments [141]. The optical properties of iron-polymer nanocomposites were studied by Ushakov *et al.* [142]. The polymer matrix chosen for the work was a linear low density polyethylene. Different iron concentrations in the polyethylene matrix were investigated at room temperature, in the visible and near-infrared regions [142]. The authors demonstrated that optical methods, such as linear optical spectroscopy, facilitate the size determination of ultra-small nanoparticles (through direct energy measurements), determination which is reportedly rather difficult to perform by other known methods [142]. The metal particles examined in this study had diameters between 2–8 nm.

Another important class of metal-containing materials is represented by the polymer/aluminum nanocomposites. Huang*et al.* focused their efforts on studying polyethylene-aluminum (PE/Al) nanocomposites [143,144,145]. In their initial study published in 2008 the authors describe the preparation of Al-polymer nanocomposites with concentrations ranging from 1 wt% to 48 wt% metal dispersed in a PE matrix, employing Al particles with diameters of 102 nm [143]. Nanoscale dispersions were observed at concentrations lower than 4 wt%, and agglomerates were detected to be present at higher concentrations. The nanocomposites obtained in this work exhibited electrical and mechanical properties believed to be good enough for practical applications [143]. In an attempt to investigate the dispersion of surface treated Al nanoparticles in the polymer matrix, the same group later prepared PE/Al nanocomposites, via solution compounding, using Al nanoparticles (100 nm diameter) with and without surface modification respectively [145]. The authors observed a strong correlation between the time and concentration dependences of DC-conductivity and the rheological properties of different nanocomposite systems. It was also noticed that the modification of the Al nanoparticles with a silane coupling agent could lead to improved dispersion of the Al particles in the polymer matrix [145]. Conclusions drawn by the researchers illustrate two mechanisms that affect the rheological properties of the Al/PE nanocomposites: while the particle–particle interaction is the dominant mechanism in the "non-surface-treated" composites, the particle-polymer interaction is the dominant one in the "surface-treated" counterparts [145]. One year later, another study from the same group focused on the effects of surface treatment of Al nanoparticles on the electrical properties of linear low density polyethylene composites [144]. It was found that, due to the surface treatment, the Al nanoparticles with non-polar octyl groups not only increased the percolation threshold and the resistivity, but also improved the dielectric properties as compared to the composites filled with non-treated Al [144]. The researchers also noticed that the surface allows to easily control the frequency and concentration dependences of dielectric constant and provided an excellent approach to considerably reduce the dielectric loss of the nanocomposites. The authors consider the findings to be very encouraging for practical applications of the polymer/Al nanocomposites in the electrical and electronic industries [143,144,145]. In conclusion, the improved electrical properties could be directly attributed to the good dispersion and special electrical features of the surface-treated nanoparticles in the polymer matrix [144].

Finally, the last important class of metal-containing nanocomposites covered in this report is the group of copper (Cu)-based materials. A novel in-situ synthetic method for the preparation of copper-polymer nanocomposite materials is presented by Mallick *et al.* [146]. The authors used o-toluidine and cupric sulfate as precursor materials. While during the chemical reaction o-toluidine was oxidized to form poly(*o*-toluidine), the cupric sulfate was reduced to form copper nanoparticles with diameters of ca. 2.5 nm [146]. This procedure demonstrates a single step synthetic route for the preparation of Al–polymer nanocomposite materials. The conductivity of the poly-(*o*-toluidine)-Cu composite material was found to be approximately 10^{-3} S cm^{-1}. As future work, the authors aim to further investigate the potential applications and the larger scale production of such composites [146]. Moreover, the recent availability of nanomaterials suitable to make conductive composites using much smaller particles than in the past served as a driving force for the study performed by Untereker *et al.* [147]. The researchers focused on answering two questions in their study: how does particle size influence the maximum conductivity, and what is the maximum achievable conductivity for these materials. The authors did not find great variations in the conductivity of materials like platinum, carbon black, and silver particles dispersed in polyurethane composites, particles with sizes from micrometer to nanometer range [147]. A surprising finding was that in all the examples studied, the highest conductivity achieved was only about 1% of that of the pure bulk conductive materials. Further experiments to emulate these conductive composites with platinum, carbon black, copper, and nickel particles, and without polymer matrix, showed similar results, leading to the conclusion that the composite systems are intrinsically limited by the contact between filler particles [147]. This finding triggered the hypothesis that small sized particles alone do not improve the making of conductive composite materials. The authors believe, however, that high conductivities might be accomplished if the filler contact density is reduced by sintering or by using high-aspect ratio fillers [147]. At last, a polymer based nanocomposite containing stabilized copper nanoparticles was proposed to serve as a biostatic coating by Cioffi *et al.*[148]. The authors investigated correlations between biological effects and material properties and showed a slow release of metal species to act as growth inhibitors for living organisms [148]. The center goal of the study was to evaluate metal-polymer nanocomposite materials for perspective applications in the preparation of antibacterial paints/coatings to be used in household, biomedical/ hospital, and aerospace industries [148]. The Cu nanoparticles described in this study were found to have diameters around 4.6–4.9 nm.

4. POLYMER/CLAY NANOCOMPOSITES

The exfoliation, intercalation and aggregation of clays in PEO nanocomposites have been extensively studied in the past. Although in complicated multi-component systems it is difficult to draw conclusions if too many parameters are varied, the reality is that the functionality of each component is often altered by addition of another. Due to this fact and for a more complete picture of how a system behaves in a real environment, all the constituents must be considered at once. Typically, to examine the polymer-clay interactions a combination of methods is advisable. Among them, microscopy and scattering are techniques for studying structure and providing a measure of size, shape and interfacial polymer conformation and are frequently employed by various researchers. Additionally, rheology and mechanical testing are much utilized techniques because they can adequately distinguish between properties of chemically versus physically cross-linked polymer-clay materials. In this chapter are reviewed and discussed only the polymer nanocomposites based on laponite and montmorillonite, since these two clays are among the most studied and researched.

4.1. Poly(Urethane Urea) Composites

Poly(urethane ureas) (PUU) are a class of copolymers that exhibit many engineering applications owing to their dual morphology. The presence of both hard and soft segments in the polymer structure contributes to the stiffness, strength and elastic nature of the polyurethane. Despite the excellent elastic properties, the utilization of PUU is limited by the melting of the hard segments at elevated temperatures. One way to improve the mechanical and thermal properties of PUU is the dispersion of inorganic particles into the polymer matrix. In their work, Sormana *et al.* reported the preparation of PUU nanocomposites filled with monoamine-modified laponite (MAML) or diamine-modified laponite (DAML) [149]. A schematic showing the chemical structure of modified laponites is presented in Figure 11. An ion exchange technique was used for the coating on the surface of the laponite clay with the organic diamine or monoamine. The authors observed that both the tensile strength and the elongation at break of the polyurethane-clay nanocomposites (PUCN) containing DAML increased when compared with the neat PUU. This behavior was the result of a better interaction of the inorganic phase with the hard segments of the PUU. DSC thermograms exhibited nearly indistinguishable differences in the glass transition temperatures of the PUCN and pure PUU, suggesting that the modified laponite did not interfere with the soft segment phase-free volume of the PUU. The nanocomposites containing DAML revealed higher tensile strength than those with MAML. Mishra *et al.* utilized the same technique for the modification of laponite clay with dodecylamine hydrochloride [150]. The PUCN (synthesized by *in situ* polymerization technique) showed an increase in the thermal stability; however, their elongation at break was decreased due to the preference of modified laponite for the more polar and soft propylene glycol-segments of the PUU. Surprisingly, when the PUCN were prepared by a solution mixing technique, Mishra *et al.*observed that the modified laponite had affinity for the hard segments of the PUU [151]. The resulting PUCN showed an increase in the storage modulus at 140 °C.

On the other hand, Korley *et al.* used a different approach to improve the thermomechanical properties of the PUCN [152]. A solvent exchange process was employed to disperse neat laponite clay in PUU polymers containing a hydrophobic hard segment, 1,6-hexamethylene diisocyanate – 1,4-butanediol (HDI-BDO), and a soft segment that was either a hydrophilic poly(ethylene oxide)-poly(propylene oxide)-poly(ethylene oxide) (PEO-PPO-PEO) domain, or a hydrophobic poly(tetra-methylene oxide) (PTMO) one. The loaded PEO-PPO-PEO-based PUU nanocomposites showed a decrease in strength and toughness when compared with the same unloaded systems or with the PTMO-based PUU loaded nanocomposites. Dynamic mechanical analysis (DMA) and differential scanning calorimetry (DSC) tests revealed that upon loading with laponite clay the mobility of the soft segments in the PEO-PPO-PEO-based PUU composites was reduced, suggesting a favored attraction of the clay platelets to the polar soft segments that negatively affected the mechanical properties of the material. In contrast, the laponite platelets in the PTMO-based PUU nanocomposites were primarily confined to the hard domains enhancing the toughness and the strength while maintaining the flexibility of the materials.

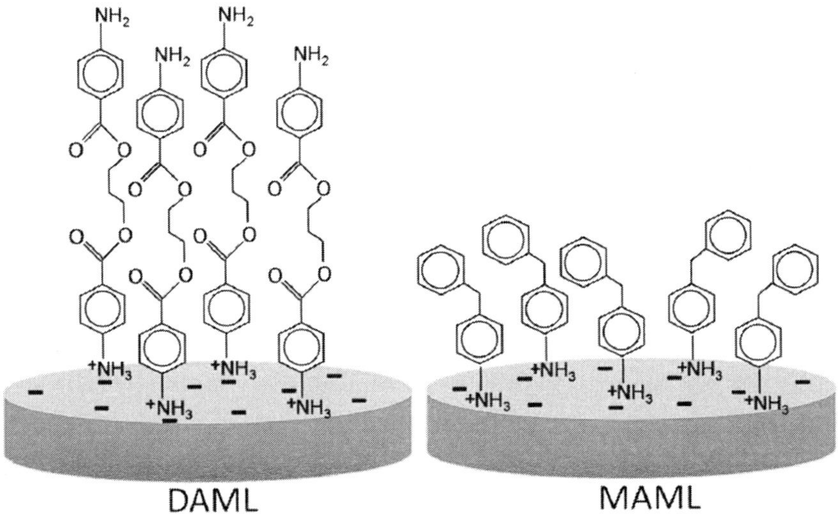

Figure 11. Schematic illustrating the chemical structure of modified laponites. For simplicity, only one side of modification is presented.

4.2. Poly(Methyl Methacrylate)/Laponite Composites

Poly(methyl methacrylate) (PMMA) is a thermoplastic polymer widely used due to its low cost and facile handling and processing. However, when subjected to stress the PMMA is predisposed to fracture. The dispersion of laponite clay platelets into the polymer matrix is employed as a way to improve the properties of PMMA. Fang *et al.* used a Pickering emulsion (an emulsion system stabilized by particles) polymerization to synthesize laponite-mediated poly (methyl methacrylate) (PMMA) nanospheres [153]. The authors prepared the emulsion by mixing a solution of laponite in distilled water with a solution of methylmetacrylate in *n*-heptane, where the laponite clay was used as a stabilizer. The polymerization of MMA was initiated by 2,2'-azobis(2-methylpropionamide) dihydrochloride. The resultant PMMA nanoparticles, wrapped in laponite clay, showed superior electrical conductivity when compared to neat PMMA. The authors hypothesized that the increase in the electrical conductivity of the nanospheres is caused by the intrinsic electrical properties of laponite clay.

Chang *et al.* used tyramine-modified laponite clay to enhance blending with the PMMA chains [154]. Tyramine hydrochloride was used for the modification of laponite clay. Due to the presence of silicate-oxygens and of the phenol groups partially covering the silicate layers, the PMMA macromolecules were absorbed onto the silicate through hydrogen bonding and the resultant nanocomposites displayed exfoliated and intercalated morphologies of the modified laponite into the PMMA matrix. A schematic of the tyramine-modified layered laponite clay is depicted in Figure 12. The authors concluded that the size of the modified clay in the polymer matrix critically impacts the structure and thermal properties of the nanocomposites. For example, when employing silicate clays with a larger diameter, the corresponding PMMA matrix consists of intercalated nanocomposites. An explanation for this behavior is that during the evaporation of the solvent the silicate layers with larger diameter tend to accumulate resulting in silicate-rich domains. Wang *et al.* employed a different technique to enhance blending and mechanical properties of the PMMA nanocomposites by chemically bonding dual-functionalized laponite clay with PMMA [155]. The dual- functionalized laponite clay was synthesized via edge functionalization (silanol condensation) of the laponite with γ-methacryloxypropyltrimethoxysilane (MPTS) followed by cation exchange reactions with cetyltrimethylammonium bromide (CTAB). The dual-functionalized laponite was then copolymerized with MMA to develop a polymer-grafted PMMA nanocomposite through a free radical polymerization. The silanol condensation was performed to avoid the linkage of the clay sheets, while the CTAB intercalation allowed the polymer to break through the clay sheets to a greater extent. The DMA results showed that the PMMA-laponite-MPTS-CTAB nanocomposites have a higher storage modulus than PMMA and PMMA-laponite-MPTS nanocomposites. Finally, Wheeler *et al.* used the same technique for the preparation of PMMA nanocomposites only that the laponite clay was covalently bonded with a different acrylate/methacrylate compound and the binding of the polymer was performed via two types of polymerization: 1) *in situ* polymerization of MMA using AIBN as a free radical initiator; 2) brush polymerization using atom transfer radical polymerization (ATRP) initiator on laponite clay [156]. The TEM analysis indicates that the CTAB treated nanocomposites clays show some separated sheets but aggregates are predominant; these results are in contradiction with those obtained by Wang *et al.* [155].

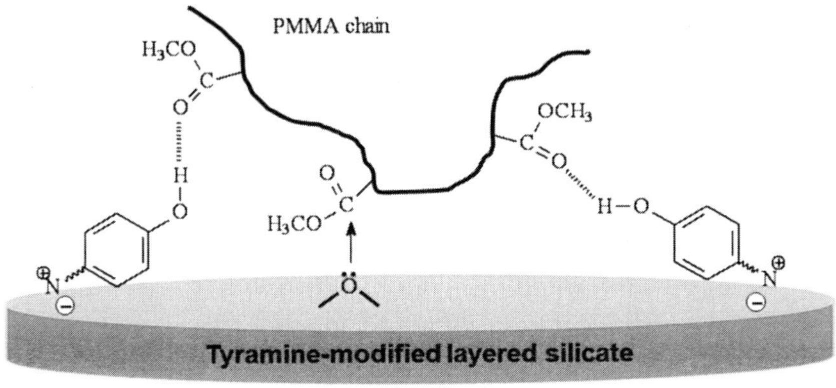

Figure 12. Schematic showing interaction and hydrogen bonding between modified clay and PMMA.

4.3. Polystyrene/Laponite Composites

Konn *et al.* used a nitroxide-mediated polymerization (NMP) of styrene-modified laponite clay to control the growth of polystyrene (PS) chains from the surface of the silicate clay, as a way to enhance dispersability of laponite platelets in polar and nonpolar monomers [157]. The laponite clay was modified by performing cation exchange with a quaternary ammonium alkoxyamine that had incorporated in its structure a *N*-tert-butyl-*N*-[1-diethylphosphono-(2,2-dimethylpropyl)] (DEPN) group. The quaternary ammonium functional group is responsible for the attachment to the laponite clay while the DEPN is used as an initiator for the NMP technique. The authors observed that when low molecular weights of PS-functionalized laponite platelets were dispersed in tetrahydrofuran (THF), the corresponding clay suspension also included some clay aggregates. After increasing the polymer chain length the clay dispersion was promoted as a result of steric barrier created by polymer chains surrounding the clay platelets. Dynamic light scattering (DLS) measurements of the suspensions showed that PS-functionalized laponite exhibited a good dispersability and stability in organic, polar, and nonpolar monomers. X-ray diffraction (XRD) measurements of films obtained from these suspensions revealed that even after casting the clay platelets were still fully exfoliated.

Furthermore, Herrera *et al.* used seeded emulsion (co)polymerization to synthesize polymer/modified-laponite composite colloids [158]. The laponite clay was functionalized with a free radical initiator, 2,2'-azobis(2-methylpropionamide) dihydrochloride, by cation exchange. Because the corresponding clay platelets were strongly aggregated in the presence of water, tetrasodium pyrophosphate (a peptizing agent) was used to accomplish a good dispersion of the clay in water. Next, the emulsion (co)polymerization was accomplished using styrene and butyl acrylate. The authors observed that increasing the amount of initiator used for the modification of laponite clay resulted in silicate aggregates with larger sizes and composite particles with improved morphologies. In a later work, Herrera *et al.* used the earlier mentioned colloids to prepare films by water evaporation and composite particles coalescence [159]. The resultant transparent films (containing 10wt% modified-laponite) exhibited an increase in the thermal stability and glass transition of 18 °C and 12 °C, respectively when compared to the properties of the copolymer in the absence of modified-laponite platelets. Ruggerone *et al.* also employed an emulsion polymerization technique to synthesize polystyrene-modified-laponite nanocomposites [160]. The laponite clay was modified with 5 wt% macromonomer (poly(ethylene oxide)1000 monomethyl ether metacrylate) while the polymerization was performed with styrene using 2,2` azobis(cyanopentanoic acid) as initiator. The authors noticed that the extent of the modified-laponite significantly affected the mechanical properties of the PS-laponite nanocomposites films. For example, films with a loading of 5wt% modified-laponite displayed an increase in the stiffness and tensile strength when compared with the unmodified films. On the other hand, films with a content of 10 and 20 wt% modified-laponite revealed reduced mechanical properties. This behavior was explained by the presence of large silicate clay agglomerates formed at high clay filling.

At last, Caruso *et al.* employed a layer-by-layer self-assembly technique for the coating of PS spheres templates with laponite clay multilayers [161]. Prior to coating with silicate clay, the PS particles were modified by the adsorption of 3 layers of positively and negatively charged polyelectrolytes [poly(diallyldimethylammonium chloride) and poly(sodium 4-styrenesulfonate)]. The benefit of this procedure was that the thickness of the multilayer wall and the diameter of the composite particles could be controlled by adjusting the initial PS diameter and the number of layers deposited on the templates. The authors demonstrated that through the use of this technique, the laponite clay is a potential candidate for building blocks utilized in hollow sphere fabrication. These laponite spheres were obtained by heating the PS colloidal templates at elevated temperatures for the removal of the organic core.

4.4. Poly(Ethylene Oxide)/Laponite Composites

Loizou *et al.* investigated the dynamic responses in poly(ethylene oxide) (PEO)/laponite nanocomposite hydrogels by means of SANS and rheology measurements [2]. The authors employed different molecular weight PEOs for the preparation of gels with the same mass composition (3% laponite and 2% PEO). All samples had a pH of 10 and a similar NaCl concentration. SANS measurements indicated that at rest no anisotropy was present for any of the systems while in the presence of an applied flow field, anisotropy developed into the gels containing the higher molecular weights of PEO. It was also observed that the anisotropy increases with the increasing molecular weight of polymer. This behavior was explained by an enhancement of polymer-laponite clay interconnections as a result of the increase of the polymer chain length. The rheological measurements conducted on the gels containing the higher molecular weight PEOs showed a frequency-dependent storage modulus suggesting the formation of a transient network between the laponite clay and the polymer. A network-like structure on the micron and nanometer length scale was confirmed by SEM and TEM images of the hydrogels [1]. The complexation of PEO oxygens with the Na^+ counterions of the silicate clay was also confirmed by solid-state NMR experiments which revealed a decrease in the segmental mobility of the PEO [162]. Stefanescu *et al.* studied the structure and the thermal properties of multilayered laponite/PEO nanocomposites prepared by shear-spreading the PEO/laponite nanocomposite hydrogels [54]. SEM and SANS measurements obtained from the films indicated that the laponite platelets oriented in the direction of spreading, resulting in well dispersed and oriented multilayers. XRD data showed an increase in the order reflection for the films with higher amounts of laponite due to a dense packing of the platelets that had a tendency to stack and orient. The authors also observed that different compositions and different polymer molecular weight affected the crystallinity of the nanocomposite films. As indicated by DSC data, the crystallinity of the films increased with the increase of the PEO molecular weight while lower melting temperatures were observed with the increase of laponite concentration. Later on, hybrid polymer-clay nanocomposites were investigated by Stefanescu *et al.* [6]. The authors prepared nanocomposite films containing two types of clay (laponite and montmorillonite) as a way to improve the properties of the films. DSC measurements revealed that higher amounts of montmorillonite in the system led to an increase of crystallinity of PEO in the nanocomposite. This behavior was the result of a much lower surface per gram of montmorillonite available for PEO coordination than laponite. As a consequence the uncoordinated PEO chains tended to move and rearrange increasing in this way the crystallinity of the films. However, the XRD showed that higher montmorillonite amounts in the nanocomposite films improved the orientation of the clay platelets. The increased size of the montmorillonite clay was responsible for the behavior where the clay platelets followed the axis induced by the spreading direction as opposed to the laponite clay that gradually lost parallelism due to their lack of high aspect-ratio. Stefanescu *et al.* also performed a mechanical study on the hybrid PEO-clay nanocomposite gels and multilayered films [55]. Rheological measurements of the hybrid nanocomposite gels showed a shear thinning behavior which indicated an overall orientation of the nanostructures in the gels. Moreover, the storage and loss modulus increased with the increasing amount of laponite into the sample due to improved distribution of surface area of the clay which elevated the viscosity of laponite rich gels. However, DMA measurements of the solid multilayered films revealed that the behavior of elastic and loss modulus was opposite to the one observed for the storage and loss modulus of the precursor nanocomposite gels. The increase in the crystalline fraction of the PEO in the films containing higher amounts of montmorillonite was found responsible for this behavior.

Loyens *et al.* prepared PEO/laponite and PEO/modified laponite nanocomposites via a melt-compounding procedure [163]. The modification of laponite with poly(ethylene glycol) components was performed in order to increase the compatibility of the clay with PEO. The authors observed that the addition of laponite clay inhibited the crystallization of PEO. This behavior was the result of the coordination of PEO with the laponite platelets which restricted the mobility of the polymer chains. TGA measurements indicated that nanocomposites based on modified laponite exhibited an increased thermal stability at low clay loadings while the nanocomposites based on pure laponite showed an increase in thermal stability only at high clay concentrations. Only a slight improvement of the mechanical properties was observed for the PEO nanocomposites based on pure or modified laponite. In a later work, Loyens *et al.* demonstrated that the presence of a salt ($NaClO_4$) in melt-compounded PEO/laponite nanocomposites did not exhibit improved mechanical properties nor increased ionic conductivity when compared to the corresponding PEO/ $NaClO_4$ complex [52]. However, when the laponite clay was replaced with Cloisite30B the nanocomposites showed excellent mechanical properties and good ionic conductivity.

4.5. Diblock and Triblock Copolymers/Laponite Composites

Gournis *et al.* investigated the morphology of two polymer-clay nanocomposites: laponite/PEO clay/homopolymer and laponite/PEO-*b*-polyisoprene (PEO-PI) clay block copolymer nanocomposites [164]. An adjustment of the layer spacing was observed for the laponite/PEO clay, a result of the elasticity provided by the intercalated polymer. In the case of laponite/PEO-PI clay, "hairy" plates were obtained, where the hydrophilic PEO was intercalated into the clay and the hydrophobic PI restrained the laponite platelets from aggregation. A schematic of the "hairy" structure in the laponite/PEO-PI is presented in Figure 13. Michell *et al.* also investigated diblock copolymer-clay nanocomposites where the utilized copolymers were synthesized using PS, PI and dioctadecyl dimethyl ammonium-modified laponite [165]. The visoeleastic data for the ordered and disordered states of the nanocomposites indicated minor changes compared to those of the unfilled polymer. This denoted that the addition of the silicates into the system had a negligible effect on the viscoelastic properties suggesting that a network structure was absent from these block copolymer-modified-laponite nanocomposites. On the contrary, Agrawal *et al.* observed that the addition of laponite platelets to an ABA triblock coplolymer, poly(lactide)-PEO-poly(lactide), had a significant impact on the rheological properties of the nanocomposites [166]. A particle-polymer network with enhanced elasticity was formed due to the PEO adsorption onto the silicate platelets.

Moreover, Liu *et al.* investigated the temperature-triggered gelation of laponite/poly(*N*-isopropyl acrylamide)/poly quaternarized *N,N*-dimethylaminoethyl methacrylate clay graft copolymer nanocomposites [167]. The addition of laponite clay to the copolymer system had as a result: 1) a decrease of the gelation temperature of the nanocomposites; 2) a decrease of the critical concentration value of the copolymer needed for the gelation process. This observation was explained by the presence of interaction between laponite and copolymer through electrostatic interactions and hydrogen bonding.

De Lisi *et al.* studied the thermodynamics of adsorption of triblock copolymers (PEO-PPO-PEO) at the interface of laponite clay/solution when varying the hydrophilicity, the architecture and the molecular weight of the copolymer [168]. Calorimetric results and modeling kinetic data indicated that the architecture of the copolymers did not have a significant influence on the interactions between the laponite platelets and the copolymers. However, keeping the ethylene oxide/propylene oxide ratio constant and varying the molecular weight of the copolymers, the authors observed that higher molecular weights resulted in an increase in the affinity for the laponite clay. Moreover, the greater the hydrophilicity of the copolymer the bigger was the affinity for the laponite surface. De Lisi *et al.* also investigated the effect of PEO-PPO-PEO triblock copolymers on aqueous laponite clay dispersions by means of small-angle neutron scattering (SANS) [169]. SANS data indicated that the addition of PEO-PPO-PEO to the laponite clay system retarded the gelation process in the time-window. Nelson *et al.* also performed SANS measurements on the adsorbed PEO-PPO-PEO triblock copolymers on laponite clay [170]. The authors observed that the hydrophobic PPO segments were preferentially adsorbed at the surface of the laponite platelets while the hydrophilic PEO were dangling in the solution. Later on, De Lisi *et al.* used the same triblock copolymer-laponite systems for thermal and structural investigation [171]. The authors noticed that the enthalpy of melting of the nanocomposites was affected by the macromolecules radiating away from the laponite clay while the macromolecules associated with the silicate did not have any contribution to the enthalpy of melting. The X-ray diffraction data of the nanocomposites indicated an exfoliated structure. Finally, De Lisi *et al.* also investigated the affinity toward an organic material, phenol, of the modified laponite with PEO-PPO-PEO triblock copolymers [172]. The authors observed that the addition of copolymers to a phenol/laponite solution led to the replacement of the organic material by the PEO-PPO-PEO. However, when the inversed copolymer (PPO-PEO-PPO) was used, the anchoring of phenol to the laponite surface was enhanced.

4.6. Polyurethane/Montmorillonite Composites

Montmorillonite (MMT), a layered silicate clay, has been the focus of extended research for the preparation of polyurethane (PU)/clay nanocomposites. Due to its high aspect ratio and stiffness, MMT was found to enhance the thermal stability, morphology [7,173,174,175], mechanical and physical properties [7,176,177,178] of polymer/clay nanocomposites. Dan *et al.* used a melt processing technique as a way to improve the morphological and physical properties of PU/MMT clay nanocomposites [173]. The PUs were synthesized using 4,4`-diphenylmethane diisocyanate and 1,4-butanediol as hard segments and poly(tetramethylene oxide) glycol or poly(butylenes adipate) glycol as soft segments. The clay added to the polymer system was either pure MMT or modified MMT clay with a quaternary ammonium salt containing two $-CH_2CH_2OH$ groups. The XRD and TEM images of the PU/clay nanocomposites indicated that the pure MMT had a very poor dispersion with thick agglomerated clay particles present in the nanocomposites while the modified MMT showed dispersion close to exfoliation. The behavior was caused by the formation of H-bonding between the carbonyl groups in PUs and the hydroxyl groups present in the modified MMT which resulted in a very good dispersion of the clay. A significant increase in the tensile properties of the nanocomposites was observed with the addition of both pure MMT and modified MMT to the system. However, the modified MMT increased the tensile modulus and stress to a greater extent than the pure MMT due to a better dispersion of the exfoliated silicate layers.

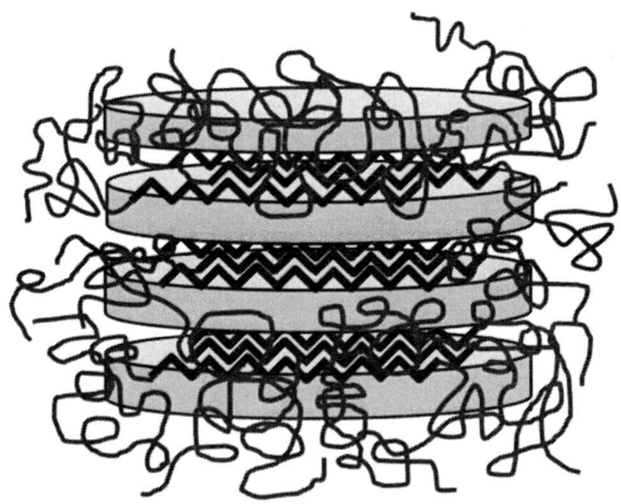

Figure 13. Schematic showing PI chains emanating from the laponite platelets filled with PEO chains ("hairy" plates).

Furthermore, Dai *et al.* also synthesized PU/MMT nanocomposites via a solution-intercalation technique [7]. The sodium cations of the MMT clay were exchanged with organophilic alkyl-ammonium ions in order to improve the dispersity of the stiff nanolayers of the silicate clay into the polymer solution. Due to the presence of intercalated or exfoliated clay layers, the nanocomposites containing the modified MMT exhibited an increase in the tensile strength and elongation at break. A cationic-exchange reaction of the MMT was also performed by Lee *et al.* [174].

The XRD and TEM images of the PU/modified MMT nanocomposites indicated that the silicate platelets were exfoliated. Because of the high thermal insulation effect of the MMT clay, the PU/clay nanocomposites showed an increase in the thermal resistance. The tensile properties of the nanocomposites were also improved with the addition of the modified MMT. In addition, Cheng-Yang *et al.* synthesized PU/clay nanocomposites with very good mechanical and thermal properties by using a de-aggregated, solvent-swollen, organic modified MMT [177]. The modified MMT with 12-aminolauric acid was first de-aggregated into smaller size particles in DMF to prepare a stable suspension. Next, the suspension was mixed with the PU and highly exfoliated PU/clay nanocomposites were obtained, as indicated by TEM micrographs. The presence of the modified MMT into the polymer system decreased the mobility of the polymer chains which resulted in the retardation of the glass transition of the nanocomposites. The TGA measurements showed a great improvement of the thermal solubility of the nanocomposites with the addition of modified MMT. Moreover, an increase in the mechanical properties was observed. Tien *et al.* also used a swelling agent-modified MMT for the preparation of PU/clay nanocomposites [178]. The authors observed that with the increase of hydroxyl groups present in the swelling agent, the extent of intercalation of the modified MMT clay also increased. As a result, the tensile strength of the nanocomposites was dramatically improved. The behavior was attributed to the increase of the interfacial bonding between the PU molecules and the modified MMT containing the highest amount of hydroxyl groups.

4.7. Poly(ε-Caprolactone)/Montmorillonite Composites

Polymer clay nanocomposites with improved properties have been reported using poly(ε-caprolactone) (PCL) and MMT. Several methods have been employed in the past for the preparation of PCL/MMT nanocomposites, such as *in situ* [179,180] or bulk [181] polymerization, melt blending [182], and solution mixing [183]. Wu *et al.* used a solution mixing technique for the preparation of PCL/MMT nanocomposites [183]. A solution of MMT dispersed in deionized water and *N,N*-dimethylformamide was mixed with a PCL/*N,N*-dimethylformamide solution. The resulting polymer/clay nanocomposites showed an increase in the melting crystallization temperature, phenomenon explained by the presence of MMT which could suppress the mobility of the PCL segments. TGA measurements indicated that an increase in the MMT content led to enhanced thermal stability of the PCL/clay nanocomposites. Tarkin-Tas prepared PCL/clay nanocomposites making use of a different technique [179]. *In situ* ring opening polymerization (ROP) of ε-caprolactone was employed in the presence of an initiator, dibutyltin dimethoxide. The clay used in the process was MMT modified with 1-decyl-2-methyl-3-(11-hydroxyundecyl) imidazolium cation via an ion-exchange reaction. The hydroxyl groups on the long alkyl chain of the alkyl-imidazolium cation acted as initiators in the ROP of the ε-caprolactone and also as chain transfer agents. Variations in the molar ratio of monomers to hydroxyl groups raised the possibility of adjusting the molecular weights of the PCL to the desired value. XRD results indicated the formation of an intercalated morphology of the nanocomposites whereas the TEM images allowed the visualization of highly exfoliated clay platelets throughout the polymer matrix. The high mobility of the hydroxyl groups at the end of the long flexible alkyl chain was found responsible for inducing the exfoliation, due to a very good availability of the functional groups for the initiation of the ROP of ε-caprolactone. TGA measurements revealed that the clay did not have any effect on the degradation temperature of the nanocomposites. DSC data showed that the clay platelets could act as nucleating agents for the PCL crystallization. As a result, an increase in the mechanical properties of the PCL/clay nanocomposites was observed. Liao *et al.* also performed an *in situ* ROP for the preparation of nanocomposites [180]. However, instead of the conventional heating the authors used microwave irradiation which proved to have a greater monomer conversion and molecular weight and to be much faster. High degrees of exfoliated and/or intercalated structure were observed by TEM images. TGA measurements showed that the thermal stability of the nanocomposites was affected by both the clay and the molecular weights of the polymer.

Kiersnowski *et al.* explored a bulk polymerization technique for the preparation of PCL/clay nanocomposites [181]. Both pure MMT and MMT modified with hexadecyltrimethylammonium bromide (HTAB) or dodecylamine hydrochloride (DAC) were used in the process. The goal of the research was to understand how the presence of clay influences the intercalation of PCL. Formation of nanocomposites was observed for PCL/MMT and PCL/MMT-HTAB systems whereas microcomposites were obtained for PCL/MMT-DAC systems. Attractive interactions between polymer and neat silicate, as well as a relatively increased gallery height in MMT-HTAB, were found responsible for the formation of nanocomposites. In the case of PCL/MMT-DAC, DAC acts as a barrier for the incoming polymer to the silicate layer. Moreover, the gallery height of the MMT-DAC is too small and not enough volume is available to accommodate macromolecules. As a result, the growth of polymer chain stopped at early stages and microcomposites were formed.

At last, Yu *et al.* employed a melt blending technique to prepare hybrid poly(L-lactide)/PCL/organically modified MMT nanocomposites [182]. The authors observed that the addition of up to 5 wt% of modified MMT to polymers led to the formation of more exfoliated individual particles, improving in this way the thermal stability and crystalline abilities of poly(L-lactide)/PCL/modified MMT nanocomposites. The tensile strength, modulus and elongation at break were also enhanced with loading up to 3% of modified MMT, whereas higher contents of clay induced an increased brittleness in the material.

4.8. Polylactic Acid and/or Poly(Lactic-co-Glycolic Acid)/Montmorillonite Composites

Preparation of polymer clay nanocomposites using polylactide has also been previously reported [184,185,186,187,188,189,190]. Polylactide is a biodegradable polymer with excellent processability and physical properties. Despite some useful characteristics, the heat resistance, melt-viscosity and mechanical properties of polylactide are not satisfactory. To improve these properties, MMT clay and/or organically modified MMT were used among other fillers for the preparation of nanocomposites. Lee *et al.* combined an electrospinning process with a salt leaching/gas foaming method as a way to prepare modified MMT reinforced polylactide nanocomposite scaffolds with potential applications in the facile transport of metabolic nutrients, cell implantation and blood vessel invasion [187]. Dimethyl dehydrogenated tallow ammonium was employed as a cation for the modification of the MMT clay. The electrospun fibers resulted from the electrospinning process provided an interconnected nano-sized porous network. However, the porosity of the scaffolds was difficult to control especially when larger-sized pores were required. To obtain micro-sized pores, a salt leaching/gas foaming method was employed. NH_4HCO_3/NaCl salt particles scattered in the interlaminar layers of electrospun fibers sheets were mechanically kneaded and molded by compression. Next, the samples were immersed in hot water to generate gaseous ammonia and CO_2 within the solidifying polymer matrixes. The resulting scaffolds presented dual porous structure with nano-sized and micro-sized pores as indicated by SEM micrographs. The incorporation of the modified

MMT clay to the polylactide fibers increased the mechanical and biodegradable characteristics of the scaffolds.

Moreover, Urbanczyk *et al.* used *in situ* ROP in supercritical CO_2 to synthesize polylactide/clay nanocomposites [189]. MMT modified with dimethyl (dehydrogenated tallowalkyl) ammonium cation (MMT1) or with bis (2-(hydroxyethyl) methyl) (tallowalkyl) ammonium cation (MMT2) were used to enhance the properties of the polylactide. The authors observed that the nature of the clay had an important role on the morphology of the nanocomposites synthesized by ROP of polylactide. The XRD and TEM images revealed that nanocomposites prepared with MMT1 had intercalated morphologies whereas those containing MMT2 revealed exfoliated morphologies. The hydroxyl groups present in the MMT2 clay could act as initiators and gave rise to polylactide chains grafted onto the clay surfaces resulting in an efficient delamination of the clay stacks. The delaminated polylactide/clay nanocomposites indicated a great improvement in both stiffness and toughness when compared with the unfilled polylactide matrix. Jiang *et al.* also obtained nanocomposites with improved properties when employing a melt extrusion technique to prepare polylactide/organically modified MMT nanocomposites [186]. The polylactide/MMT nanocomposites containing low concentration of MMT showed higher tensile strength and strain-at-break when compared with pure polylactide. This behavior was attributed to the presence of high degrees of chain orientation in the nanocomposites due to the shear amplification effect caused by the MMT platelets. Additionally, Gu *et al.*studied the linear and nonlinear shear rheological behaviors of polylactide/organophilic-MMT nanocomposites [185]. A master batch method was adopted for the preparation of the nanocomposites and dioctadecyl dimethyl ammonium bromide was employed for the modification of MMT clay. The authors observed that the linear viscoelastic properties of the melted nanocomposites displayed a solid-like rheological response (a result of strong friction between clay platelets) with increasing the clay content in the polylactide matrix. An increase of storage moduli, loss moduli and dynamic viscosities of the nanocomposites was also detected with the increase of the MMT content. The nonlinear viscoelastic results of the nanocomposites indicated that at low shear rates, the steady shear viscosities were much higher than the one of the pure polylactide and that the viscosity increased dramatically with the MMT content, whereas at higher shear rates the steady shear viscosities were lower than that of polylactide, as a result of the formation of shear-induced alignment of the dispersed MMT platelets.

Drug-loaded polymer/MMT nanoparticles have also been reported in the past [184,188,190]. For example, Dong *et al.* and Sun *et al.* prepared poly (D,L-lactide-co-glycolide)/MMT (PLGA/MMT) nanoparticles for oral delivery of paclitaxel, an anticancer drug [188,190]. An emulsion/ solvent evaporation method was employed for the preparation of paclitaxel-loaded PLGA/MMT nanoparticles. The authors observed that the presence of MMT clay into the system could significantly increase the interactions between nanoparticles and cells due to the formation of London-van-der Vaals forces and H-bonds [190]. The presence of the strong interactions was essential to enhance the absorption of the drug-loaded nanoparticles. In vitro drug release studies of the loaded nanoparticles revealed an initial burst in the first day followed by a release of paclitaxel at a slow constant rate. It was observed that for the nanoparticles containing MMT, the initial burst was slightly reduced and the speed of drug release was increased when compared to the MMT free nanoparticles. Feng *et al.* also synthesized nanoparticles for oral delivery of anticancer drugs with docetaxel as a model drug [184]. Docetaxel-loaded polylactide-vitamin E d-α-tocopheryl polyethylene glycol 1000 succinate (TPGS) /MMT nanoparticles were prepared by a modified solvent extraction/evaporation method where TPGS was used due to its high emulsification effect as well as the high drug encapsulation efficiency.

4.9. Poly(Ethylene Oxide)/Montmorillonite Composites

Nanocomposites of poly(ethylene oxide) with montmorillonite have been extensively studied in the past decade [5,191,192,193,194]. It has been recently shown that the ionic strength in PEO-MMT nanocomposite gels, dictated by addition of various Li^+ and Na^+ salts, affects the extent of internal orientation of such systems when subjected to elongational flow [194]. For example, Li based salts triggered the formation of stronger polymer clay networks, owing to the ability of Li^+ ions to screen water/PEO hydrogen bonding to a greater extent than Na^+ ions. The disruption of hydrogen bonding promoted the coordination of PEO oxygens to the metals from the surface of the MMT platelets. Additionally, it was found that for salts of the same metal (e.g. Li^+), the bulkiness of the counter ion could also affect the extent of orientation in such processes. For example gels containing the bulky SO_4^- counterion could orient easier than gels containing the relatively small Cl^- anion [194].

Furthermore, in films prepared from such PEO-MMT gels it was observed that the nanoscopic MMT platelets oriented with their surface parallel to the plane of the film [193]. The strength of the network in PEO-MMT gels appeared to have a significant impact on the morphology of films prepared via a layer-by-layer approach, where a lower gel viscosity resulted in finer polymer-clay layers in the films. Additionally, it was observed that the strength of the network in the nanocomposite gels containing various salts was inversely proportional to the fraction of crystalline polymer, and ultimately to the storage modulus, in the resultant nanocomposite films [193]. Figure 14 presents a SEM micrograph obtained from a PEO-MMT film containing Li_2SO_4. Two 2D SAXS patterns are additionally presented for the cases when the X-ray beam was projected perpendicular to the edge and to the top surface of the film. Highly oriented layers resulted from the layer-by-layer preparation method can be observed in the SEM micrograph. In addition, the anisotropic SAXS pattern observed from the edge of the film coupled with the isotropic pattern seen when the X-ray beam is projected from the top clearly indicate that the MMT platelets are oriented with the surface parallel to the plane of the film.

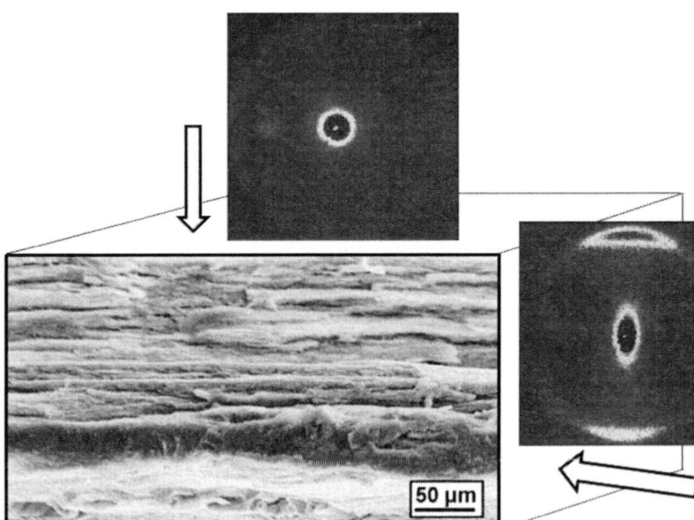

Figure 14. Schematic of a PEO-MMT film, containing Li_2SO_4, that was prepared through a layer-by-layer approach. The exposed freeze-fractured edge was analyzed via SEM and a micrograph is displayed. Additionally 2D SAXS patterns are presented for the cases when the X-ray beam is projected on the edge and on the top surface of the film.

4.10. Polypropylene/Montmorillonite Composites

Polypropylene (PP) is one of the most widely used polyolefin polymers because of its low density, low cost and excellent chemical resistance. When organically modified MMT is dispersed in the polymer matrix an enhancement in mechanical properties, heat distortion temperature and a decrease in flammability of the corresponding nanocomposites is typically observed. Sun *et al.* synthesized such nanocomposites through a melt-mixing method by using MMT modified with tetraethylene pentamine and vinyl triethoxy silane as a coupling agent [195]. The modification of the hydrophilic MMT to a hydrophobic MMT was necessary to increase the compatibility of the clay with the PP while the coupling agent was used to enhance the interlayer spacing of the MMT. The authors observed an improvement in the mechanical and thermal properties of the nanocomposites, as a result of a good dispersion of organomodified MMT in the polymer matrix, as indicated by TEM micrographs and XRD patterns.

Rohlmann *et al.* investigated the linear viscoelasticity behavior of PP/organophilic MMT nanocomposites prepared via melt-mixing [196]. The MMT was modified with dimethyldialkyl ammonium halide to ensure a better compatibility with the polymer which facilitated the exfoliation and dispersion of the clay platelets in the polymer. However, a complete exfoliation was not achieved for the PP/ organophilic MMT systems. The addition of PP grafted with maleic anhydride (a coupling agent) resulted in a complete exfoliation of the clay platelets. Dynamic strain sweep measurements of the nanocomposites indicated a linear viscoelastic behavior that was found to depend on the concentration of the clay. The maximum strain of linear viscoelastic behavior is reduced with the increase in the amount of the exfoliated platelets suggesting that the linear viscoelastic region was sensitive to the composite structure. Moreover, the annealed nanocomposites indicated a solid-like rheological behavior, a result of percolated network superstructure formed due to physical interaction between the MMT platelets.

Unlike Rohlmann *et al.* [196], Qin *et al.* observed a nearly exfoliated dispersion of clay platelets in the polymer matrix for the nanocomposites containing PP/ maleated PP/ modified MMT with diocta-decyldimethyl ammonium chloride [197]. The photo-oxidative behavior of these nanocomposites was investigated and it was detected that an exposure to UV radiation resulted in a faster photo-oxidative degradation than that of pure PP. Photoresponsive groups introduced in the system with the addition of modified MMT and maleated PP were responsible for the increase of photo-oxidative degradation of the PP matrix. Nowacki *et al.* prepared a similar system of isotactic PP/ maleated PP/ organically modified MMT, with the difference that the modification of MMT was performed with octadecyl amine [198]. The authors investigated the nucleation of spherulites in the nanocomposites under shear. It was observed that shear-induced crystallization of the films obtained by compression molding augmented the nucleation density when compared with samples crystallized in static conditions. A decrease in the spherulite sizes was also noted. A possible explanation for the behavior could be the effect of increased local stress and orientation of polymer chains under shearing. Higher shear rates also resulted in an increase of the nucleation intensity.

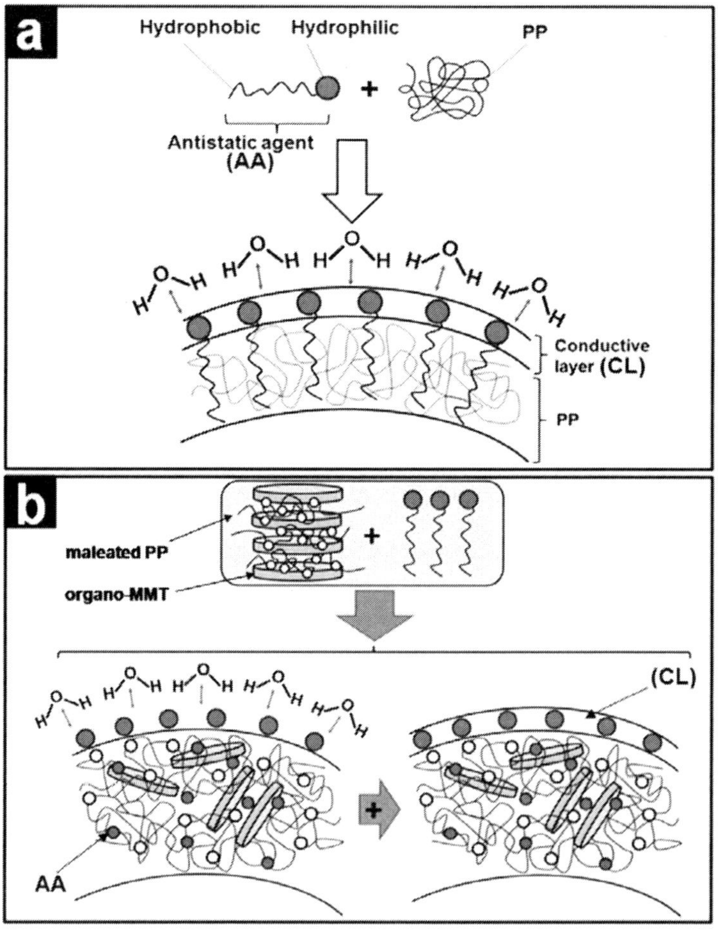

Figure 15. Schematic depicting possible interaction mechanisms between either PP and AA (a), or between PP, AA and organo-MMT (b).

Finally, Chow *et al.* incorporated in the PP/ maleated PP/ organo-MMT nanocomposites an antistatic agent (AA) based on polyamide/polyether block amide [199]. The authors observed that the presence of the AA had minor effects on the tensile strength, tensile modulus, elongation at break and impact strength of the nanocomposites. However, a better wettability and interfacial interaction between organo-MMT and PP matrix was achieved with the addition of the AA. The surface resistivity was also investigated and it was observed that at lower concentration of AA (3–6 wt%) the nanocomposites exhibited an insulator behavior while at higher concentration of AA (9 wt%) an antistatic behavior was achieved. A possible mechanism for the interaction and intercalation of PP/ maleated PP/ organo-MMT nanocomposites is depicted in Figure 15.

5. CONCLUSIONS

Metal-polymer nanocomposite materials are becoming more popular for applications requiring low cost, high metal surface areas. Catalytic systems seem to be the most prevalent application for a wide range of metals used in polymer nanocomposites, particularly for metals like Pt, Ni, Co, and Au, with known catalytic activities. Electrically active devices obtained by embedding metal nanoparticles in conductive polymer matrices are also a popular set of applications for these types of materials. Worthy to be mentioned are optically active devices, along with a wide variety of Fe, Co and Ni based magnetically active products. From a muscle like structure to bioactive nanocomposites, these materials deliver enhanced properties over the original components, allowing for a great expansion their usage envelope.

From a synthetic perspective, today's researchers have developed highly specific pathways for obtaining very well tailored properties. In addition, very cost effective and generally applicable synthesis pathways have been designed. Such methods are well suited for a large variety of metals and deliver nanoscale metal-polymer composite materials with significantly improved properties over the starting materials. Some of these methods are based on unique chemical bonds and/or treatments that help compatibilize metal surfaces with the organic matrixes; others simply utilize steric effects like mechanical wrapping of the polymers around metal nanospheres to obtain a metal-polymer nanocomposite product, while other yet use compounding methods to mechanically insert metal particles in polymer substrates.

There seems to be a trend of preparing these nanocomposites "tailor-made" for a specific application, rather than customizing a universal product. Even though the great majority of the research described here presents a synthetic method of some sort, there are some papers published on modifications on premade metal-polymer nanocomposite. Overall, metal-embedded polymer nanocomposites bring a whole new level of possibilities to the world of engineered materials, enhancing already existing properties within the starting materials, as well as creating a new set of properties found only in the composite structures.

Furthermore, following the recent developments in the fields of nanotechnology, CNT and clay reinforced polymer composites received tremendous attention in the past decades, and currently occupy elite places among the most studied polymer composites. Some of the most frequently utilized techniques to prepare polymer/CNT and/or polymer/clay nanocomposites include melt mixing, solution casting, electrospinning and solid-state shear pulverization. Literature reports indicate that similar to the solution casting method, electrospinning and solid-state shear pulverization are techniques completely benign to the various polymer matrixes. On the other hand, prolonged mixing times, which are often needed to achieve good filler dispersion when the melt compounding technique is used, regularly lead to extensive polymer degradation. Regardless of the technique utilized, some reports suggest that a good dispersion of CNTs and clays in polymer matrixes and formation of percolated networks sometimes may not be synonymous, especially when the dispersed fillers are aligned predominantly in one direction. When not functionalized, often times clays and CNTs can only be dispersed in polymer matrixes as microscale aggregates because of their poor affinity for the polymers and increased tendency for agglomeration. The interfacial adhesion between clays and/or CNTs and the polymer matrixes is often times improved by functionalizing the filler, fact that allows for a homogeneous nanoscale-dispersion of the reinforcing agent within the matrixes. Other times various surfactants (such as anionic surfactants) may be employed to achieve an extensive dispersion of the fillers. Some of the chemicals (covered in this review) utilized to functionalize CNTs include maleic anhydride, octadecylamine, polyethylene glycol, poly(hexamethylene adipamide), poly(bisphenol-A-co-epichlorohydrin), and various acids. On the other hand, some of the chemicals utilized here to functionalize clays include γ-methacryloxypropyltrimethoxysilane, cetyltrimethyl-ammonium bromide, PEO-PPO-PEO copolymers, 12-aminolauric acid, poly(ε-caprolactone), dimethyl (dehydrogenated tallowalkyl) ammonium and/or bis (2-(hydroxyethyl) methyl) (tallowalkyl) ammonium cations, etc. When properly dispersed within the matrixes, CNTs and clays have the ability to improve the properties of the resultant materials several orders of magnitude relative to the unfilled polymers. The enhanced properties may include tensile behavior, strength, toughness, stiffness, electrical and thermal conductivity and crystallization kinetics. An example of the latter case is that ordered nanotubes can act as nucleating templates for the growth of oriented crystalline superstructures. The presence of CNTs often leads to an increase in the melting/crystallization temperature of the polymer matrix, as it is the case for isotactic polypropylene. However, for some other polymers, such as poly(ethylene oxide), the presence of nanotubes leads to a decrease of the melting/crystallization temperature. In addition, some studies suggest that the presence of CNTs do not strongly impede the chain motion of polymer matrix.

ACKNOWLEDGEMENTS

The authors thank William H. Daly and Ioan I. Negulescu of LSU for their tremendous guiding and supporting efforts throughout the years.

REFERENCES

1. Loizou, E.; Butler, P.; Porcar, L.; Kesselman, E.; Talmon, Y.; Dundigalla, A.; Schmidt, G. Large scale structures in nanocomposite hydrogels. Macromolecules 2005, 38, 2047–2049.

2. Loizou, E.; Butler, P.; Porcar, L.; Schmidt, G. Dynamic responses in nanocomposite hydrogels. Macromolecules 2006, 39, 1614–1619.

3. Schmidt, G.; Nakatani, A.I.; Butler, P.D.; Karim, A.; Han, C.C. Shear orientation of viscoelastic polymer-clay solutions probed by flow birefringence and SANS. Macromolecules 2000, 33, 7219–7222.

4. Schmidt, G.; Nakatani, A.I.; Han, C.C. Rheology and flow-birefringence from viscoelastic polymer-clay solutions. Rheol. Acta 2002, 41, 45–54.

5. Stefanescu, E.A.; Dundigalla, A.; Ferreiro, V.; Loizou, E.; Porcar, L.; Negulescu, I.; Garno, J.; Schmidt, G. Supramolecular structures in nanocomposite multilayered films. Phys. Chem. Chem. Phys. 2006, 8, 1739–1746.

6. Stefanescu, E.A.; Daly, W.H.; Negulescu, I.I. Hybrid polymer/clay nanocomposites : Effect of clay size on the structure of multilayered films. Macromol. Mater. Eng. 2008, 293, 651–656.

7. Dai, X.; Xu, J.; Guo, X.; Lu, Y.; Shen, D.; Zhao, N.; Luo, X.; Zhang, X. Study on structure and orientation action of polyurethane nanocomposites. Macromolecules 2004, 37, 5615–5623.

8. Chatterjee, T.; Mitchell, C.A.; Hadjiev, V.G.; Krishnamoorti, R. Hierarchical polymer-nanotube composites. Adv. Mater. 2007, 19, 3850–3853.

9. Jiang, H.; Moon, K.; Li, Y.; Wong, C.P. Surface functionalized silver nanoparticles for ultrahigh conductive polymer composites. Chem. Mater. 2006, 18, 2969–2973.

10. Keng, P.Y.; Shim, I.; Korth, B.D.; Douglas, J.F.; Pyun, J. Synthesis and self-assembly of polymer-coated ferromagnetic nanoparticles. ACS Nano 2007, 1, 279–292.

11. Kwong, H.Y.; Wong, Y.W.; Wong, K.H. Temperature dependence of magnetoresistivity of cobalt-polytetrafluoroethylene granular composite films. J. Appl. Phys. 2007, 102, 114303.

12. Pirkkalainen, K.; Leppänen, K.; Vainio, U.; Webb, M.A.; Elbra, T.; Kohout, T.; Nykänen, A.; Ruokolainen, J.; Kotelnikova, N.; Serimaa, R. Nanocomposites of magnetic cobalt nanoparticles and cellulose. Eur. Phys. J. D 2008, 49, 333–342.

13. Mohanraj, G.T.; Dey, P.K.; Chaki, T.K.; Chakraborty, A.; Khastgir, D. Effect of temperature, pressure, and composition on DC resistivity and AC conductivity of conductive styrene-butadiene rubber-particulate metal alloy nanocomposites. Polym. Compos. 2007, 28, 696–704.

14. Panda, M.; Srinivas, V.; Thakur, A.K. Surface and interfacial effect of filler particle on electrical properties of polyvinyledene fluoride/nickel composites. Appl. Phys. Lett. 2008, 93, 242908.

15. Valmikanathan, O.P.; Ostroverkhova, O.; Mulla, I.S.; Vijayamohanan, K.; Atre, S.V. The effect of synthesis procedure on the structure and properties of palladium/polycarbonate nanocomposites. Polymer 2008, 49, 3413–3418.

16. Tang, Z.Y.; Kotov, N.A. One-dimensional assemblies of nanoparticles: Preparation, properties, and promise. Adv. Mater. 2005, 17, 951–962.

17. Tripp, S.L.; Dunin-Borkowski, R.E.; Wei, A. Flux closure in self-assembled cobalt nanoparticle rings. Angew. Chem., Int. Ed. 2003, 42, 5591–5593.

18. Farrell, D.; Ding, Y.; Majetich, S.A.; Sanchez-Hanke, C.; Kao, C.C. Structural ordering effects in Fe nanoparticle two- and three-dimensional arrays. J. Appl. Phys. 2004, 95, 6636–6638.

19. Hilgendorff, M.; Tesche, B.; Giersig, M. Creation of 3-D crystals from single cobalt nanoparticles in external magnetic fields. Aust. J. Chem. 2001, 54, 497–501.

20. Carotenuto, G.; Martorana, B.; Perlo, P.; Nicolais, L. A universal method for the synthesis of metal and metal sulfide clusters embedded in polymer matrices. J. Mater. Chem. 2003, 13, 2927–2930.

21. Jose, M.V.; Steinert, B.W.; Thomas, V.; Dean, D.R.; Abdalla, M.A.; Price, G.; Janowski, G.M. Morphology and mechanical properties of Nylon 6/MWNT nanofibers. Polymer 2007, 48, 1096–1104.

22. Chrissafis, K.; Antoniadis, G.; Paraskevopoulos, K.M.; Vassiliou, A.; Bikiaris, D.N. Comparative study of the effect of different nanoparticles on the mechanical properties and thermal degradation mechanism of in situ prepared poly(E-caprolactone) nanocomposites. Compos. Sci. Technol. 2007, 67, 2165–2174.

23. Thomassin, J.M.; Lou, X.; Pagnoulle, C.; Saib, A.; Bednarz, L.; Huynen, I.; Jerome, R.; Detrembleur, c. multiwalled carbon nanotube/poly(epsilon-caprolactone) nanocomposites with exceptional electromagnetic interference shielding properties. J. Phys. Chem. C 2007, 111, 11186–11192.

24. Abraham, T.N.; Debdatta, R.; Siengchin, S.; Karger-Kocsis, J. Rheological and thermal properties of poly(ethylene oxide)/multiwall carbon nanotube composites. J. Appl. Polym. Sci. 2008, 110, 2094–2101.

25. Narh, A.K.; Jallo, L.; Rhee, K., Y. The effect of carbon nanotube agglomeration on the thermal and mechanical properties of polyethylene oxide. Polym. Compos. 2008, 29, 809–817.

26. Song, Y.S. Effect of surface treatment for carbon nanotubes on morphological and rheological properties of poly(ethylene oxide) nanocomposites. Polym. Eng. Sci. 2006, 46, 1350–1357.

27. Averett, R.D.; Realff, M.L.; Jacob, K.I. The effects of fatigue and residual strain on the mechanical behavior of poly(ethylene terephthalate) unreinforced and nanocomposite fibers. Compos. Part A Appl. Sci. Manuf. 2009, 40, 709–723.

28. Steinert, B.W.; Dean, D.R. Magnetic field alignment and electrical properties of solution cast PET-carbon nanotube composite films. Polymer 2009, 50, 898–904.

29. Brosse, A.C.; Tence-Girault, S.; Piccione, P.M.; Leibler, L. Effect of multi-walled carbon nanotubes on the lamellae morphology of polyamide-6. Polymer 2008, 49, 4680–4686.

30. Li, Y.; Shimizu, H. Conductive PVDF/PA6/CNTs nanocomposites fabricated by dual formation of cocontinuous and nanodispersion structures. Macromolecules 2008, 41, 5339–5344.

31. Mitchell, C.A.; Krishnamoorti, R. Dispersion of single-walled carbon nanotubes in poly(epsilon-caprolactone). Macromolecules 2007, 40, 1538–1545.

32. Chatterjee, T.; Krishnamoorti, R. Steady Shear response of carbon nanotube networks dispersed in poly(ethylene oxide). Macromolecules 2008, 41, 5333–5338.

33. Chatterjee, T.; Yurekli, K.; Hadjiev, V.G.; Krishnamoorti, R. Single-Walled carbon nanotube dispersions in poly(ethylene oxide). Adv. Funct. Mater. 2005, 15, 1832–1838.

34. Yoo, H.J.; Jung, Y.C.; Cho, J.W. Effect of interaction between poly(ethylene terephthalate) and carbon nanotubes on the morphology and properties of their nanocomposites. J. Polym. Sci., Part B: Polym. Phys. 2008, 46, 900–910.

35. Ahn, B.W.; Chi, Y.S.; Kang, T.J. Preparation and characterization of multi-walled carbon nanotube/poly(ethylene terephthalate) nanoweb. J. Appl. Polym. Sci. 2008, 110, 4055–4063.

36. Wang, K.; Li, W.W.; Gao, C. Poly(epsilon-caprolactone)-functionalized carbon nanofibers by surface-initiated ring-opening polymerization. J. Appl. Polym. Sci. 2007, 105, 629–640.

37. Wakamatsu, N.; Takamori, H.; Fujigaya, T.; Nakashima, N. Self-Organized single-walled carbon nanotube conducting thin films with honeycomb structures on flexible plastic films. Adv. Funct. Mater. 2009, 19, 311–316.

38. Chen, H.; Liu, Z.; Cebe, P. Chain confinement in electrospun nanofibers of PET with carbon nanotubes. Polymer 2009, 50, 872–880.

39. Hu, G.J.; Feng, X.Y.; Zhang, S.M.; Yang, M.S. Crystallization behavior of poly(ethylene terephthalate)/multiwalled carbon nanotubes composites. J. Appl. Polym. Sci. 2008, 108, 4080–4089.

40. Nyczyk, A.; Hasik, M.; Turek, W.; Sniechota, A. Nanocomposites of polyaniline, its derivatives and platinum prepared using aqueous Pt sol. Synth. Met. 2009, 159, 561–567.

41. Zotti, G.; Vercelli, B.; Berlin, A. Gold nanoparticle linking to polypyrrole and polythiophene: monolayers and multilayers. Chem. Mater. 2008, 20, 6509–6516.

42. Huang, S.W.; Neoh, K.G.; Kang, E.T.; Han, H.S.; Tan, K.L. Palladium-containing polyaniline and polypyrrole microparticles. J. Mater. Chem. 1998, 8, 1743–1748.

43. Wilson, J.L.; Poddar, P.; Frey, N.A.; Srikanth, H.; Mohomed, K.; Harmon, J.P.; Kotha, S.; Wachsmuth, J. Synthesis and magnetic properties of polymer nanocomposites with embedded iron nanoparticles. J. Appl. Phys. 2004, 95, 1439.

44. Yurkov, G.; Fionov, A.; Koksharov, Y.; Koleso, V.; Gubin, S. Electrical and magnetic properties of nanomaterials containing iron or cobalt nanoparticles. Inorg. Mater. 2007, 43, 834–844.

45. Sarkar, A.; Kapoor, S.; Yashwant, G.; Salunke, H.G.; Mukherjee, T. Preparation and characterization of ultrafine Co and Ni particles in a polymer matrix. J. Phys. Chem. B 2005, 109, 7203–7207.

46. Sun, Y.; Sun, J.; Liu, M.; Chen, Q. Mechanical strength of carbon nanotube-nickel nanocomposites. Nanotechnology 2007, 18, 505704–505704.

47. Balan, L.; Jin, M.; Malval, J.P.; Chaumeil, H.; Defoin, A.; Vidal, L. Fabrication of silver nanoparticle-embedded polymer promoted by combined photochemical properties of a 2,7-diaminofluorene derivative dye. Macromolecules 2008, 41, 9359–9365.

48. Dundigalla, A.; Lin Gibson, S.; Ferreiro, V.; Malwitz, M.M.; Schmidt, G. Unusual multilayered structures in PEO/laponite nanocomposite films. Macromol. Rapid Commun. 2005, 26, 143–149.

49. Elmahdy, M.M.; Chrissopoulou, K.; Afratis, A.; Floudas, G.; Anastasiadis, S.H. Effect of confinement on polymer segmental

motion and ion mobility in PEO/layered silicate nanocomposites. Macromolecules 2006, 39, 5170–5173.

50. Inyang, H.I.; Bae, S.; Mbamalu, G.; Park, S.-W. Aqueous polymer effects on volumetric swelling of Na-montmorillonite. J. Mater. Civil Eng. 2007, 19:1, 84–90.

51. Loiseau, A.; Tassin, J.F. Model nanocomposites based on laponite and poly(elhylene oxide): preparation and rheology. Macromolecules 2006, 39, 9185–9191.

52. Loyens, W.; Maurer, F.H.J.; Jannasch, P. Melt-compounded salt-containing poly(ethylene oxide)/clay nanocomposites for polymer electrolyte membranes. Polymer 2005, 46, 7334–7345.

53. Qiu, W.L.; Pyda, M.; Nowak-Pyda, E.; Habenschuss, A.; Wunderlich, B. Reversibility between glass and melting transitions of poly(oxyethylene). Macromolecules 2005, 38, 8454–8467.

54. Stefanescu, E.A.; Schexnailder, P.J.; Dundigalla, A.; Negulescu, I.I.; Schmidt, G. Structure and thermal properties of multilayered Laponite/PEO nanocomposite films. Polymer 2006, 47, 7339–7348.

55. Stefanescu, E.A.; Stefanescu, C.; Daly, W.H.; Schmidt, G.; Negulescu, I.I. Hybrid polymer-clay nanocomposites: A mechanical study on gels and multilayered films. Polymer 2008, 49, 3785–3794.

56. Arias, C.B.; Zaman, A.A.; Talton, J. Rheological behavior and wear abrasion resistanceof polyethylene oxide/laponie nanocomposites. J. Dispersion Sci. Technol. 2007, 28, 247–254.

57. Xu, Y.; Higgins, B.; Brittain, W.J. Bottom-up synthesis of PS–CNF nanocomposites. Polymer 2005, 46, 799–810.

58. Wang, C.; Huang, C.-L.; Chen, Y.-C.; Hwang, G.-L.; Tsai, S.-J. Carbon nanocapsules-reinforced syndiotactic polystyrene nanocomposites: Crystallization and morphological features. Polymer 2008, 49, 5564–5574.

59. Shen, J.; Zeng, C.; Lee, L.J. Synthesis of polystyrene–carbon nanofibers nanocomposite foams. Polymer 2005, 46, 5218–5224.

60. Mu, M.; Walker, A.M.; Torkelson, J.M.; Winey, K.I. Cellular structures of carbon nanotubes in a polymer matrix improve properties relative to composites with dispersed nanotubes. Polymer 2008, 49, 1332–1337.

61. Chang; Kisliuk, A.; Rhodes, S.M.; Brittain, W.J.; Sokolov, A.P. Conductivity and mechanical properties of well-dispersed single-wall carbon nanotube/polystyrene composite. Polymer 2006, 47, 7740–7746.

62. Nyden, M.R.; Stoliarov, S.I. Calculations of the energy of mixing carbon nanotubes with polymers. Polymer 2007, 49, 635–641.

63. Xie, L.; Xu, F.; Qiu, F.; Lu, H.; Yang, Y. Single-walled carbon nanotubes functionalized with high bonding density of polymer layers and enhanced mechanical properties of composites. Macromolecules 2007, 40, 3296–3305.

64. Putz, K.; Krishnamoorti, R.; Green, P.F. The role of interfacial interactions in the dynamic mechanical response of functionalized SWNTePS nanocomposites. Polymer 2007, 48, 3540–3545.

65. Cipriano, B.H.; Kota, A.K.; Gershon, A.L.; Laskowski, C.J.; Kashiwagi, T.; Bruck, H.A.; Raghavan, S.R. Conductivity enhancement of carbon nanotube and nanofiber-based polymer nanocomposites by melt annealing. Polymer 2008, 49, 4846–4851.

66. Kota, A.K.; Cipriano, B.H.; Duesterberg, M.K.; Gershon, A.L.; Powell, D.; Raghavan, S.R.; Bruck, H.A. Electrical and rheological percolation in polystyrene/MWCNT nanocomposites. Macromolecules 2007, 40, 7400–7406.

67. Cipiriano, B.H.; Kashiwagi, T.; Raghavan, S.R.; Yang, Y.; Grulke, E.A.; Yamamoto, K.; Shields, J.R.; Douglas, J.F. Effects of aspect ratio of MWNT on the flammability properties of polymer nanocomposites. Polymer 2007, 48, 6086–6096.

68. Cui, J.; Wang, W.P.; You, Y.; Liu, C.; Wang, P. Functionalization of multiwalled carbon nanotubes by reversible addition fragmentation chain-transfer polymerization. Polymer 2004, 45, 8717–8721.

69. Xu, G.; Wu, W.-T.; Wang, Y.; Pang, W.; Zhu, Q.; Wang, P.; You, Y. Constructing polymer brushes on multiwalled carbon nanotubes by in situ reversible addition fragmentation chain transfer polymerization. Polymer 2006, 47, 5909–5918.

70. Jinqi, X.; Jeremy, W.B.; David, A.B.; Tzu-Chia, T.; Michael, E.M.; Karen, L.W. Hierarchical inorganic-organic nanocomposites possessing amphiphilic and morphological complexities: influence of nanofiller dispersion on mechanical performance. Adv. Funct. Mater. 2008, 18, 2733–2744.

71. Yang, B.-X.; Shi, J.-H.; Pramoda, K.P.; Goh, S.H. Enhancement of stiffness, strength, ductility and toughness of poly(ethylene oxide) using phenoxy-grafted multiwalled carbon nanotubes. Nanotechnology 2007, 18, 125606.

72. Jin, J.; Song, M.; Pan, F. A DSC study of effect of carbon nanotubes on crystallisation behaviour of poly(ethylene oxide). Thermochim. Acta 2007, 456, 25–31.

73. Wang, Z.; Meng, X.; Li, J.; Du, X.; Li, S.; Jiang, Z.; Tang, T. A simple method for preparing carbon nanotubes/clay hybrids in water. J. Phys. Chem. C 2009, 113, 8058–8064.

74. Song, Y.S. Rheological characterization of carbon nanotubes/ poly(ethylene oxide) composites. Rheol. Acta 2006, 46, 231–238.

75. Priftis, D.; Petzetakis, N.; Sakelllariou, G.; Pitsikalis, M.; Baskaran, D.; Mays, J.W.; Hadjichristidis, N. Surface-Initiated titanium-mediated coordination polymerization from catalyst-functionalized single and multiwalled carbon nanotubes. Macromolecules 2009, 42, 3340–3346.

76. Hua, C.W.; Chen, Z.M.; Xu, Q.; He, L.H. Ring-Banded spherulites in PCL and PCL/MWCNT solution-casting films and effect of compressed CO2 on them. J. Polym. Sci. Pol. Phys. 2009, 47, 784–792.

77. Ruelle, B.; Peeterbroeck, S.; Gouttebaron, R.; Godfroid, T.; Monteverde, F.; Dauchot, J.P.; Alexandre, M.; Hecq, M.; Dubois, P. Functionalization of carbon nanotubes by atomic nitrogen formed in a microwave plasma Ar+N-2 and subsequent poly(epsilon-caprolactone) grafting. J. Mater. Chem. 2007, 17, 157–159.

78. Wu, D.F.; Zhang, Y.S.; Zhang, M.; Yu, W. Selective localization of multiwalled carbon nanotubes in Poly(epsilon-caprolactone)/ Polylactide Blend. Biomacromolecules 2009, 10, 417–424.

79. Wu, T.M.; Chen, E.C. Isothermal and nonisothermal crystallization kinetics of poly(epsilon-caprolactone)/multi-walled carbon nanotube composites. Polym. Eng. Sci. 2006, 46, 1309–1317.

80. Xu, G.Y.; Du, L.C.; Wang, H.; Xia, R.; Meng, X.C.; Zhu, Q.R. Nonisothermal crystallization kinetics and thermomechanical properties of multiwalled carbon nanotube-reinforced poly(epsilon-caprolactone) composites. Polym. Int. 2008, 57, 1052–1066.

81. Castro, M.; Lu, J.B.; Bruzaud, S.; Kumar, B.; Feller, J.F. Carbon nanotubes/poly(epsilon-caprolactone) composite vapour sensors. Carbon 2009, 47, 1930–1942.

82. Kai, W.H.; Hua, L.; Dong, T.L.; Pan, P.J.; Zhu, B.; Inoue, Y. Synthesis and characterization of fullerene grafted poly(epsilon-caprolactone). J. Appl. Polym. Sci. 2008, 107, 4029–4035.

83. Kai, W.H.; Hua, L.; Dong, T.L.; Pan, P.J.; Zhu, B.; Inoue, Y. Fullerene end-capped biodegradable poly(epsilon-caprolactone). Macromol. Chem. Phys. 2008, 209, 104–111.

84. Kai, W.H.; Hua, L.; Zhao, L.; Inoue, Y. Synthesis of novel star shaped poly (epsilon-caprolactone) utilizing fullerene as the molecular core. Macromol. Rapid Commun. 2006, 27, 1702–1706.

85. Liu, M.X.; Guo, B.C.; Du, M.L.; Chen, F.; Jia, D.M. Halloysite nanotubes as a novel beta-nucleating agent for isotactic polypropylene. Polymer 2009, 50, 3022–3030.

86. Lu, K.B.; Grossiord, N.; Koning, C.E.; Miltner, H.E.; van Mele, B.; Loos, J. Carbon nanotube/isotactic polypropylene composites prepared by LaTex technology: Morphology analysis of CNT-induced nucleation. Macromolecules 2008, 41, 8081–8085.

87. Hou, Z.C.; Wang, K.; Zhao, P.; Zhang, Q.; Yang, C.Y.; Chen, D.Q.; Du, R.N.; Fu, Q. Structural orientation and tensile behavior in the extrusion-stretched sheets of polyp ropylene/multi-walled carbon nanotubes' composite. Polymer 2008, 49, 3582–3589.

88. Pujari, S.; Ramanathan, T.; Kasimatis, K.; Masuda, J.; Andrews, R.; Torkelson, J.M.; Brinson, L.C.; Burghardt, W.R. Preparation and characterization of multiwalled carbon nanotube dispersions in polypropylene: melt mixing versus solid-state shear pulverization. J. Polym. Sci. Pol. Phys. 2009, 47, 1426–1436.

89. Wu, D.F.; Sun, Y.R.; Zhang, M. Kinetics study on melt compounding of carbon nanotube/polypropylene nanocomposites. J. Polym. Sci. Pol. Phys. 2009, 47, 608–618.

90. Xu, D.H.; Wang, Z.G.; Douglas, J.F. Influence of carbon nanotube aspect ratio on normat stress differences in isotactic polypropylene nanocomposite melts. Macromolecules 2008, 41, 815–825.

91. Haggenmueller, R.; Du, F.M.; Fischer, J.E.; Winey, K.I. Interfacial in situ polymerization of single wall carbon nanotube/nylon-6,6 nanocomposites. Polymer 2006, 47, 2381–2388.

92. Li, L.Y.; Li, C.Y.; Ni, C.Y.; Rong, L.X.; Hsiao, B. Structure and crystallization behavior of Nylon 66/multi-walled carbon nanotube nanocomposites at low carbon nanotube contents. Polymer 2007, 48, 3452–3460.

93. Kang, M.; Myung, S.J.; Jin, H.J. Nylon 610 and carbon nanotube composite by in situ interfacial polymerization. Polymer 2006, 47, 3961–3966.

94. Zeng, H.L.; Gao, C.; Wang, Y.P.; Watts, P.C.P.; Kong, H.; Cui, X.W.; Yan, D.Y. In situ polymerization approach to multiwalled carbon nanotubes-reinforced nylon 1010 composites: Mechanical properties and crystallization behavior. Polymer 2006, 47, 113–122.

95. Sandler, J.K.W.; Pegel, S.; Cadek, M.; Gojny, F.; van Es, M.; Lohmar, J.; Blau, W.J.; Schulte, K.; Windle, A.H.; Shaffer, M.S.P. A comparative study of melt spun polyamide-12 fibres reinforced with carbon nanotubes and nanofibres. Polymer 2004, 45, 2001–2015.

96. Antoniadis, G.; Paraskevopoulos, K.M.; Bikiaris, D.; Chrissafis, K. Melt-crystallization mechanism of poly(ethylene terephthalate)/multi-walled carbon nanotubes prepared by in situ polymerization. J. Polym. Sci. Pol. Phys. 2009, 47, 1452–1466.

97. Mun, S.J.; Jung, Y.M.; Kim, J.C.; Chang, J.H. Poly(ethylene terephthalate) nanocomposite fibers with functionalized multiwalled carbon nanotubes via in-situ polymerization. J. Appl. Polym. Sci. 2008, 109, 638–646.

98. Liu, F.J.; Huang, L.M.; Wen, T.C.; Gopalan, A. Large-area network of polyaniline nanowires supported platinum nanocatalysts for methanol oxidation. Synth. Met. 2007, 157, 651–658.

99. Palmero, S.; Colina, A.; Muñoz, E.; Heras, A.; Ruiz, V.; López-Palacios, J. Layer-by-layer electrosynthesis of Pt–Polyaniline nanocomposites for the catalytic oxidation of methanol. Electrochem. Commun. 2009, 11, 122–125.

100. Tristany, M.; Moreno-Mañas, M.; Pleixats, R.; Chaudret, B.; Philippot, K.; Dieudonné, P.; Lecante, P. Formation of nanocomposites of platinum nanoparticles embedded into heavily fluorinated aniline and displaying long range organization. J. Mater. Chem. 2008, 18, 660–666.

101. Okamoto, M.; Fujigaya, T.; Nakashima, N. Design of an assembly of poly (benzimidazole), carbon nanotubes, and pt nanoparticles for

a fuel-cell electrocatalyst with an ideal interfacial nanostructure. Small 2009, 5, 735–740.

102. Sapurina, I.Y.; Kompan, M.E.; Zabrodskii, A.G.; Stejskal, J.; Trchova, M. Nanocomposites with mixed electronic and protonic conduction for electrocatalysis. Russ. J. Electrochem. 2007, 43, 528–536.

103. Compagnini, G. Noble metal particles for polymer-based nanostructured thin films. Appl. Surf. Sci. 2004, 226, 216–225.

104. Zhu, Y.; Kockrick, E.; Kaskel, S.; Ikoma, T.; Hanagata, N. Nanocasting route to ordered mesoporous carbon with FePt nanoparticles and its phenol adsorption property. J. Phys. Chem. C 2009, 113, 5998–6002.

105. Fuhrer, R.; Athanassiou, E.K.; Luechinger, N.A.; Stark, W.J. Crosslinking metal nanoparticles into the polymer backbone of hydrogels enables preparation of soft, magnetic field-driven actuators with muscle-like flexibility. Small 2009, 5, 383–388.

106. Brayner, R.; Vaulay, M.J.; Fievet, F.; Coradin, T. Alginate-Mediated growth of Co, Ni, and CoNi nanoparticles: Influence of the biopolymer structure. Chem. Mater. 2007, 19, 1190–1198.

107. Burke, N.A.D.; Stover, H.D.H.; Dawson, F.P. Magnetic nanocomposites: Preparation and characterization of polymer-coated iron nanoparticles. Chem. Mater. 2002, 14, 4752–4761.

108. Forster, S.; Plantenberg, T. From self-organizing polymers to nanohybrid and biomaterials. Angew. Chem. Int. Edit. 2002, 41, 688–714.

109. Das, D.; Basu, S.; Paul, A. PVC-Cu composites with chemically deposited ultrafine copper particles. J. Mater. Sci. 1980, 15, 1719–1723.

110. Fekete, E.; Molnar, S.; Kim, G.M.; Michler, G.H.; Pukanszky, B. Aggregation, fracture initiation, and strength of PP/CaCO 3 composites. J. Macromol. Sci. B 1999, 38, 885–899.

111. King, S.; Hyunh, K.; Tannenbaum, R. Kinetics of nucleation, growth, and stabilization of cobalt oxide nanoclusters. J. Phys. Chem. B 2003, 107, 12097–12104.

112. Kiss, A.; Fekete, E.; Pukanszky, B. Aggregation of CaCO3 particles in PP composites: Effect of surface coating. Compos. Sci. Technol. 2007, 67, 1574–1583.

113. Osman, M.A.; Suter, U.W. Surface treatment of calcite with fatty acids: Structure and properties of the organic monolayer. Chem. Mater. 2002, 14, 4408–4415.

114. Raj, R.G.; Kokta, B.V.; Dembele, F.; Sanschagrain, B. Compounding of cellulose fibers with polypropylene: Effect of fiber treatment on dispersion in the polymer matirx. J. Appl. Polym. Sci. 1989, 38, 1987–1996.

115. Luechinger, N.A.; Booth, N.; Heness, G.; Bandyopadhyay, S.; Grass, R.N.; Stark, W.J. Surfactant-Free, Melt-Processable metal-polymer hybrid materials: Use of graphene as a dispersing agent. Adv. Mater. 2008, 20, 3044–3049.

116. Ghose, S.; Watson, K.A.; Working, D.C.; Connell, J.W.; Smith, J.G.; Sun, Y.P. Thermal conductivity of ethylene vinyl acetate copolymer/nanofiller blends. Compos. Sci. Technol. 2008, 68, 1843–1853.

117. Sahiner, N. In situ metal particle preparation in cross-linked poly (2-acrylamido-2-methyl-1-propansulfonic acid) hydrogel networks. Colloid Polym. Sci. 2006, 285, 283–292.

118. Chelebaeva, E.; Guari, Y.; Larionova, J.; Trifonov, A.; Guérin, C. Soluble ligand-stabilized cyano-bridged coordination polymer nanoparticles. Chem. Mater. 2008, 20, 1367–1375.

119. Zhao, W.; Zhang, Q.; Chen, T.; Lu, T. Preparation and thermal decomposition of PS/Ni microspheres. Mater. Chem. Phys. 2009, 113, 428–434.

120. Compton, J.; Thompson, D.; Kranbuehl, D.; Ohl, S.; Gain, O.; David, L.; Espuche, E. Hybrid films of polyimide containing in situ generated silver or palladium nanoparticles: Effect of the particle precursor and of the processing conditions on the morphology and the gas permeability. Polymer 2006, 47, 5303–5313.

121. Wada, Y.; Kobayashi, T.; Yamasaki, H.; Sakata, T.; Hasegawa, N.; Mori, H.; Tsukahara, Y. Nanohybrid polymer prepared by successive polymerization of methacrylate monomer containing silver nanoparticles in situ prepared under microwave irradiation. Polymer 2007, 48, 1441–1444.

122. Gao, C.; Vo, C.D.; Jin, Y.Z.; Li, W.; Armes, S.P. Multihydroxy polymer-functionalized carbon nanotubes: Synthesis, derivatization, and metal loading. Macromolecules 2005, 38, 8634–8648.

123. Deshmukh, R.D.; Composto, R.J. Surface segregation and formation of silver nanoparticles created in situ in poly (methyl methacrylate) films. Chem. Mater. 2007, 19, 745–754.

124. Mbhele, Z.H.; Salemane, M.G.; Van Sittert, C.; Nedeljkovic, J.M.; Djokovic, V.; Luyt, A.S. Fabrication and characterization of silver polyvinyl alcohol nanocomposites. Chem. Mater. 2003, 15, 5019–5024.

125. Mukherjee, P.; Nandi, A.K. Electronic properties of poly (o-methoxy aniline)-silver nanocomposite thin films: Influence of nanoparticle size and density. J. Mater. Chem. 2009, 19, 781–786.

126. Lee, J.Y.; Liao, Y.; Nagahata, R.; Horiuchi, S. Effect of metal nanoparticles on thermal stabilization of polymer/metal nanocomposites prepared by a one-step dry process. Polymer 2006, 47, 7970–7979.

127. Wang, H.L.; Gao, J.; Sansinena, J.M.; McCarthy, P. Fabrication and characterization of polyaniline monolithic actuators based on a novel configuration: integrally skinned asymmetric membrane. Chem. Mater. 2002, 14, 2546–2552.

128. Mei, Y.; Lu, Y.; Polzer, F.; Ballauff, M.; Drechsler, M. Catalytic activity of palladium nanoparticles encapsulated in spherical polyelectrolyte brushes and core shell microgels. Chem. Mater. 2007, 19, 1062–1069.

129. French, B.L.; Davis, L.M.; Munzinger, E.S.; Slavin, J.W.J.; Christy, P.C.; Thompson, D.W.; Southward, R.E. Palladium polyimide nanocomposite membranes: Synthesis and characterization of reflective and electrically conductive surface-metallized films. Chem. Mater. 2005, 17, 2091–2100.

130. Krebs, F.C.; Nyberg, R.B.; Jorgensen, M. Influence of residual catalyst on the properties of conjugated polyphenylenevinylene materials: Palladium nanoparticles and poor electrical performance. Chem. Mater. 2004, 16, 1313–1318.

131. Goldenberg, L.M.; Sakhno, O.V.; Smirnova, T.N.; Helliwell, P.; Chechik, V.; Stumpe, J. Holographic composites with gold nanoparticles: Nanoparticles promote polymer segregation. Chem. Mater. 2008, 20, 4619–4627.

132. Hata, H.; Kobayashi, Y.; Salama, M.; Malek, R.; Mallouk, T.E. pH-dependent intercalation of gold nanoparticles into a synthetic

fluoromica modified with poly (allylamine). Chem. Mater. 2007, 19, 6588–6596.

133. Li, D.; He, Q.; Cui, Y.; Li, J. Fabrication of pH-responsive nanocomposites of gold nanoparticles/poly (4-vinylpyridine). Chem. Mater. 2007, 19, 412–417.

134. Uhlenhaut, D.I.; Smith, P.; Caseri, W. Color Switching in Gold-Polysiloxane Elastomeric Nanocomposites. Adv. Mater. 2006, 18, 1653–1656.

135. Marinakos, S.M.; Shultz, D.A.; Feldheim, D.L. Gold nanoparticles as templates for the synthesis of hollow nanometer-sized conductive polymer capsules. Adv. Mater. 1999, 11, 34–37.

136. Park, J.H.; Lim, Y.T.; Park, O.O.; Kim, J.K.; Yu, J.W.; Kim, Y.C. Polymer/Gold nanoparticle nanocomposite light-emitting diodes: Enhancement of electroluminescence stability and quantum efficiency of blue-light-emitting polymers. Chem. Mater. 2004, 16, 688–692.

137. Tanaka, M.; Fujita, R.; Nishide, H. Alternate network film of thiol group-terminated polythiophene and gold nanoparticle. Polymer 2007, 48, 5884–5888.

138. Cioffi, N.; Losito, I.; Torsi, L.; Farella, I.; Valentini, A.; Sabbatini, L.; Zambonin, P.G.; Bleve-Zacheos, T. Analysis of the Surface Chemical Composition and Morphological Structure of Vapor-Sensing Gold Fluoropolymer Nanocomposites. Chem. Mater. 2002, 14, 804–811.

139. Laicer, C.S.T.; Mrozek, R.A.; Taton, T.A. Domain nucleation dictates overall nanostructure in composites of block copolymers and model nanorods. Polymer 2007, 48, 1316–1328.

140. Guo, Z.; Lin, H.; Karki, A.B.; Wei, S.; Young, D.P.; Park, S.; Willis, J.; Hahn, T.H. Facile monomer stabilization approach to fabricate iron/vinyl ester resin nanocomposites. Compos. Sci. Technol. 2008, 68, 2551–2556.

141. Baker, C.; Ismat Shah, S.; Hasanain, S.K. Magnetic behavior of iron and iron-oxide nanoparticle/polymer composites. J. Magn. Magn. Mater. 2004, 280, 412–418.

142. Ushakov, N.M.; Kochubei, V.I.; Zapsis, K.V.; Kosobudskii, I.D. Optical properties of metal-polymer nanocomposites based on iron and high-pressure polyethylene. Opt. Spectrosc. 2004, 96, 798–803.

143. Huang, X.; Jiang, P.; Kim, C.; Ke, Q.; Wang, G. Preparation, microstructure and properties of polyethylene aluminum nanocomposite dielectrics. Compos. Sci. Technol. 2008, 68, 2134–2140.

144. Huang, X.; Kim, C.; Jiang, P.; Yin, Y.; Li, Z. Influence of aluminum nanoparticle surface treatment on the electrical properties of polyethylene composites. J. Appl. Phys. 2009, 105, 014105.

145. Huang, X.; Kim, C.; Ma, Z.; Jiang, P.; Yin, Y.; Li, Z. Correlation between rheological, electrical, and microstructure characteristics in polyethylene/aluminum nanocomposites. J. Polym. Sci. Pol. Phys. 2008, 46, 2143–2154.

146. Mallick, K.; Witcomb, M.J.; Scurrell, M.S. In situ synthesis of copper nanoparticles and poly (o-toluidine): A metal–polymer composite material. Eur. Polym. J. 2006, 42, 670–675.

147. Untereker, D.; Lyu, S.; Schley, J.; Martinez, G.; Lohstreter, L. Maximum conductivity of packed nanoparticles and their polymer composites. ACS Appl. Mater. Interfaces 2009, 1, 97–101.

148. Cioffi, N.; Torsi, L.; Ditaranto, N.; Tantillo, G.; Ghibelli, L.; Sabbatini, L.; Bleve-Zacheo, T.; D'Alessio, M.; Zambonin, P.G.; Traversa, E. Copper nanoparticle/polymer composites with antifungal and bacteriostatic properties. Chem. Mater. 2005, 17, 5255–5262.

149. Sormana, J.-L.; Chattopadhyay, S.; Meredith, J.C. Mechanical and thermal properties of poly(urethane urea) nanocomposites prepared with diamine-modified Laponite. J. Nanomater. 2008, 2008, 1–9.

150. Mishra, A.K.; Chattopadhyay, S.; Nando, G.B.; Devadoss, E. Synthesis and characterization of elastomeric polyurethane-laponite nanocomposite. Des. Monomers Polym. 2008, 11, 394–407.

151. Mishra, A.K.; Nando, G.B.; Chattopadhyay, S. Exploring preferential association of laponite and cloisite with soft and hard segments in TPU-Clay nanocomposite prepared by solution mixing technique. J. Polym. Sci. Pol. Phys. 2008, 46, 2341–2354.

152. Korley, L.T.J.; Liff, S.M.; Kumar, N.; McKinley, G.H.; Hammond, P.T. Preferential Association of Segment Blocks in Polyurethane Nanocomposites. Macromolecules 2006, 39, 7030–7036.

153. Fang, F.; Kim, J.; Choi, H.; Kim, C. Synthesis and electrorheological response of nano-sized laponite stabilized poly(methyl methacrylate) spheres. Colloid Polym. Sci. 2009, 287, 745–749.

154. Chang, C.-C.; Hou, S.-S. Intercalation of poly(methyl methacrylate) into tyramine-modified layered silicates through hydrogen-bonding interaction. Eur. Polym. J. 2008, 44, 1337–1345.

155. Wang, J.; Wheeler, P.; Jarrett, W.; Mathias, L. Synthesis and characterization of dual-functionalized laponite clay for acrylic nanocomposites. J. Appl. Polym. Sci. 2007, 106, 1497–1506.

156. Wheeler, P.A.; Wang, J.; Mathias, L.J. Poly(methyl methacrylate)/ Laponite Nanocomposites: Exploring Covalent and Ionic Clay Modifications. Chem. Mater. 2006, 18, 3937–3945.

157. Konn, C.; Morel, F.; Beyou, E.; Chaumont, P.; Bourgeat-Lami, E. Nitroxide-Mediated Polymerization of Styrene Initiated from the Surface of Laponite Clay Platelets. Macromolecules 2007, 40, 7464–7472.

158. Negrete-Herrera, N.; Putaux, J.-L.; David, L.; Bourgeat-Lami, E. Polymer/Laponite composite colloids through emulsion polymerization: Influence of the clay modification level on particle morphology. Macromolecules 2006, 39, 9177–9184.

159. Negrete-Herrera, N.; Putaux, J.-L.; David, L.; De Haas, F.; Bourgeat-Lami, E. Polymer/Laponite composite latexes: Particle morphology, film microstructure, and properties. Macromol. Rapid Commun. 2007, 28, 1567–1573.

160. Ruggerone, R.; Plummer, C.J.G.; Herrera, N.N.; Bourgeat-Lami, E.; Månson, J.-A.E. Highly filled polystyrene-laponite nanocomposites prepared by emulsion polymerization. Eur. Polym. J. 2009, 45, 621–629.

161. Caruso, R.A.; Susha, A.; Caruso, F. Multilayered titania, silica, and laponite nanoparticle coatings on polystyrene colloidal templates and resulting inorganic hollow spheres. Chem. Mater. 2001, 13, 400–409.

162. Lorthioir, C.; Lauprêtre, F.; Soulestin, J.; Lefebvre, J.-M. Segmental dynamics of poly(ethylene oxide) chains in a model polymer/clay intercalated phase: Solid-State NMR investigation. Macromolecules 2009, 42, 218–230.

163. Loyens, W.; Jannasch, P.; Maurer, F.H.J. Poly(ethylene oxide)/Laponite nanocomposites via melt-compounding: Effect of clay modification and matrix molar mass. Polymer 2005, 46, 915–928.

164. Gournis, D.; Floudas, G. Hairy Plates: Poly(ethylene oxide)-b-polyisoprene Copolymers in the Presence of Laponite Clay. Chem. Mater. 2004, 16, 1686–1692.

165. Mitchell, C.A.; Krishnamoorti, R. Rheological properties of diblock copolymer/layered-silicate nanocomposites. J. Polym. Sci. Pol. Phys. 2002, 40, 1434–1443.

166. Agrawal, S.K.; Sanabria-DeLong, N.; Tew, G.N.; Bhatia, S.R. Nanoparticle-reinforced associative network hydrogels. Langmuir 2008, 24, 13148–13154.

167. Liu, R.; Tirelli, N.; Cellesi, F.; Saunders, B.R. Temperature-triggered gelation of aqueous laponite dispersions containing a cationic poly(n-isopropyl acrylamide) graft copolymer. Langmuir 2009, 25, 490–496.

168. De Lisi, R.; Lazzara, G.; Lombardo, R.; Milioto, S.; Muratore, N.; Turco Liveri, M.L. Adsorption of triblock copolymers and their homopolymers at laponite clay/solution interface. Role played by the copolymer nature. Phys. Chem. Chem. Phys. 2005, 7, 3994–4001.

169. De Lisi, R.; Gradzielski, M.; Lazzara, G.; Milioto, S.; Muratore, N.; Prevost, S. Aqueous laponite clay dispersions in the presence of poly(ethylene oxide) or poly(propylene oxide) oligomers and their triblock copolymers. J.Phys.Chem. B 2008, 112, 9328–9336.

170. Nelson, A.; Cosgrove, T. Small-angle neutron scattering study of adsorbed pluronic tri-block copolymers on laponite. Langmuir 2005, 21, 9176–9182.

171. De Lisi, R.; Lazzara, G.; Milioto, S.; Muratore, N. Laponite clay in homopolymer and tri-block copolymer matrices. J. Therm. Anal. Calorim. 2007, 87, 61–67.

172. De Lisi, R.; Lazzara, G.; Milioto, S.; Muratore, N. Aqueous nonionic copolymer-functionalized laponite clay. a thermodynamic and spectrophotometric study to characterize its behavior toward an organic material. Langmuir 2006, 22, 8056–8062.

173. Dan, C.H.; Lee, M.H.; Kim, Y.D.; Min, B.H.; Kim, J.H. Effect of clay modifiers on the morphology and physical properties of

thermoplastic polyurethane/clay nanocomposites. Polymer 2006, 47, 6718–6730.

174. Lee, H.-T.; Lin, L.-H. Waterborne polyurethane/clay nanocomposites: Novel effects of the clay and its interlayer ions on the morphology and physical and electrical properties. Macromolecules 2006, 39, 6133–6141.

175. Osman, M.A.; Mittal, V.; Morbidelli, M.; Suter, U.W. Polyurethane adhesive nanocomposites as gas permeation barrier. Macromolecules 2003, 36, 9851–9858.

176. Plummer, C.J.G.; Rodlert, M.; Bucaille, J.-L.; Grunbauer, H.J.M.; Månson, J.-A.E. Correlating the rheological and mechanical response of polyurethane nanocomposites containing hyperbranched polymers. Polymer 2005, 46, 6543–6553.

177. Chen-Yang, Y.W.; Lee, Y.K.; Chen, Y.T.; Wu, J.C. High improvement in the properties of exfoliated PU/clay nanocomposites by the alternative swelling process. Polymer 2007, 48, 2969–2979.

178. Tien, Y.I.; Wei, K.H. High-tensile-property layered silicates/polyurethane nanocomposites by using reactive silicates as pseudo chain extenders. Macromolecules 2001, 34, 9045–9052.

179. Tarkin-Tas, E.; Goswami, S.K.; Nayak, B.R.; Mathias, L.J. Highly exfoliated poly(ε-caprolactone)/organomontmorillonite nanocomposites prepared in situ polymerization. J. Appl. Polym. Sci. 2008, 107, 976–984.

180. Liao, L.; Zhang, C.; Gong, S. Preparation of poly(-caprolactone)/clay nanocomposites by microwave-assisted in situ ring-opening polymerization. Macromol. Rapid Comm. 2007, 28, 1148–1154.

181. Kiersnowski, A.; Gutmann, J.S.; Piglowski, J. Influence of organic modifiers on morphology and crystallization of poly(ε-caprolactone)/synthetic clay intercalated nanocomposites. J. Polym. Sci. Pol. Phys. 2007, 45, 2350–2367.

182. Yu, Z.; Yin, J.; Yan, S.; Xie, Y.; Ma, J.; Chen, X. Biodegradable poly(l-lactide)/poly([var epsilon]-caprolactone)-modified montmorillonite nanocomposites: Preparation and characterization. Polymer 2007, 48, 6439–6447.

183. Wu, T.; Xie, T.; Yang, G. Preparation and characterization of poly([epsilon]-caprolactone)/Na+-MMT nanocomposites. Appl. Clay Sci. 2009, 45, 105–110.

184. Feng, S.-S.; Mei, L.; Anitha, P.; Gan, C.W.; Zhou, W. Poly(lactide)-vitamin E derivative/montmorillonite nanoparticle formulations for the oral delivery of Docetaxel. Biomaterials 2009, 30, 3297–3306.

185. Gu, S.-Y.; Ren, J.; Dong, B. Melt rheology of polylactide/montmorillonite nanocomposites. J. Polym. Sci. Pol. Phys. 2007, 45, 3189–3196.

186. Jiang, L.; Zhang, J.; Wolcott, M.P. Comparison of polylactide/nano-sized calcium carbonate and polylactide/montmorillonite composites: Reinforcing effects and toughening mechanisms. Polymer 2007, 48, 7632–7644.

187. Lee, Y.H.; Lee, J.H.; An, I.-G.; Kim, C.; Lee, D.S.; Lee, Y.K.; Nam, J.-D. Electrospun dual-porosity structure and biodegradation morphology of Montmorillonite reinforced PLLA nanocomposite scaffolds. Biomaterials 2005, 26, 3165–3172.

188. Sun, B.; Ranganathan, B.; Feng, S.-S. Multifunctional poly(d,l-lactide-co-glycolide)/montmorillonite (PLGA/MMT) nanoparticles decorated by Trastuzumab for targeted chemotherapy of breast cancer. Biomaterials 2008, 29, 475–486.

189. Urbanczyk, L.; Ngoundjo, F.; Alexandre, M.; Jérôme, C.; Detrembleur, C.; Calberg, C. Synthesis of polylactide/clay nanocomposites by in situ intercalative polymerization in supercritical carbon dioxide. Eur. Polym. J. 2009, 45, 643–648.

190. Dong, Y.; Feng, S.-S. Poly(d,l-lactide-co-glycolide)/montmorillonite nanoparticles for oral delivery of anticancer drugs. Biomaterials 2005, 26, 6068–6076.

191. Loyens, W.; Maurer, H.J.F.; Jannasch, P. Melt-compounded salt-containing poly(ethylene oxide)/clay nanocomposites for polymer electrolyte membranes. Polymer 2005, 46, 7334–7345.

192. Malwitz, M.M.; Dundigalla, A.; Ferreiro, V.; Butler, P.D.; Henk, M.C.; Schmidt, G. Layered structures of shear-oriented and multilayered PEO/silicate nanocomposite films. Phys. Chem. Chem. Phys. 2004, 6, 2977–2982.

193. Stefanescu, E.A.; Stefanescu, C.; Negulescu, I.I.; Daly, W.H. Effect of ionic species on the structure and properties of salt-containing

PEO-montmorillonite nanocomposites. Macromol. Chem. Phys. 2008, 209, 2320–2330.

194. Stefanescu, E.A.; Petrovan, S.; Daly, W.H.; Negulescu, I.I. Elongational rheology of polymer/clay dispersions: determination of orientational extent in elongational flow processes. Macromol. Mater. Eng. 2008, 293, 303–309.

195. Sun, Y.-h.; Luo, Y.-F.; Jia, D.-M. Preparation and characterization of polypropylene/solid-state organomodified montmorillonite nanocomposites. Polym. Compos. 2008, 29, 357–363.

196. Rohlmann, C.O.; Failla, M.D.; Quinzani, L.M. Linear viscoelasticity and structure of polypropylene-montmorillonite nanocomposites. Polymer 2006, 47, 7795–7804.

197. Qin, H.; Zhang, S.; Liu, H.; Xie, S.; Yang, M.; Shen, D. Photo-oxidative degradation of polypropylene/montmorillonite nanocomposites. Polymer 2005, 46, 3149–3156.

198. Nowacki, R.; Monasse, B.; Piorkowska, E.; Galeski, A.; Haudin, J.M. Spherulite nucleation in isotactic polypropylene based nanocomposites with montmorillonite under shear. Polymer 2004, 45, 4877–4892.

199. Chow, W.S.; Tham, W.L. Effects of antistatic agent on the mechanical, morphological and antistatic properties of polypropylene/organo-montmorillonite nanocomposites. Express Polym. Lett. 2009, 3, 116–125.

INDEX